Studies in History and Philosophy of Science

Volume 43

More information about this series at http://www.springer.com/series/5671

Delphine Antoine-Mahut • Stephen Gaukroger
Editors

Descartes' *Treatise on Man* and its Reception

Springer

Editors
Delphine Antoine-Mahut
Département de Sciences Humaines
CERPHI, IHRIM, CNRS, UMR 5317. ENS de Lyon
Lyon Cédex 07, France

Stephen Gaukroger
Unit for the History and Philosophy of Science
University of Sydney
Sydney, NSW, Australia

ISSN 0929-6425 ISSN 2215-1958 (electronic)
Studies in History and Philosophy of Science
ISBN 978-3-319-46987-4 ISBN 978-3-319-46989-8 (eBook)
DOI 10.1007/978-3-319-46989-8

Library of Congress Control Number: 2016963413

Printed on acid-free paper

This Springer imprint is published by Springer Nature
The registered company is Springer International Publishing AG
The registered company address is: Gewerbestrasse 11, 6330 Cham, Switzerland

Contents

Chapter 1
The Story of *L'Homme*

Delphine Antoine-Mahut

Abstract The story of *L'Homme* is a true novel. Its plot weaves itself along three intersecting points: an *unfinished* text, a *copied or plagiarised* text, and a *corrected* text. Telling this whole and complex story helps us to understand the true place of Descartes in the history of modern anthropology and in the contemporary attempt to explain cognition, memory, sensation and human health. In a nutshell, it is probably the best way to understand the Cartesian contribution to the vast philosophical programme of seeking to know oneself.

1.1 *L'Homme* Sidelined

Descartes' *L'Homme* has long been forgotten or marginalised by commentators on Cartesianism, the Classical Age and the history of the sciences. The latest developments in critical editions of the text, compared to Descartes' other writings, are a symptom that establishes this diagnosis. The *Meditations* and the *Discourse on the Method* have been translated and re-translated time and again, re-edited, and made part of the examination programme in France and in various universities all over the world. By contrast with *L'Homme*, other similarly unfinished and posthumous texts have been and are currently the subject of important work: this is particularly the case with the *Regulae ad directionem ingenii.*[1] Descartes the natural philosopher has, over the past 30 years, been the subject of an unquestionable resurgence of interest and the focus of major pieces of current Cartesian research, reassessing the significance of the *Principia philosophiae* and the scientific essays.[2] In the work on

[1] A new manuscript has recently been found at Cambridge by Richard Serjeantson. This discovery has reignited lively discussion on this text.

[2] For a selection of the most important, see Frédéric de Buzon, *La science cartésienne et son objet: "Mathesis" et phénomène*. Paris: Champion, 2013; Desmond Clarke, *Descartes. A Biography*, New York: Cambridge University Press, 2007; Daniel Garber, *Descartes Embodied. Reading*

D. Antoine-Mahut (✉)
Département de Sciences Humaines, CERPHI, IHRIM, CNRS, UMR 5317. ENS de Lyon, 15 Parvis René Descartes, BP 7000, 69342 Lyon Cédex 07, France
e-mail: delphine.antoine-mahut@ens-lyon.fr

D. Antoine-Mahut, S. Gaukroger (eds.), Descartes' *Treatise on Man* and its Reception, Studies in History and Philosophy of Science 43,
DOI 10.1007/978-3-319-46989-8_1

physical questions, particular attention has been paid to physiology.[3] Despite this, there are only three modern available editions of *L'Homme*: the text in volume XI of the complete edition of the *Oeuvres de Descartes*, edited by Charles Adam and Paul Tannery,[4] the edition of Annie Bitbol-Hespériès and Jean-Pierre Verdet, published by Seuil in 1996 (in which *L'Homme* is placed in continuity with *The World*, of which it constitutes Chapter 18[5]) and that of Thierry Gontier, published by Fayard

Cartesian Philosophy through Cartesian Science. Chicago, Cambridge University Press: 2000; Daniel Garber, *Descartes's Metaphysical Physics*. Chicago, University of Chicago Press: 1992; Daniel Garber and Sophie Roux (eds.), *The Mechanization of Natural Philosophy*, Dordrecht: Springer, 2013, 237–262.; Stephen Gaukroger, *Descartes: An Intellectual Biography*. Oxford: Oxford University Press, 1995; Stephen Gaukroger, ed. *Descartes' Natural Philosophy* Routledge, 2000; Stephen Gaukroger, John Schuster & John Sutton eds., *Descartes' Natural Philosophy*, London and New York: Routledge, 2000; Stephen Voss, ed., *Essays on the Philosophy of Science of René Descartes*, New York, Oxford: Oxford University Press, 2003. Concerning works not yet published, see notably Delphine Bellis' thesis, presented by Zittel, C., "Delphine Bellis, Le visible et l'invisible dans la pensée cartésienne. Figuration, imagination et vision dans la philosophie naturelle de René Descartes (2010)". *Studium*, 2011, 4 (3), 183–185.

[3] For a selection of the most important, see Vincent Aucante, *La philosophie médicale de Descartes*. Paris: PUF, 2006; Annie Bitbol-Hespériès, *Le principe de vie chez Descartes*. Paris: Vrin, 1996; Géraldine Caps, *Les "médecins cartésiens". Héritage et diffusion de la représentation mécaniste du corps humain (1646–1696)*. Georg Olms Verlag; Hildeshein. Zürich. New York: 2010; Denis Des Chene, *Physiologia. Natural Philosophy in Late Aristotelian and Cartesian Thought*. Ithaca and London: Cornell University Press, 1996; Denis Des Chene, *Spirits and Clocks, Machine and Organism in Descartes*. Ithaca and London: Cornell University Press, 2001; François Duchesneau, *Les modèles du vivant de Descartes à Leibniz*. Paris: Vrin, 1998; Ronan De Calan, *Généalogie de la sensation. Physique, physiologie et psychologie en Europe, de Fernel à Locke*. Paris: Honoré Champion, 2012; Delphine Kolesnik-Antoine (Antoine-Mahut), *L' Homme cartésien. La "force qu'a l'âme de movoir le corps": Descartes, Malebranche*, PUR, 2009; Mirko, D. Grmek, *La première révolution biologique. Réflexions sur la physiologie et la médecine du XVIIe siècle*. Paris: Payot, 1990; Franco Aurelio Meschini, *Neurofisiologia cartesiana*. Firenze: Leo Olschki, 1998; and Emanuela Scribano, *Macchine con la mente. Fisologia e metafisica tra Cartesio e Spinoza*. Carocci editore e Frecce: Roma, 2015. Concerning works in progress not yet published, see Barnaby R. Hutchins, *Obscurity And Confusion: Nonreductionism in Descartes's Biology and Philosophy*. PhD dissertation, Department of Philosophy and Moral Sciences, Ghent University, 2016.

[4] René Descartes, *Œuvres*. Published by Charles Adam and Paul Tannery. New presentation by Bernard Rochot and Pierre Costabel, 11 volumes. Paris: Vrin-CNRS, 1964–1974, re-ed. 1996.

[5] Two other editions are currently in publication. The first is that of Annie Bitbol-Hespériès, to be published in volume 3 of the new edition of the *Complete Works* of Descartes directed by Denis Kambouchner (Paris: Tel, Gallimard). Annie Bitbol-Hespériès continues with and is completing the work in which she was the first to engage with the medical sources of the treatise in the 1980s under the direction of Geneviève Rodis-Lewis, which resulted in the 1990 publication of her thesis: *Le principe de vie chez Descartes* (Paris: Vrin). In it she consolidates the reintegration of *L'Homme* in the continuity of *The World*, and was also the first to popularise it in France on the occasion of her edition of *L'Homme* published by Seuil in 1996. The second edition, at my initiative, will be published by Garnier Flammarion in 2017 in paperback. The apparatus criticus has a simple function of explaining allusions and striking the most important theoretical points. The special feature of this edition lies in the adoption of a different perspective from the previous work. By reproducing the full text of the Clerselier edition, it effectively assumes the role of a reading of the text from its receptions. My introduction places here the foundations of this methodological option and the

in 1999 (without any apparatus criticus at all). Only a few translations are to be found, which are, moreover, exceptional.[6] And even when critical studies of the life sciences raise general questions about mechanism and its role in the history of the life sciences,[7] *L'Homme* itself is never made the subject of a comprehensive critical study or even an issue of a journal.

Two kinds of explanations—complementary and exclusive of one another—can be proposed for this surprising lacuna. The first concentrates on its reception. It shows how very different theoretical motivations have, paradoxically, similar effects and blur the place of the text in Descartes' work. The second focuses on features of and the nature of the text itself. It shows that Descartes and his principal readers have, if not deliberately, also played a role in this blurring.

One can start by pleading the progress of science and using the topos of the obsolescence of the older in relation to the newer. Descartes' works on physiology, understood as a branch of mechanistic physics, fall under this rubric in two ways: they would be invalidated by later anatomical discoveries, and they would be intrinsically unable to account for the specificity of the living in relation to natural phenomena in general. Wherever one turns in the history of science, whether from the perspective of the eyes and the hands in dissection or what modern technologies reveal of the body without opening it up, to that of the different theoretical tools used to account for the dynamism and complexity of the living, Descartes and his *L'Homme* look like 'has beens', that is, at best outlandish and at worst obsolete means of clarifying present thought about the human being.

Paradoxically, the upshot of this reading is similar to that produced on the basis of diametrically opposed motivations. Those who wish to rehabilitate the rationalism of Cartesian psychology and to free themselves from any determinant biological anchorage are also led to leave *L'Homme* to one side. The exemplary case of this is the man who institutionalised the history of philosophy in France in the nineteenth century by invoking a spiritualist Descartes: Victor Cousin.[8] In his edition of

---

contribution of Annie Bitbol-Hespériès to this volume introduces us to her own project. The reader can thus complement one approach with the other.

[6] For the most important, see Descartes. *The World and Other Writings*. Ed. Stephen Gaukroger, Cambridge Texts in the History of Philosophy: Cambridge University Press, 2004; in Dutch: Descartes. *De wereld. De mens. Het zoeken naar de waarheid.* Redactie & annotatie Erik-Jan. Bos, Han Van Ruler. Inleidingen Erik-Jan. Bos; Vertaling Jeanne Holierhoek. Boom: Amsterdam, 2011; in German: René Descartes, *Die Welt: Abhandlung über das Licht. Der Mensch* (Philosophische Bibliothek), transl. Christian Wolhers, Hamburg, Meiner: 2015; and in Italian: René Descartes, *Opere postume 1650–2009*, a cura di Giulia Belgioioso, con la collaborazione di Igor Agostini, Francesco Marrone, Massimiliano Savini, Milano: Bompiani, 2009 (translation of *L'Homme* by Siegrid Agostini). It should be added that a team from the University of Lisbon, directed by Adelino Cardoso, is currently working on a Portuguese translation of *L'Homme* in the Latin edition by Florent Schuyl.

[7] For the latest developments, see Claire Crignon, Delphine Antoine-Mahut, "Etude bibliographique", in *Gesnerus, Swiss Journal of the History of Medicine and Sciences*, "Teleology and mechanism in Early Modern Medicine", Vol. 71 (2014), N°2, 187–203.

[8] Victor Cousin (1792–1867) is considered the leader of a spiritualist school founded on a rationalist interpretation of the Cartesian cogito and as the first French person to have truly proposed the

Descartes' *Complete Works* (1824–1828), he separated the *Discourse on the Method* and the scientific essays, and in his courses and his publications, he always took care to esteem the *Meditations* and the *Discourse*, evoking as little as possible the physiology texts that might provide succour to his positivist phrenologist adversaries.

In addition to being designated the father of a mechanism that underestimated the complexity of the living in the physical universe, Descartes has thus also come to play a historiographical role as the founder of 'modern subjectivity', and by extension of a 'dualism' of the body and soul. Ignorance of *L'Homme* would in this way be the logical correlate of 'the ghost in the machine', set out by Gilbert Ryle in the 1950s[9]: not only would Descartes hold back the history of science with his far-fetched hypotheses about the pineal gland and animal spirits, but in addition—and this sometimes explains it—he corrupted from the inside what should have been saved from physiology, through the metaphysical postulates of an immortal and immaterial soul, which had nothing to do with what even his contemporaries considered to be natural philosophy. What these current critiques pass over in silence needs to be stressed: all these arguments have been used since the seventeenth century, and with an acuity that is often missing in contemporary debates because of the omnipotence of 'scientism'.[10]

Finally, to the extent that precise considerations on organic life are integrated into the science of man, or on the contrary to the extent that one seeks to show how the rational soul can be emancipated from it, *L'Homme* is set aside.

---

history of philosophy. He served as, successively or jointly, all institutional powers: president of the Board of Education, president of the philosophy *Agrégation* and the *Académie des Sciences Morales et Politiques*, director of the *École normale*. He determined the curricula, trained and appointed teachers and, in short, founded a state philosophy. On this point, see the excellent work of Patrice Vermeren: Victor Cousin. *Le jeu de la philosophie et de l'Etat*. Paris: L'Harmattan, 1995.

[9] Gilbert Ryle, *The Concept of Mind*, Hutchinsons University Library (particularly chapter 1: "Descartes' Myth"), re-ed. University of Chicago Press, with an introduction by D. Dennett, 2002. See also Desmond Clarke, "Exorcising Ryle's Ghost from Cartesian Metaphysics", *Philosophical Inquiry* 23 (3–4): 27–36 (2001);. Dubois (1970), "Ryle et Merleau-Ponty: faut-il exorciser le fantôme qui se cache dans la machine humaine?", *Revue Philosophique de la France et de l'Etranger*, t. 160, 1970, 299–317; H.P. Rickmann, "Exorcising the Ghost in the Machine", *Philosophy*, 63, 1988 (246), 487–499; Desh Raj Sirswal, "Gilbert Ryle on Descartes' Myth", K.U. Research Journal of Arts and Humanities, 2007, 81–86 and Antti Revonsuo, "Is there a Ghost in the Cognitive Machinery?", *Philosophical Psychology*, 1993, 6, (4), 387–405. The issue of these debates is interestingly synthesised by Daniel C. Dennett in *Consciousness Explained* (Little: Brown and Company, 1991) through a reference, rare enough in these corpora to be highlighted, to the *Treatise on Man*: "The embarrassment of dualists on this point is actually simpler than the mention of the alleged laws of physics would suggest. It is the same inconsistency as that which arises from children – which they joyfully tolerate in order to laugh – during the stories of Casper the Friendly Ghost. How can Casper both pass through walls and catch a falling towel? How can mental substance both escape any physical measurement and control the body? A ghost in the machine does not help us with our theories if it cannot move things around him - like a violent and noisy spirit that can tip over a lamp and slam a door. But all that can move a physical thing is itself a physical thing (albeit perhaps a kind of strange and so far not studied physical thing)".

[10] On the historical nature of these debates, see Delphine Antoine-Mahut, "Esprit, es-tu là ? Eléments pour une hantologie de Casper, le gentil fantôme", in *Le corps et l'esprit. Problèmes cartésiens, problèmes contemporains*, ed. Sandrine Roux, Paris: Editions des archives contemporaines, 2015, 129–146.

The second explanation proceeds from features of the text itself in the overall work of Descartes. Contemporary commentators have been increasingly perplexed and have engaged in lively critiques over the authenticity of the text that we call 'Descartes' *L'Homme* and that which we use as a reference, namely Clerselier's 1664 edition (reproduced in the Adam and Tannery edition). Recent research has brought to light the existence of important differences between Clerselier's French edition and Schuyl's Latin edition (1662), it being understood that Clerselier's text is not a retranslation into French of the French text that Schuyl had himself translated into Latin.[11] This in addition underlines the importance of the fact that in Descartes' lifetime there were several different copies of an 'original' in circulation, and we have no reason to believe that this was the one that Clerselier possessed, or that of Schuyl, or that of either of them. Not publishing *L'Homme*, or waiting to do it, derives in this case from an intellectual and philological probity requirement of the historian of philosophy: as long as nothing assures us that the text that we are working on is really that of Descartes, the interpretations that we are able to propose on this basis remain in the realm of conjecture.

And when we go back to Descartes' lifetime, we observe that, even though it was unpublished, it is the text that brought a scandal with it. On the one hand, there are the public lectures on physiology of the Dutch physician Henricus Regius[12] at the University of Utrecht from 1638, which sparked the 'Querelle d'Utrecht'.[13] The arguments were both physical and metaphysical: the eradication of substantial

---

[11] Claude Clerselier (1614–1684) was a lawyer in the Parisian parliament. He was the stepbrother of Pierre Chanut and the stepfather of Jacques Rohault. Notably, he was the editor and translator of several of Descartes' works, especially of the latter's *Letters*, published in three volumes in Paris, 1657, 1659 and 1667, and of the 1664 edition of *L'Homme*. He published a new edition of *L'Homme*, as chapter 18 of *The World*, in 1667 and another from the *Principles of Philosophy* in 1681. He is a self proclaimed faithful disciple of Descartes', conscious of restoring the true meaning of the master's writings and proving their orthodoxy in a posthumous context that challenges them.

Florentius Schuyl (1619–1669)? Here I will partially reproduce note 4, p. 24 of the correspondence edition between Descartes and Regius by Erik-Jan. Bos (*The Correspondence between Descartes and Henricus Regius.* Zeno: The Leiden-Utrecht Research Institute of Philosophy, 2002): "After his graduation in the Utrecht Faculty of Arts, (he) studied philosophy and theology in Leiden for a short while, before being appointed professor of philosophy at the illustrious School at's-Hertenbosch un 1640. In the 1640s, Schuyl embraced Cartesianism. Eventually he became interested in the philosopher's posthumous works, and in 1662 he published Descartes' *Traité de l'homme* in a Latin translation (…) In 1664, Schuyl graduated in medicine and he was immediately appointed professor of medicine in Leiden".

[12] Henricus Regius (1598–1679) was the first chair of medicine and botany at the University of Utrecht (1638). There he taught Cartesian physiology and the circulation of the blood and maintained that the consistency of Cartesianism implies excluding all that relates to metaphysics from the limits of human reason.

[13] On the Querelle, see especially Theo Verbeek, *La Querelle d'Utrecht*, Paris: Les impressions nouvelles, 1988 and Theo Verbeek, *Descartes and the Dutch: early reactions to Cartesian philosophy, 1637–1650*, Carbondale: Southern Illinois University Press, 1992, and Theo Verbeek, "Ens per accidens: le origini della Querelle di Utrecht," *Giornale critico della filosofia italiana*, 71 (1992), 276–288.

forms from the explanation of living things and the improper extension, according to Regius, of the eradication of the human soul itself could not fail to irritate the theologians and Aristotelians of the University.

But above all—and we generally neglect to properly address this in light of the one study of the controversy on the nature of the soul[14]—it is the accusation made by Descartes against Regius, on the question of plagiarism, in the *Fundamenta Physices*, which appeared in 1646,[15] of what he considered the 'best part' of

[14] On Regius, see *Lettres à Regius et Remarques sur l'explication de l'esprit humain*, ed. Geneviève Rodis-Lewis Paris: Vrin, 1959; Delphine Kolesnik-Antoine (Antoine-Mahut), "La question des passions chez Regius et Descartes. Premiers éléments d'interprétation", in *Azimuth*, Storia e Letteratura, "The Domain of the Human. Anthropological Frontiers in Modern and Contemporary Thought"; "Il dominio dell'umano. Frontiere antropologiche tra moderno e contemporaneo". Dir. Simone Guidi, 2013, 13–32; Delphine Kolesnik-Antoine (Antoine-Mahut), "Les expériences physiologiques chez Henricus Regius: les pierres lydiennes du cartésianisme ?", in *Journal of Early Modern Studies*, II, "The creative role of experimentation in Early Modern Science". Dir. Dana Jalobeanu, April 2013,125–145; Delphine Antoine-Mahut, "Peut-on à la fois être cartésien et sceptique?" in *Pour et contre le scepticisme. Théories et Pratiques de l'Antiquité aux Lumières.* Elodie Argaud, Nawal el Yadari, Sébastien Charles and Gianni Paganini eds. Paris: Champion, 2015, 55–70; Delphine Bellis, "Empiricism without Metaphysics: Regius' Cartesian Natural Philosophy", Mihnea Dobre and Tammy Nyden (eds.), *Cartesian Empiricisms*, Springer: Studies in History and Philosophy of Science, 2013, 151–183; Roberto Bordoli (a cura di), *René Descartes. Henricus Regius. Il carteggio. Le polemiche*, Naples: Cronopio, 1997; Erik-Jan. Bos in *The Correspondence between Descartes and Henricus Regius*, Zeno, The Leiden-Utrecht Research Institute of Philosophy, 2002; Erik-Jan. Bos, "Henricus Regius et les limites de la philosophie cartésienne", in: D. Kolesnik-Antoine (Antoine-Mahut) (ed.), *Qu'est.-ce qu'être cartésien?*, Lyon: ENS editions, 2013, 53–68; Robin Buning, "Henricus Regius and the earliest teaching of Cartesian philosophy at Utrecht University", Delphine Kolesnik-Antoine (Antoine-Mahut) and Catherine Secrétan (eds.), *Les Pays-Bas aux XVIIe et XVIIIe siècles*, Paris: Champion, 2015; Desmond Clarke, "The Physics and the Metaphysics of the Mind: Descartes and Regius", John Cottingham and P. Hacker (eds.), *Mind, Method and Morality: Essays in Honour of Anthony Kenny*, Oxford: Oxford University Press, 2010, 187–207; Paolo Farina, "Sulla formazione scientifica di Henricus Regius: Santorio Santorii e il *De statica medicina*," *Rivista Critica di Storia della Filosofia*, 30 (1975), 363–399; "Il corpuscolarismo di Henricus Regius: materialismo e medicina in un cartesiano olandese del seicento," in Ugo Baldini (ed.), *Ricerche sull'atomismo del Seicento* (Firenze, 1977); T.P. Gariepy, "Mechanism without Metaphysics: Henricus Regius and the Establishment of Cartesian Medicine", Yale University, 1990; Gideon Manning, "Naturalism and Un-naturalism Among the Cartesian Physicians," *Inquiry*, 51/5 (2008), 441–463; Karl E. Rothschuh, "Henricus Regius und Descartes. Neue Einblicke in die frühe Physiologie (1640–1641) des Regius", *Archives internationales d'histoire des sciences*, 21, 1968, 39–66; Theo Verbeek (ed.) *Descartes et Regius. Autour de l'Explication de l'esprit humain*, Amsterdam/Atlanta, Rodopi, coll. "Studies in the History of Ideas in the Low Countries", 2, 1993; Theo Verbeek, "The Invention of Nature. Descartes and Regius," in Stephen Gaukroger, John Schuster and John Sutton, eds., *Descartes' Natural Philosophy* (London, 2000), 149–167; and Catherine Wilson, "Descartes and the corporeal mind. Some implications of the Regius affair", in *Descartes'Natural Philosophy.* Edited by S. Gaukroger, J. Schuster and J. Sutton, Routledge: London and NY, 2000, 659–679.

[15] *Fundamenta Physices*, Amsterlodami, Ludovicum Elzevirium, 1646. The text is revised and completed in 1654 in the *Philosophia naturalis, editio secunda, Priore multo locupletor, & emendatior*, Amsterdam: Ludovic Elzevier; then in 1661: *Philosophia naturalis; in qua tota rerum universitas, per clare & facilia Principia, explanatur*, Amsterlodam, Ludovic & Daniel Elzevier. There is a known French translation attributed to Claude Rouxel: *Philosophie naturelle*, Utrecht, Rodolphe van Zyll, 1687.

L'Homme , namely the explanation of muscular contraction, which led Descartes to make a public statement, in the letter-preface to the *Principles of Philosophy* (1647), on what he considered to be authentically 'Cartesian'.

Starting from these two criteria—the precise restoration of the mechanism of the movement of the muscles and the relation between this physical explanation and the metaphysical principles ensuring the separation of mind and body—he hopes to give his "nephews" the key to understand it. But he also makes of the invisible text of *L'Homme* a plot line from which it will be hard to escape.

The framework of this plot line can be described through a threefold mediation: the mediation of *copies* (how many are there? Through whose hands did they pass and when? Where is the original?); the mediation of *discourse* (who to believe? The accusers or those defending themselves? What are the respective roles of those who entered into the battle and took part on one side or the other?); finally, the mediation of posthumous *editions* and especially Clerselier's edition which is itself derived from the preceding meditations.

From Descartes himself up to the most recent receptions, passing through long-term university institutionalisation, from a spiritualist model of Cartesianism, *L'Homme* is thus at the heart of a blurring which mixes the real and the imaginary. It was important to ask up to what point the epistemological project of restoring the true history of this text was plausible and what weight the historian of Cartesian ideas should place, in this history, on the effects of the real produced by the imaginary. For despite the deformations and about-turns, etc., to which it is sometimes necessary to submit our initial theses,[16] these effects contribute to attesting to the power and the topicality of these theses.

The discussion was opened during the international colloquium that I organised at the ENS de Lyon on 16 and 17 January 2014 as part of the ANR research programme 'Anthropos'[17] while Stephen Gaukroger was visiting professor at LABEX (*laboratoire d'excellence*) 'COMOD' (*Constitution et Origines de la Modernité*, ENS de Lyon).[18] It was subsequently pursued via different channels and resulted in this collective volume, which is a much larger re-worked version of the original programme.[19]

In adopting a perspective internal to Cartesian philosophy or one studying reception, each of the contributions aims, in an original way, to write a chapter of this story.[20] It remains for us, in this introduction, to lay the theoretical framework, together with those factual elements needed to explain as clearly as possible the voice of *L'Homme*.

---

[16] For a theorisation and application of the history of philosophical ideas taking charge of the Descartes "ad hocs" produced by these receptions, see F. Azouvi, *Descartes et la France. Histoire d'une passion nationale*. Paris: Fayard, 2002.

[17] anthropos.ens-lyon.fr

[18] www.labexcomod.eu

[19] We would like to thank Springer for their remarkable hospitality.

[20] When we refer to one of this volume's contributions, we will give, in brackets, the name of the author in italics followed by an asterisk.

The story of *L'Homme* is a true novel. Its plot weaves itself along three intersecting points: an *unfinished* text, a *copied or plagiarised* text, and a *corrected* text.

1.2 *L'Homme*: An Unfinished Text

The writing of the text is generally placed in the early 1630s. Descartes nevertheless abandoned the project of finishing and publishing it on learning of the condemnation of Galileo in 1633. In fact, *L'Homme* constitutes Chapter 18 of *The World* or the *Treatise on Light (Gabriel Alban-Zapata*)*. It is the application, to the living body, and particularly to the human body, of the same explanatory principle of the whole as that which has been demonstrated heuristically in accounting for the principal changes in the physical universe.[21] In Descartes' work, the extension of the mechanisation of physics to physiology proceeds from an application of the laws of nature and its rules of collision—theorised in *The World* and later developed in *Principia Philosophiae*—to the study of the living body, especially to large warm-bloodied animals with lungs, in which the regular motion of the blood marks each of them out as *this* body, circling in rings in the way of the earth's water. What then sets the human body apart is the joining of a rational soul to this machine. The end of the treatise specifies that it will examine the modalities of this union, and this is the point at which the text ends.[22] Finally, that human reason is open to registering various sensible effects up to their insensible causes signifies that, acting on a living body, it can and must undergo an embryogenesis allowing it to follow the evolution of the body in time, from the beginning of the mechanism up until its rupture, by resorting solely to the explanatory principles of extension, shape, and motion. This objective is not that of *L'Homme* but will be that of *The Description of the Human Body* written from 1648[23] and which Clerselier published under the title of *A Treatise on the Formation of the Foetus* in the 1664 edition.

The reconstruction of the story (in the sense here of any first narrative) of the text thus begins with what Descartes himself recounts about himself in 1638 in his first public work, as it happens, in Part V of the *Discourse on the Method*. After having

---

[21] We detail the conditions of this application to the human body in "La machine du corps", in *Lectures de Descartes*, eds. Frédéric de Buzon, Elodie Cassan and Denis Kambouchner, Paris: Ellipses, 2015, 229–252.

[22] "Now before I pass to the description of the rational soul, I want you once again to reflect a little on all that I have just said about this machine." (AT XI, 200, Gaukroger, 168).

[23] To explain why he didn't take care of questions relative to generation earlier, Descartes often alleges a lack of experience. See, for example, *Discourse on the Method*, V, AT VI, 45–46: "From the description of inanimate bodies and plants I went on to describe animals, and in particular men. But I did not yet have sufficient knowledge to speak of them in the same manner as I did of the other things - that is, by demonstrating effects from causes and showing from what seeds and in what manner nature must produce them. So I contented myself with supposing that God formed the body of a man exactly like our own."

rallied behind Harvey's discovery of the blood[24] and tied it to 'the rules of mechanics' and to the respiratory function; after having then set out in a synthetic manner the principal characteristics of 'the fabric of the nerves and muscles'; after then having distinguished animal machines and 'true men' which are distinguished by a rational soul which, 'by the way', specifies the actions of the body; after all these he tells us that in *L'Homme* he has examined in what way the soul comes to be joined to and united with the body-machine:

> After that I described the rational soul, and showed that, unlike the other things of which I have spoken, it cannot be derived in any way from the potentiality of matter, but must be specially created; and I showed how it is not sufficient for it to be lodged in the human body like a helmsman in his ship, except perhaps to move its limbs, but that it must be more closely joined and united with the body in order to have, besides this power of movement, feeling and appetites like ours and so constitute a real man.[25]

Descartes thus presents to his reader the project of *L'Homme* as an unpublished but completed project. For this *a posteriori* reconstruction, clearly unverifiable by the reader, has a very strong theoretical benefit: it reinserts the description of the principal functions of the machine in a *historia* of the mind which begins by doubting everything, discovering at the same time the foundational existence of his thought and the distinction between it and his body, and it makes, from the examination of the modalities of the union, a subsequent step, in the order of reasons, with the discovery of this distinction and the autonomous functioning of the body which results from it: autonomous in that there is no recourse to a soul which would account for its life.

This story then continues with other texts, most often also incomplete and unpublished, which Descartes tells us—or which commentators tell us (the coincidence of the two is not always precise)—constitute a rewriting or continuation of the aborted project of *L'Homme*.

The correspondence with Elizabeth of 1648 mentions a treatise 'on the generation of animals',[26] and it is often concluded that this refers to the same text as that which Clerselier published under the title of A *Treatise on the Formation of the Foetus*. Here Descartes will correct or at least complete the perspective of *L'Homme*, abandoning the synthetic point of view of description of parts in favour of a more

---

[24] The work *Exercitatio anatomica de motu cordis et Sanguinis in Animalibus* was published in Frankfurt in 1628. Descartes reproduces Harvey's overall explanation but inverts the diastolic and systolic phases. On this point, see notably Annie Bitbol-Hespériès, *Le principe de vie chez Descartes*, *op.cit.*, and François Duchesneau, *Les modèles du vivant de Descartes à Leibniz*, *op.cit.*

[25] AT VI, 46.

[26] To Elizabeth, 25th January 1648, AT V, 112: "I am now working on another manuscript, which I hope Your Highness may find more agreeable. This is a description of the functions of animals and of man. The draft I made about twelve or thirteen years ago (which Your Highness has seen) fell into the hands of some people who transcribed it badly, and I thought I ought to put it in order - that is to say, rewrite it. Just in the last eight or ten days I have even ventured to try to explain the manner in which animals develop from the very beginning of their existence. I say 'animals' in general, for I would not be so bold as to undertake such a thing in the particular case of man, because I simply do not have a sufficient number of observations for such an undertaking."

analytic approach, which highlights their formation and is made possible by the time elapsed and the observations carried out between the two versions of the text. In this way, the mechanism was absolved of any blame for failing to provide a dynamic description of the generation of living things.

In some cases, it is a different treatise, according to him, completed, published and renowned, which is taken as having completed *L'Homme* as well as the *Principles of Philosophy*: the treatise on the *Passions of the Soul* (1649). It will complete it, in the dual sense that it will synthesise the major physiological achievements in the first part, and will further develop the unifying dimension that *L'Homme* lacks, which Descartes claims to have completed in his account in the *Discourse on the Method*, but which the *Principles of Philosophy* themselves (1647) still present as being what will furnish the material for Parts V and VI, which are still to come. The most successful work on the comparison of the two texts to date is the two volumes of Denis Kambouchner, *L'Homme des passions* (Paris: Albin Michel, 1996).

Each of these hypotheses indicates that, from the perspective internal to Descartes' work, *L'Homme* constitutes an important part of the history that the public and official version will nevertheless come to correct and complete.

It must be added to the complexity of these different reconstructions, which began with Descartes himself, that *L'Homme* is also a text that is largely *copied or plagiarised*.

1.3 *L'Homme* Copied and Plagiarised

What is known thus far about the history of the copies? The first reference work on this question is the impressive bibliography of Descartes' works (1637–1704) edited by Matthijs Van Otegem in 2002 (Zeno, The Leiden-Utrecht Research Institute of Philosophy). Chapter 9 of volume 2 is devoted to *L'Homme*. It provides both precious information and important motifs of perplexity.

In collating the texts where Descartes mentions the manuscript[27] and the copies of *L'Homme*, and testimonials such as the letter of 8 April 1642 from Heereboord to Colvius, or the letter of 9 November 1662 from Chapelain to Bernier, and the prefaces to the Latin and French editions of 1664, we obtain the following: three copies (from Van Zurck, Pollot and Heereboord[28]) were circulated, one from Descartes'

[27] Notably the letter to Mersenne in November 1646, in which Descartes states: "It is now twelve or thirteen years since I described all the functions of the human or animal body; but the manuscript is in such a mess that I would be hard put to it to read it myself. Nevertheless, four or five years ago I could not avoid lending it to a close friend, who made a copy which was then recopied by two more people, with my permission but without my rereading or correcting the transcripts" (AT IV, 566–577).

[28] Here we reproduce the very helpful notes of the Biographical Lexicon by Erik-Jan. Bos, *The Correspondence between Descartes and Henricus Regius*, *op.cit.*, 249–256. They give us some idea of the variety of Descartes' visits. Anthony Studler van Van Surck ou Zurck (1608–1666) "was a correspondent of a close friend of Descartes; his acquaintance with the Frenchman dates from

manuscript and two copies of this copy. And if one takes seriously the accusations of plagiarism made by Descartes against Regius (we will return to these shortly), the existence of a fourth copy cannot be excluded. Finally, it is known, from the letters addressed to her, that Princess Elizabeth had at least seen a copy or held one. This could have come from Descartes, or even Pollot, who we know served as an intermediary between Descartes and her.

Schuyl used the copies of Van Zurck and Pollot. He compared the two versions and made a Latin translation of the text as it was in 1642. And he revised this translation in a 1664 edition (which Adam and Tannery do not mention), notably using a copy (therefore the third) provided by Heereboord.

However, the 1664 edition revealed a text that was different from that of 1662. The possibility cannot be excluded that Descartes reworked *L'Homme* after the early 1630s and that the composition of the original itself should itself placed in a story that can now be understood as a chronology. These different stages would explain the differences between the two Latin editions and the French edition of 1664 (Clerselier's) and the Latin edition of 1662 (Schuyl's); a new rivalry could even be imagined between the two 1664 editions. But at the same time, passages in the correspondence with Elizabeth (1648), in which Descartes explains that he is working on a new treatise on the generation of animals, could refer equally to the *Treatise on the Formation of the Foetus* as to a reworking of the original text of *L'Homme*, the publication project of the first which would then compensate for abandoning the latter for a radically different text.[29]

Finally, let us turn to Clerselier's text. We cannot establish definitively that Clerselier had a copy of *L'Homme* before 1657, the date of the appearance of the

---

1633, when they both lived in Amsterdam (AT I 268–269). In October 1633, he went to Leiden to study law, and he matriculated again at Leiden University in 1636 and 1639. He was Lord of Sweyburg and Bergen (from 1640), Knight of Holland, and 'Hoogheemraad van de Uitwaterende Sluizen'. He acted as Descartes' banker in Holland".

Jean-Alphonse Pollot (1603–1668) "Pollot joined the Dutch States' army at the age of 17. Despite the loss of his right arm during the siege of 's-Hertogenbosch in 1629, he stayed in the service of the Dutch army. In 1633 he was appointed a captain. He was also known as 'Monsieur Alphonse" to distinguish him from his brother Jean-Baptiste Pollot (who died in 1641), whom he succeeded as chamberlain of Frederik-Hendrik in 1642. After the Stadholder's death, Amalia van Solms appointed him her personal steward in 1648. He returned to Geneva in 1659. He was a close friend of Descartes', who profited from his connections in The Hague when the Utrecht Vroedschap intended to put him on trial in 1643″.

Adriaan Heereboord (1613–1661) "studied from 1631 until 1637 at the Statencollege in Leiden. In 1640, he was appointed associate professor of logic at Leiden University, receiving the degree of *magister philosophiae* from Golius in February 1641. In 1643 and 1644, he held disputations both pro and contra Descartes and Regius (…), but Descartes records his pro-Cartesianism in a letter to Pollot of 8 January 1644, adding that in his most recent disputations '(Heereboord) s'y declare plus ouvertement pour moi, et me cite avec beaucoup (plus) d'éloges, que n'a jamais fait Mr. de Roy (Regius)' (AT IV, 77). Heereboord's staunch defence of the New Philosophy met with serious opposition from the professors of theology Jacob Trigland (1583–1654) and Jacob Revius (1586–1658), and he professor of philosophy Adam Stuart (1591–1654), which battle resulted in the so-called Leiden Crisis in 1647".

[29] For his part, M. Van Otegem does not clear up this ambiguity.

last volume of the correspondence. The Elzeviers had planned to publish a preview of the text,[30] but we do not know why this project was abandoned. Clerselier says nothing about these events after 1660, whether the difficulty lay in finding people competent to produce the figures (on the figures, see *Steven Nadler**). He says he sent a copy to Huyberts and one to Gutschoven, and maintains, particularly in criticism of Schuyl's edition, that he holds the original manuscript of the text (the same claim on Schuyl's part becomes invalidated at a stroke). But what does this mean, if what he says is true?[31] Is it the 'draft' mentioned by Descartes in the November 1646 letter to Mersenne? All we can conclude with certainty is that Clerselier's copy contains the chapter numbers of the *Treatise on Light*, that it thus refers to a state of the text anterior to the 1642 version used by Schuyl. M. Van Otegem's conclusion then makes sense: 'It can be concluded that the authority of Clerselier's *Treatise on Man* (1664) should not be taken for granted and that for the study of the text one cannot confine oneself to this edition alone' (op. cit., 491).

Continuing the work of F.A. Meschini (*Franco-Aurelio Meschini** explores other angles), who himself makes use of A. Bortolotti's work,[32] who insists that, if we are to establish a reliable edition of *L'Homme*, we need to compare the 1662 and 1664 versions but also to integrate what we can learn from the Dutch translations, which were able to use editions other than Clerselier's. The aim is to provide sanction for the idea, raised but not definitively assumed as such by Van Otegem, of a reworking of the text of *L'Homme* by Descartes himself with a view to publication. F-A. Meschini interprets the letters to Elizabeth of 6 October 1645 and 31 January 1648, and the 23 November 1646 letter to Mersenne as referring to *L'Homme*, which Descartes took up again with a view to publishing it. This would explain the difference at the beginning of the text between the two Schuyl editions: Schuyl's copy integrates this later stage of drafting, whereas Clerselier maintains that this is pure reconstruction.

As far as the copies are concerned, Meschini relies on Descartes' texts on Regius to defend the existence of a copy consulted by Regius in 1646, the date of publication of the *Fundamenta Physices*. And he stresses the importance of Elizabeth in the eventual dissemination of the text, referring us to a note published by Sylvain

[30] The Elzeviers are a renowned family of Dutch typographs of Brabant origin (from Leuven), active during all of the seventeenth century, principally in Leyde and Amsterdam.

[31] Concerning a "false letter" composed by Clerselier, see Giulia Belgioioso, "Un faux de Clerselier", *Bulletin cartésien*, XXXIII, 2005, in *Archives de Philosophie*, 2005, 68, 1, 148–158, also consultable at www.cartesius.net; and her article "Les 'correspondances' de Descartes" in *DesCartes et DesLettres. 'Epistolari' e filosofia in Descartes e nei cartesiani*, cura di Francesco Marrone, Le Monier, 2008, 8–32.

[32] F.-A. Meschni, "Filologia e scienza. Note per un' edizione critica de *L'Homme* di Descartes", *Le Opere dei Filosofi e degli scienziati. Filosofia e scienza tra testo, libro e biblioteche*. A cura di F-A. Meschini. Con la collaborazione di F. Puccini. Leo S. Olsschki, 2011, 165–204 and A. Bortolotti, "I manoscritto di Descartes nella seconda metà del' 600", *Rivista du storia della filosofia*, IV/1987, 675–695.

Matton in the *Bulletin Cartésien* XXXVI.[33] This testimony, attributed to Urbain Chevreau, precisely states the following:

> At this time, I said to [Chanut] that I had a *Treatise on Man* by M. Descartes, which he had passionately sought in vain for the works of this philosopher, whom he preferred to the ancients and the moderns. He asked me more than once how this treasure had come into my hands; and I replied (…) that I had had it from M. de la Voyette, gentleman to the Queen, who had been Page to the Prince of Orange, and that he had it from Madame Princess Elizabeth, famous pupil of M. Descartes. When he had looked at the first leaves, he begged me earnestly to lend it to him, with the promise of returning it to me from Hamburg, where he would have it copied with the greatest diligence. Being in Paris, he shared it with M. de Clerselier, his brother in law, who put this very badly copied manuscript in order, and he has since communicated this to others and in the preface to the treatise which was printed, whence one can see the fate of this manuscript, although there is no mention of me or of M. de la Voyette.

Taking our starting point from this, and distancing ourselves from Meschini's hypothesis, we can still imagine that Elizabeth herself was able, when meeting with Regius in 1643 or a little after,[34] to show him this copy. This would be a supplementary scenario covering the story understood as a chronology of facts with the story envisaged as a narration.

But the reopening of the plagiarism dossier leads to another surprise. For it allows us to propose the opposite hypothesis, which invalidates the thesis of Regius' plagiarism of *L'Homme*.

On Descartes' side, the dossier comprises the letters to Mersenne of 5 October and 23 November 1646 (AT IV, 510–511); to Huygens of 5 October 1646 (AT IV, 517–518); and to Elizabeth of March 1647 (AT IV, 626–627); as well as the letter-preface to the French edition of the *Principles of Philosophy*, which publicly confirms the rupture with Regius.

The terms of the accusation are these: Regius procured, without the knowledge and against the wishes of Descartes, a copy of *L'Homme*. He plagiarised, in his *Fundamenta Physices*, the thesis which Descartes held most strongly because he considered it to be paradigmatic of the functioning of the global machine: that of the correspondence between antagonistic muscles, but mistaking the role of the valves (in preventing the depletion of animal spirits), or even missing the very existence of the valves. Finally, Regius cut the tree of knowledge from its metaphysical roots, by reducing what the human mind is able to know rationally in the domain of natural philosophy, and relegating any other form of metaphysical knowledge to the register of supernatural revelation.

These accusations then become established readings posing as 'characters' in this Cartesian novel. Beginning with Clerselier's insinuations in the preface to his

[33] "Un témoignage oublié sur le manuscrit du *Traité de L'Homme* de Descartes", *Archives de Philosophie*, volume 71, book 1, Spring 2008, 148–149.

[34] On the theoretical affinities between Elizabeth and Regius, see Delphine Antoine-Mahut, "Élisabeth philosophe: un cartésianisme empirique?", in *Élisabeth face à Descartes: deux philosophes*? Marie-Frédérique Pellegrin and Delphine Kolesnik-Antoine (Antoine-Mahut) eds., Paris: Vrin, 2014, 119–138.

1664 edition, proceeding through Baillet up to T-P. Gariepy, in his PhD thesis devoted to Regius, incorporates the plagiarism thesis without questioning,[35] Regius has become designated the renegade looter, when he is not being described as too stupid to understand the subtleties of Cartesian metaphysics. From reading to reading, the saga of a plagiarised Descartes is amplified, distorted and widely reported.

The work of T. Verbeek and E-J. Bos allows us to go back to these clichés. In his introduction to his edition of the correspondence between Descartes and Regius (Zeno, 2002), Bos particularly notes the words of a certain Carolus Fabricius, of whom we know almost nothing, and there is nothing, according to Bos, that prevents us from identifying him with Regius himself. This text is the preface to the second edition of the *Brevis explication mentis humanae*,[36] which is a reply to the *Notae in programma parues* in 1648. Its interest lies in the fact that it reverses all the terms of the accusation of plagiarism, in order to direct them at Descartes himself, thus constructing for posterity a wholly different history of *L'Homme*:

> Those who use their eyes in examining the construction of the muscles as published by Regius in his Physics (*Fundamenta Physices*, 1646, *Philosophia Naturalis*, 1654), and read his description, will see clearly that Regius has taken fully into consideration and sufficiently distinguished the different pores, valves, turning first from the brain to the muscles, then from antagonistic pairs of muscles, preventing the return of spirits into the brain, and alternating between one muscle and another and functioning by reciprocal movement. This is why, just as these slanderers would like to say that Regius had plagiarised the writings of Descartes on animals, even though he had never been aware of them (as Regius has in any case always said), there is every right to say the same thing not of Regius but of Descartes, namely that he has appropriated the entire physiology of Regius (i.e. the theses of the *Physiologia*, published in 1643), which he has known for years, and put its contents to his own purposes.

Whatever the identity of the author of this text, it is of interest to note here that he has never been reported in Cartesian historiography. We draw on it as supplementary information for our novel: the authority of its narrator is a key element in what has survived of this or these stories. From this point of view, Descartes has, incontestably, been considered a better scriptwriter than Regius.

There remains one final point to outline that concerns the various processes surrounding and supporting the successive editions of *L'Homme*. For *L'Homme* is a *corrected* text or at least one shaped by reading instructions from its editors, whose interpretation of its meaning is maximally prescribed by them.

[35] *Mechanism without Metaphysics: Henricus Regius and the Establishment of Cartesian Medicine*, *op.cit.*

[36] The full title is: *Brevis explicatio mentis humanae sive animae raitonalis; antea publico examini proposita, et deinde opêra Henricii regii Ultrajectini nonihil dilucidate, et à notis Cartesii vindicata*, Utrecht, G. van Zijl, 1657.

1.4 *L'Homme* Corrected

The first of these frameworks of interpretation, and to date still the perennial one, is that of the Clerselier edition. One can thereby synthesise the hermeneutic mechanisms put in play by each of the texts framing *L'Homme* in this edition: Clerselier's preface, the French translation of Schuyl's Latin preface by Clerselier's son, and the *Remarques* of Louis de La Forge.[37]

Clerselier's approach is contemporaneous with the placing of the works of Descartes on the Index, *donec corrigantur*, starting in 1662 (on the reception of *L'Homme* in this context, see *Domenico Collacciani**) particularly because of the critique of substantial forms and the assimilation of the essence of body to extension.[38] He therefore addresses himself to two sorts of readers: 'All those who believe that the soul of man is corporeal, or who hold that it is just a disposition or even a temperament of the body', and 'those who want the soul to be corporeal, but who deny that it is an assembly or temperament of the body, although they speak a little differently from the others, yet nevertheless fall back on this belief.' It is particularly a question of readers of the *Meditations* being stopped at the beginning of the traversal of the mind, i.e. at the moment when it is imagined to be 'air, or fire, or some other body'. Clerselier addresses himself particularly to these latter readers, because he judges that it is 'principally' these who Descartes targets in *L'Homme*: physicians, and those physicians who, while profiting immensely from the eradication of all occult qualities from the human body, do not hesitate to stress the mismatch between more recent observations and Descartes' principles. It is also the reason why he has chosen other physicians in order to respond to these physicians (on the reception of *L'Homme* by Dutch physicians, see *Tad Schmaltz**).

Given this readership, the crucial thing is to forcefully call to mind these principles, so as to pre-empt wrong interpretations of the text of *L'Homme*, i.e. interpretations that we would today count as materialist. Clerselier's whole aim consists in basing the physiological developments of the treatise on the thesis, which comes later in Descartes' work, on the substantial distinction, and mobilising texts of

---

[37] I began the analysis of these processes in "Les voies du corps. Schuyl, Clerselier et La Forge lecteurs du traité de *L'Homme* de Descartes", in *Consecutio temporum* (consecutiotemporeum.org), *Rivista critica della postmodernità*, « Corpo, desiderio, lavoro: per un nuovo materialismo », eds. Francesco Toto and Roberto Finelli, n°2, February 2012. I synthesise, complete and reorganise them here.

Louis de La Forge was born in 1632 in La Flèche and died in 1666 in Saumur. Son of a doctor and himself a medical doctor, he was a grand admirer of Descartes and took upon himself the task of completed the master's work from principles theorised by the latter. He is thus the commentator and one of the illustrators of *L'Homme* in Clerselier's edition of 1664. He is also the author of the *Treatise on the Human Mind* (1666), in which he is especially at pains to develop the considerations relative to the union of the body and soul.

[38] On this dossier of condemnations, see John Cottingham, Roger Ariew and Tom Sorell in *Descartes' Meditations background source materials*, Cambridge: Cambridge Philosophical Texts in Context, 1998.

Augustine to establish a consensual authority.[39] Here Clerselier very clearly follows the instructions required by the reconstruction of the project of *L'Homme* in the fifth part of the *Discourse on the Method* (the substantial distinction between mind and body, then the descriptions of the principal functions of the soul, and then the description of the union) and then, through subsequent editing, in reply to the assaults of the 'Querelle d'Utrecht', of the *Meditations*. Consequently it makes *L'Homme* a text that not only reveals its meaning only once the distinction of later texts is integrated into it, but which forms a path towards the more correct notion of the union of the later texts. By this it sets out to correct the errors that lie in Schuyl's Latin preface, and it prescribes La Forge's *Remarques* as a more precise response to physicians in developing the unitary thesis.

Schuyl had in fact tried to reconstruct the final argument of the treatise, which led on his account to the *Meditations*. He wanted to show the reader the orderly way going from the rejection of the animal soul and the vegetative soul as explanations of life, to the promotion in man alone of a rational intellective soul actually distinct from body. To do this he began a new development, based on the progress of knowledge on the functioning of plants, between the 1630s and the end of the 1660s. Then he considered animals, and finally man. In Schuyl's economy of arguments, the focus on animals and on the absence of a soul in them (The argument on the soul of animals had equally been made public in the fifth part of the *Discourse* and in the correspondence), serves thus to stem any temptation to make a direct leap from physics to man. It protects and encloses the exceptional character of the 'true man' in Cartesian ontology. But according to Clerselier, a *historia* of the mind taking physics as its point of departure, and not the substantial distinction, has already historically revealed its dangers; the condemnations at the beginning of the 1660s constitute a direct result of it. The temptation of this *historia* had been very clearly formulated by friends of Mersenne:

> However much we ponder on the question of whether the idea that we have of our mind, i.e. the notion or the concept of the human mind, contains nothing of the corporeal in it, we do not dare nevertheless to assert that thought cannot in any way belong to a body subject to some sort of motion. (*6th Objections*, AT IX-1, 223-224

And what did Descartes reply to Mersenne? He repeats the opposite *historia* 'on the basis of arguments set out in the *Meditations*', which, starting from the discovery of the mind and its modes, step by step definitively establishes, metaphysically, the principle of the real distinction between substances 'by following the rules of logic' (idem, 238).

Clerselier next thought to pursue the Cartesian project by inserting[40] a number of details insisting on the substantial distinction into the French translation of Schuyl's Latin preface. In this way, he places the critique of animals' souls in continuity with and not as the principle of this substantial distinction. But at the same time this

---

[39] Several works have been dedicated to this question on the relations between Descartes and Augustin, in particular Emmanuel Faye's collective on "Cartésianisme et augustinisme" (*Corpus*, 2000), that provides important bibliographical indications.

[40] Through his son at least. I give examples of these insertions in "Les voies du corps", *op.cit.*

allows us to understand better the originality and relevance of Schuyl's reading: it is precisely because *L'Homme* shows that the bodies of animals function very well without a soul that it is necessary to restore the missing link of the critique of animal souls, in order to stress the specificity of the human body in Cartesianism and the subsequent joining of the soul to this machine, without wrongly joining it to bodies other than human ones. In other words, what Schuyl is seeking to relate to us is what Descartes has truly done, without imputing to him what he hasn't done, in any case in the text before our eyes, and that only an *a posteriori* reconstruction, based on another context and informed by other readings, is able to envisage it thus:

> always trying to abridge his best work as much as possible, as in the his *Méditations métaphysiques*, he is content to show clearly that out soul and its thoughts can never follow from and be dependent on matter, without having fun refuting the absurdities that would follow from the contrary view. Similarly in this treatise (*L'Homme*), without stopping to uncover the errors into which many have fallen, due to their not distinguishing the body and the soul, he is content to explain how corporeal motions arise in our body; and how animals, who have no knowledge, can nevertheless, solely by the disposition of their organization, do things which many, without taking care, and misled by a false and deceptive appearance, have thought cannot be achieved except by the intermediary of a knowing soul.

It remains to analyse La Forge's framing of and corrections to *L'Homme* (*Philippe Drieux**, *Emanuela Scribano**). He presents himself in the role of disciple with a fourfold objective: *to supplement, correct, link, and apply*.

Whereas Clerselier focused on the distinction, and because Schuyl followed a dangerous path in reconstructing the aims of the treatise, La Forge's primary task is to make good 'Descartes' failure' to do so on the question of the union of mind and body. This will be the particular subject of another text that he wrote at the same time as his *Remarques* but which was published 2 years later: the *Treatise on the Human Mind.*[41] But the single preoccupation is reflected in his *Remarques* in the defence of certain nodal points of Cartesian physiology. This is above all the case with the pineal gland, whose principal function in Cartesian anthropology is to assure the transmission or passage between the soul and the body. La Forge here reproduces in the case of the gland the reverse symmetrical gesture that Clerselier makes with regard to the substantial distinction in the case of the bodily mechanism: he reads the developments in *L'Homme* in the light of those in the first part of the *Passions of the Soul* and passages from the correspondence made public in Clerselier's edition, importing *a posteriori* the single theory that is lacking in the physiology of the treatise. Far from being contrary to the anatomical facts, as the physician Bartholin and his followers had maintained,[42] the unobservable nature of

[41] Desmond Clarke offers an excellent presentation and translation in English: *Treatise on the Human Mind*, ed. and trans. by Desmond Clarke (Dordrecht: Springer 1997).

[42] Caspar Bartholin (the Ancient) was born in February 1585, 12th in Danemark (Sweden today) and he died in July, 1629, the 13rd. He publishes the *Anatomicae Intitutiones Corporis Humani* (1611). He has two sons: Thomas Bartholin (the Ancient), 1616-1680-, Danish physician, mathematician and theologian; and Erasme Bartholin (1625–1698), Danish mathematician. Thomas Bartholin has two sons: Caspar Bartholin the Young (1655–1738) and Thomas Bartholin the Young, historian. Thomas Bartholin the Young (Caspar's Son) publishes *Anatomia, ex Caspari*

the gland now becomes the condition of possibility of its functioning: the more 'immaterial' this gland is, the better it is able to convey a variety of sensory information from the body to the soul, thereby accounting for the complexity of play of the human machine.[43]

Secondly, it is a question of *correcting* the passages or expressions that could lead to confusion, in a polemical context different from that of the elaboration of the treatise, but always bearing in mind later developments in Descartes' thought. Two examples can be given: a metaphysical one, in that it engages the soul in the Cartesian sense, and a purely physiological one, in that the new physics corrects the scholastic metaphysics of entities. Firstly, the word 'idea' is used in *L'Homme* in the physiological sense of impressions tied to animal spirits on the gland. This meaning is clearly incompatible with the purely psychological definition given later in the *Meditations*. And above all, it gives food for thought for those who, reading Schuyl too quickly, would be tempted to jump directly from physics to physiology, and to make an unjustified extension of the physiological meaning to ideas in man. The upshot of La Forge's instructions here, asking for a renunciation of the use of the term as far as *L'Homme* is concerned, is thus identical to that of Clerselier: the term 'idea' in Descartes is reserved for the psychological idea, to avoid establishing relations of paternity between the Descartes of *Man* and the writings of Hobbes, for example[44] (on the Hobbes/Descartes relationship, see *Arnaud Milanese**). Now to an example of the purely physiological meaning: the commentary of article 5, devoted to the explanation of the rarefaction of the blood in the heart. The text of *L'Homme* uses for this the verb 'to exhale' and the word 'vapours'. However, these terms suggest, on the one hand, that a 'virtue' or 'power' of this heart could initiate this exhalation (so that a metaphysics, this time a faulty one, could come to vitiate physiology) and, on the other hand, that Descartes would support the thesis of the

---

Bartholini parentis Institutionibus, omniumque recentiorum et propiis observationibus tertiùm ad sanguinis circulationem reformata (1641 for the first edition, then 1651, 1655, 1666, etc). It was translated in French in 1647, under the title: *Institutions anatomiques de Gasp. Bartholin, augmentées et enrichies (...) par Thomas Bartholin, docteur en médecine et fils de l'auteur et traduites en français par Abr. Du Prat, docteur en médecine* (this translation was established considering the second edition).

[43] The most illuminating passage on this point is *a fortiori* the argument that concludes the response to Bartholin's objections and is anchored in the most recent anatomical observations: "Someone might add, to support Bartholin, that Sylvius found sand and a stone in this gland. It does not matter. Because even if his body became stone, provided he has sufficiently large pores to let in spirits, and he is not big enough to make it sag, and still hold it down, so that it does not cease to be suspended, it will not become the seat of the soul (...) I say more, when in place of the gland there would be no particular body, such as it is of the very gland, and that it would be nothing else than the place of the discharge of the arteries and the lassis choroid, as may have happened in the beginning of the formation of the brain, and in the heads of those where there was said to have been found nothing but water (assuming that these observations are true), this place would not become the seat of the soul".

[44] I develop this point in "Reintroducing Descartes in the history of materialism: The effects of the Descartes/Hobbes debate on the first reception of Cartesianism" in *Cartesian Mind and Nature*, Stephen Gaukroger and Catherine Wilson eds., New York: OUP, forthcoming.

role of the air in cooling the blood in the lungs before it enters the heart. Descartes had explicitly argued the opposite in his correspondence with Plempius at the beginning of 1638.[45] This is once again a case of La Forge correcting Descartes himself, using the letters to Plempius, the fifth part of the *Discourse* and the *Passions of the Soul*, but equally the explanations of rarefaction and fermentation offered in *The World*, later developed in the *Principles*, in order to reconstruct his 'true thought':

> the two drops of blood entering the heart there are rarefied, expanding and heating them, and in this way spread one into the arterial vein and the other into the aorta, just like the way in which milk when heated ferments and spreads over the edges of the vessel (...) Those who have read the author's subsequent writings will not doubt what I say (... or that) the word 'exhale' should be explained as a rushing or spreading, and 'vapours' by a boiling of the blood. Therefore, so that the words in the treatise do not mislead, I believe that we should substitute the expressions I have just used for them.

These explanations are developed in article 6 to account for the case of a child leaving his mother's womb, when the air in the lungs is coarser for some time afterwards. La Forge's third stated objective in his *Remarques* is that of *binding* together the explanations of the treatise with those of the two other treatises, *L'Homme* and the *la Description du corps humain* (*Description of the Human Body*). In short, what is at issue is restoring the consistency and continuity of the synthetic and analytic approaches of description of the human body and its development, exhibiting the heuristic function of mechanism. This is the meaning of the dynamic definition of the machine that he proposes at the beginning of the *Remarques*, which restores the language of the organism in order to specify the living body:

> by the word Machine, one must never understand anything other than a body composed of many organic parts which are united, working together to produce movements, which they would not be able to do if they were separated. I use the term organic parts for all kinds of simple or composite bodies which, being united together, are able by their arrangement, shape, motion, rest, and position to produce the motions and functions of the machine of which they are parts.

Finally, more generally (and this process synthesises the three others), La Forge *applies* what Descartes has done to conjecture 'what he could have done', allowing his followers to 'finish what remains to be described in the formation of the animal.' If it happens that the limits of the explanation of the beginnings of life have not been

---

[45] Plempius to Descartes, January 1638, AT I, 498; Descartes to Plempius, 15 February 1638, AT I, 530 (translation) and Tel VIII-2, 414 and Plempius to Descartes, March 1638, AT II, 54. I analyse the terms and the issues of this exchange in "La machine du corps", *op. cit.*, 246–251. Vopiscus Fortunatus Plemp(ius) (1601–1671) was "born in Amsterdam of a Catholic family. (He) was educated at a Jesuit college in Ghent. He was a student at Leuven, Leiden, Padua and Bologna, where he graduated in 1624. He practised in Amsterdam from 1624 to 1633. In 1634 he was appointed professor of medicine at Leuven University. Descartes and Plemp knew each other personally, and it seems likely that they dissected animals together in Amsterdam in the early 1630s. Plemp's initial rejection of the theory of the circulation of the blood disappointed Descartes, and his apparently unauthorised publication of a shortened version of Descartes' letters to Plemp on the subject in *De fundamentis medicinae* (1638) put an end to their relationship. In the second edition of his work (1644) Plemp accepted the circulation of the blood, be it along Harveyan lines » (Erik-Jan. Bos, *The Correspondence between Descartes and Henricus Regius*, op.cit, 251).

pushed far enough by Descartes, the explanatory principles that he left us with, together with his deductive method, will give his 'nephews' all the means to complete the advance of the sciences and keeping their foundations.

But to correct *L'Homme* in a context other than his own, whether it be that of Descartes himself, raises important methodological questions for the historian of philosophy. For by adding novels to novels and in giving in to the ease of a teleological reading, do we not end up stifling the voice of Descartes? It would in this case be tantamount to renouncing the 1664 edition as a reference edition: it is certainly necessary to do work there, but in order to underline the eventual errors of interpretation and to bring to light the similarities between the different translations or corrections and the processes of genuine betrayal. In short, we would need to return to a version of *L'Homme* that was purified and placed again in the context of the problems of the 1630s, and to establish it in continuity with the Cartesian project of explaining the world (*Annie-Bitbol-Hespériès**).

Yet there are also posthumous readings that correct other teleological illusions, of which we are less aware because they continue to be ours. When he worked on the physiological meaning of the term 'idea', La Forge reminded us that Descartes was first (or at least, also) identified, in a Cartesian context, by this physiological sense,[46] and not at all as a promoter of the innate ideas that remained incomprehensible to an assiduous reader such as Regius. The Descartes of the *Meditations*, or the primarily spiritualist Descartes who reaches us is thus very different, or at least reduced, compared to the Descartes as understood by his contemporaries. And there are details in La Forge on this question that allow us to understand it. We have seen the feedback effect of the criticism of Schuyl by Clerselier, which consisted in bringing to light what the specificity of the 'true' man was for Descartes; from an invisible falsehood mixed with matter ('vegetative' and 'sensitive' souls), we progress towards the sole true immaterial thing which for this reason, can be united with a body, the rational soul. Finally, we have seen that by revaluing the material tenuousness of the pineal gland in the light of the explanation of its role in the functioning of the human union of mind and body, La Forge subtly articulated these two invisibles, the material and the immaterial.

Thus the specificity of this link between the invisible material and the invisible immaterial, i.e. between physiology and metaphysics, in which lies the unsurpassable contribution of these first receptions of *L'Homme*, for the restitution of the authenticity of the Cartesian project and the understanding of the different ways in which it will then be understood, distorted, and criticised. It is at this point that we finish this introduction of interpretative frameworks.

---

[46] See in particular the explanations addressed to the "scrupulous Cartesians" in the preface of the *Treatise on the human mind:* "I also used words which I thought were the most meaningful and intelligible, even specifying the meaning of words such as 'mind' or 'idea' which could have been equivocal. If any overly scrupulous Cartesian is shocked to see that I limited the second term exclusively to the forms of thought of the mind, although Mr. Descartes uses it also to signify the forms of animal spirits (with which these other forms of our thoughts are linked), I ask them to consider that, in a subject as obscure as this and in which misunderstanding was very much to be feared, I could not be too careful in the choice of words nor try too hard to avoid equivocations and disputes over words" (Clarke, op. cit., 5, 6).

1.5 The Visible and the Invisible in *L'Homme*

The scientific context of the early 1660s can be briefly described as that of a delimitation attempt in the field of anatomy, understood as showing the internal structure of the human body by a combinations of eyes and hands in properly conducted dissections,[47] by contrast with a physiology[48] seen as a product of pure rationality and at every moment risking abstraction and invention, and thus error. Physiological reconstructions could be accompanied by metaphysical considerations, although this was not always the case. Metaphysics in itself is not necessarily crucial. But it is understood as the study of realities whose nature goes beyond the physical register, and thus the competence of science. The new anatomy is thus constituted by the double exclusion of domains that may have common elements, or even merge: the domain of 'physiology' and the domain of 'metaphysics'.

From this point of view, and in mapping once more the evolution that we are interested in, we can thematise a reversal of the context of the development of *L'Homme* and that of its posthumous edition. Descartes takes up his pen in a context in which Fernel's distinction between anatomy, considered as the study of the parts of the body 'apparent to the senses', and physiology, conceived as the study of those parts 'known solely by thought', prevails. According to Fernel, it is the study of the 'elements' and the 'faculties' or 'hidden principles' of those parts invisible to the senses, which provide the intelligible key to the global functioning of body, or of physiology as a whole. But these faculties or principles are not necessarily of a material nature: what Fernel calls 'physiology' also deploys considerations of a metaphysical nature, and metaphysics of Aristotelian and Galenist inspiration.

The 'anatomists' to whom Clerselier and La Forge address themselves in the early 1660s reinvest in this Fernellian sense of anatomy. But this is to dismiss physiology understood as a privileged domain of scholastic entities and reclaim the radical homology of the functioning, and thus the explanations, of the visible material and the invisible material. They pursue in this sense, in physiology, the Cartesian gesture of eradicating forms from the physical universe (on Stenon's exemplary case in this history, see *Raphaële Andrault**).

It is here that the essential role of commentators such as La Forge appears, if we are to understand the specificity of the Cartesian contribution, and in particular that of *L'Homme*, in this story. Indeed, anatomists, especially those who were retrospec-

---

[47] On this iatro-mechanistic context, see the excellent edition *Discours sur l'anatomie du cerveau* of Niels Sténon by Raphaële Andrault. Paris: Classiques Garnier, 2009.

[48] The full text of the *Universa Medicina* (3rd edition, the text was published in Frankfurt in 1574) reads: "(…) the human body having been divided by anatomy into its parts that are apparent to the senses, we must move on to those that are only known by the mind, and by continuing further, see of what elements each of these are composed, what is the mixture of elements, what is their temperament, which virtues and faculties are hidden in them, and by which mind and heat they are preserved. And when the resolution of these things will be known, we will then know by their composition what are the efficient causes of all things, and which moods are engendered, what are the functions of each of them, and what is the office of all of them. And we will thus understand all physiology, which teaches by demonstration the knowledge of the natural composition of man".

tively designated 'iatro-mechanists' in the nineteenth century, are the true 'nephews' of Descartes. They used the up-to-date knowledge gained through perfecting optical instruments and the techniques of dissection to go continually deeper into the detail of the explanation of the beginning of life and its development. They pursued the battle against the faculties and occult qualities, and rid the scientific field of its obsolete metaphysical dross. More precisely: this gesture is not completely justified by Cartesian anthropology, or by the exceptional characteristic that reveals human nature in Descartes' natural philosophy, and which the establishment of a strict continuity between *L'Homme* and *Le Monde* could prevent us from understanding. What the first readers of Descartes show is that in some way the complete project of *L'Homme* deploys an invisible which is *at the same time* physical and metaphysical. It helps us to problematise the reasons why *L'Homme* could not be reduced to a treatise on anatomy, without that being any longer a treatise on physiology in Fernel's sense. They make room for the real contribution of Descartes to the scientific 'revolution' that will lead to biology: a conception of physiology, pushing to the furthest limits of human reason, which reveals the autonomous functioning of the human body, thus integrating the progress of anatomy by his 'nephews'; but still leaving an empty space for the variety and complexity of the actions that the human soul can initiate in this machine, or can receive from it, in the exceptional case of the human mind/body compound. This is why posterity has been able to isolate each of these viewpoints, playing off the one against the other. An extreme avatar of the first leads to contemporary reductionism, or to a relegation of Cartesian science towards obsolescence; and a hardened descendent of the second leads to different forms of 'the ghost in the machine'.

In order to specify one last time what Descartes understood by 'man' and by the rational study of this man in *L'Homme*, we will return to two successive receptions: that of the function of the illustrative figures in the 1664 edition, and that of the beginnings of life in the subsequent commentary which Malebranche proposes about the *Treatise on the Formation of the Foetus* in the *Entretiens sur la métaphysique et la religion* (*Dialogues on Metaphysics and Religion*)X and XI (the first edition published in 1688).

Clerselier reports difficulties in finding people willing and able to illustrate *L'Homme*. But what would be a 'good' illustration for *L'Homme*?

The first answer derives directly from the explanations offered by Descartes in article 203 of the fourth part of the *Principles*. Insofar as the material invisible, whose effects can be observed through sensation, is homologous in its functioning with those effects explicable by extension, shape, and motion, these represent this invisible to the reader's imagination allowing him to see what is not seen, thus showing him the causes of these effects. As for the illustrations of the brain, La Forge explains that they are 'presented, not in the same way that they appear, but in the way in which we would see them if our senses were subtle enough to discover them (article 1).

This feature of the illustrations can also help to hierarchise them among one another and to distinguish 'good' figures from even better ones. For what makes the reader 'see' better is the choice to schematise the invisible by mobilising as much as possible elements that are already visible, facilitating the transition from one to the other. In article 24, to distinguish himself from Gutschoven (the other illustrator of *L'Homme*), La Forge justifies his choice of chest muscles rather than those of the eye for the sake of explaining the function of respiration, by the desire to make more visible the difficult but nevertheless heuristic thesis of the correspondence between antagonistic muscles:

> I have not used the preceding figure to explain the way in which respiration operates, although I could, like M. Gutchoven, have done this, because it is good to see that it is not just in the case of the eye muscles, which is what appears from the communication that the author speaks of. This is why I preferred to take the two chest muscles, i.e. the *inferior serratus posticus* and the *serratus major*, whose tendons are clearly antagonistic, and so are better suited to showing the same thing occurs in muscles which we do not see.

But the figures also have another function: that of covering all possible experience, or schematising that which we will never see, at least not in any human body, yet which is needed to explain the various actions of the latter. This is especially the case with the representation of the pineal gland. If La Forge 'has represented it significantly larger than it is in nature, and as anatomists are accustomed to make it in their illustrations', this is in fact

> not only because the reader will be able to understand the text better, but also because it is much larger in a living animal, as represented in the figure, than it is in the head of an animal that is dead, since even Bartholin notes that it seems bigger in cadavers when they are opened promptly, while this cavity is still warm.

The figuration thus integrates three related things: the unobservable nature of the gland in the human body is connected to what one really observes in large warm-blooded animals, and to the differences that we see when we dissect the corpse of a dead animal and a living animal, which leads to them being represented very differently from what is 'natural' in man, that is to say, with a magnifying mirror. But this ideal microscope serves above all to render the *function* of this gland intelligible. It serves to distinguish the Cartesian project from that of 'anatomists' such as Bartholin, who allege that the unobservable nature of the gland in 'natural' man puts the whole of Cartesian anthropology in peril. It serves to make visible in Cartesian physics what the specificity of the human being consists in, namely the union of a soul to a body functioning without this soul but able to enter into a process of interaction with it, something which one cannot find in any other created being. Thus it is precisely the fact that the illustrations in *L'Homme* are not classical anatomical plates that allows them to attain a complete 'intelligence' of the Cartesian project:

> if the figures does not resemble what is in nature, one should not be surprised, since the aim has not been to produce a work of Anatomy, which shows exactly how the parts of bodies lie, and the relation or proportion that they have to one another, but only to explain by these means what M. Descartes sets out in his book, where he speaks most often of things that are not sensible, which have to be made sensible in order for them to become more intelligible. But there is nothing easier than to put them in the natural way, and to conceive of them as they are, once we have considered them otherwise than they are.[49]

---

[49] Sténon perfectly understood the significance of this approach when he writes, in his *Discours sur l'anatomie du cerveau* (*op.cit.*, 92–93): "No one else (other than Descartes) has explained mechan-

But our account on this intimate entanglement of physiology and metaphysics would not be complete without a final detour through its theological reception. This reception allows us to turn to Malebranche.

Why choose Malebranche? In an anecdotal way, but one that has become conical through Fontenelle's eulogy on him, because Malebranche turned to philosophy as a result of having discovered *L'Homme*, the 1664 edition, in a bookshop on the Rue Saint Jacques.[50] If this was a real surprise in the Oratorian's journey, it was because he saw in *L'Homme* and in the *Treatise on the Formation of the Foetus*, in an Augustinian framework, the best natural demonstration of the simplicity and goodness of God's work.

Next, on a deeper level (and this follows from it), because Malebranche was a defender of Cartesian mechanism against all odds. We see him, for example, restoring the Cartesian explanation of muscular antagonism, building on Borelli (on the Italian reception of *L'Homme*, see *Raffaele Carbone**) and in a context where chemical explanations still prevailed,[51] or again defending to the end the function assigned by Descartes to this centre of the brain to which the soul is united, and which anatomists can call whatever they like, so long as this function is preserved.[52] For the mechanist account of the human body demonstrated the simplicity and fruitfulness of the laws, instituted by God within our body, of bodies with one another, and of the relation between souls and bodies.

---

ically all the actions of man, and primarily those of the brain; others will discover man himself; Mr. Descartes only tells us of a machine that nevertheless shows us the inadequacy of what others teach us (...). He knew too well the shortcomings of history that we have of man to begin to explain its true composition. Hence he undertakes not to do it in his *Treatise on Man*, but to explain a machine that makes all actions of which men are capable. Some of his friends explain here a little differently than he; yet we see at the beginning of his book that he heard it that way".

[50] Bernard le Bovier de Fontenelle, *Œuvres complètes*, Volume VI, 1694–1727 (ed. Alain Niderst) Paris Corpus: Fayard, 1994. Malebranche's eulogy is found on pages 337–360.

[51] The Cartesian explanation of muscle contraction serves as a paradigm for the application of the correct method to follow to seek truth, in book VI, Part II, Chapter VIII of *The Search After Truth*. The *De motu animalium* (1680–1681) by Jean-Alphonse Borelli is notably cited in a very positive way in the *Dialogues on Metaphysics and Religion*, X, article VI. Theodore states: "I recently read a book, *Du mouvement des animaux*, that merits examination. The author considers with care the game of the machine necessary to change its place. He explains precisely the force of the muscles, and the reasons for their situation, all by the principles of geometry and mechanics. But what he does not stop at is what is easiest to find in the machine of the animal, he makes known so much art and wisdom in he who formed him and fills the mind of the reader with admiration and surprise".

[52] See notably *The Search After Truth*, II, Part I, chapter 1, §II: "Whether therefore, in the sense of Willis, the two small bodies, he called corpora striata, lies the common sense that brain convolutions conserve species of memory, and the corpus callosum is the seat of the imagination; whether, following the feeling of Fernel in the pia mater, which surrounds the brain substance; whether in the pineal gland of Descartes; or finally in some other unknown part hitherto, that our soul has its main functions, one does not put at a loss. It is enough that a main part; and it is even absolutely necessary, as also that the bottom of Descartes system remains. For it must be remarked that when he was mistaken, as are many of appearance, when he assured that it is in the pineal gland the soul is immediately united, however it could do harm deep in his system, which we still draw all the utility that can be expected of real, to advance in the knowledge of man. "

Malebranche is thus led to unprecedented extensions of the strict Cartesian system. The model of impact is applied to action at a distance between bodies in space, to account for contagion of the air and means and in time, to explain the propagation, via the uterus of the mother, of original sin. But at the same time, these successive extensions became unprecedented in the Malebranchean regime. In an occasionalist system that is primarily concerned with combatting the pride of souls always trying to compete with God, and denouncing the dispersal of divine power in a nature populated by forms and entities, *bodies* become invested with a 'power' that can make them independent of the theocentrism which underlies them, thus making it possible to turn against the system itself.[53]

These difficulties are most manifest in the interpretation of the beginnings of life. The challenge for Malebranche can be summed up as follows: how does one keep the theoretical gains made by the eradication of forms and qualities while preserving the creative act of God of the first principle of metaphysics that he is? The first point becomes clear in the deepening of Descartes' criticisms of Harvey in the *Treatise on the Formation of the Foetus*. The second is represented, from *The Search after Truth*,[54] to the eleventh *Dialogue on Metaphysics and Religion*, through a progressive elaboration of the hypothesis of preformation.

In the fifth part of the *Discourse on the Method*, Descartes publicly supported Harvey's discovery of the circulation of the blood. Nevertheless, when it came to explaining the beginnings of life, he was obliged to disagree with the English physician and his theories. These essentially turned on the order and definition of the phases of systole and diastole. According to Harvey, the active phase of the beat matches the systole, i.e. the contraction of the cardiac muscle, which expels blood into the aorta and into the pulmonary artery. The release phase which follows is the passive phase or diastole, during which the chambers of the heart are enlarged and collect the blood coming from the veins. For Descartes, the opposite is the case:

> Harvey was not so successful, it seems to me, on the question of the heart's movement. For he imagined … that when the heart lengthens, its cavities increase in size, and that when it shortens, they become narrower. I claim, instead, to demonstrate that they become even larger when the heart shortens. (*DCH*, AT XI, 241).

Descartes' aim is to explain the beating of the heart solely through the laws of physics, thus without making an appeal to 'pulsatic or pulsific force':

---

[53] On this point, see Delphine Antoine-Mahut (ed.), *Les malebranchismes des Lumières. Etudes sur les réceptions contrastées de la philosophie de Malebranche, fin XVIIe-XVIIIe siècles*. Paris: Champion, 2014; and Angela Ferraro's thesis, *La réception de la philosophie de Malebranche au XVIIIe siècle. Métaphysique et épistémologie*, defended the 15th April 2016 at the Sapienza University (Rome), under the joint supervision of Carlo Borghero (Sapienza University) and Denis Moreau (University of Nantes).

[54] The reference text on this point, and in which Malebranche links cosmogenesis and embryogenesis to show how Descartes has given natural reason and religion, is *The Search after Truth*, VI, II, IV, OC II (Vrin, Robinet), 343–344. Malebranche reads Descartes by himself, thus employing the same hermeneutic method as La Forge or Schuyl. But he is interested for his part in other texts. Here, in the rare passages in which Descartes refers to religion.

> If we suppose that the heart moves in the way that Harvey describes, not only must we imagine some faculty which causes this movement, but the nature of this faculty is much harder to conceive of than whatever Harvey purports to explain by invoking it. … If instead we restrict our consideration to the expansion of the blood which must follow necessarily from the heat which everyone recognises is greater in the heart than in all other parts of the body, it will be plain to see that this expansion alone is sufficient to move the heart in the way that I have described. (*DCH*, AT XI, 243–244).

One can understand at once what is at stake in the controversy with Harvey on the order of the systole and diastole: Descartes needs the active phase to be the diastole or rarefaction, because it is this, *from the start*, that explains the consecutive propulsion of the blood into the rest of the body. Heat is the principle of life because without it neither the formation nor the regularity or constancy of bodies would be thinkable. Of course, one could seek to go further, even as far as mixing seeds. But the mixing of these two seeds does not allow us to identify that without which life would be unthinkable. What prevails in the mechanist explanation, and takes us back at once to the impossibility and the comparative uselessness of taking things back to the mixing of seeds, is thus the thought that the point of departure of the lawfulness which prevails in the organism. It is here that human reason, or the domain of philosophy, ceases, and all the later discoveries will attest to the heuristic power of an explanation which, as such, will remain unchanged in its principles.

It is here that Malebranche sees an opportunity to intervene. For, in the meanwhile, the context has been reshaped by two new events. Firstly, in 1651 Harvey published his work on the generation of animals.[55] In it he defended an epigenetic theory, i.e. a theory according to which each living organism is formed anew, under the impulse of an auto-generative capacity, which forcefully recalled the faculties with which Plumpish opposed the inadequacy of mechanism. This epigenetic theory was very successful in the 1670s.[56] At the same time, Descartes had been accused of temerity for having believed that the world and the organism were formed by themselves, by the free play of the laws of motion and the impacts of bodies in the wake of creation,[57] and perhaps even worse, for having claimed to know in what way these bodies had been created by God.[58]

[55] *Exercitationes de Generatione Animalium. Quibus accedunt quaedam de partu; de membranis ac humoribus uteri; & de conceptione*. Londini: Typis Du-Gardianis; Impensis O. Pulleyn, 1651.

[56] On this point, see Jacques Roger, *Les sciences de la vie dans la pensée française du XVIIIe siècle: la génération des animaux de Descartes à l'Encyclopédie*. Paris: Armand Colin, 1963. See also André Robinet, *Malebranche de l'académie des sciences*. Paris: Vrin, 1970.

[57] Schematically, these readings are rooted in Article 47 of Part III of the *Principia philosophiae*, in which Descartes states that matter "can successively take all forms". On this point, see André Charrak, "Descartes au principe des cosmogenèses matérialistes", *Corpus. Revue de philosophie*, n° 61: "Matérialisme et cartésianisme", 2011.

[58] The text most clearly stating the need to defend Descartes from such accusations, after his blacklisting in the early 1660s, is the *Lettre écrite à un sçavant Religieux de la Compagnie de Jesus: Pour montrer, I. Que le Systeme de Monsieur Descartes, et son opinion touchant les bestes, n'ont rien de dangereux. II. Et que tout ce qu'ilen a écrit, semble estre tiré du premier Chapitre de la Genese*… (1668), by Gérauld de Cordemoy.

We must begin by recalling that this was not what Descartes intended to do, and that it was wrong to blame him for it. It is necessary to remember this because it circumscribes once again the domain of what it is possible to know for human reason, without deluding oneself. It demonstrates the ability of the latter to know the effects of divine laws and the manner in which they are particularised in secondary causes. The text that is clearest and most synthesising on this point is to be found in the *Dialogues on Metaphysics and Religion* XI, §VIII:

> Aristes. I have heard however that M. Descartes had started a treatise on the formation of the foetus in which he claims to explain how an animal can be formed from the mixture of the seed of the two sexes.
>
> Theodore. The philosopher's unfinished work can help us comprehend how the laws of motion suffice to make the parts of an animal grow little by little. But that these laws can form them and bind them all together is what no on will ever prove. Apparently M. Descartes recognised this himself; for he did not pursue his ingenious conjectures any further.
>
> Aristes. His enterprise as a little rash.
>
> Theodore. Very rash if his aim was to explain the construction of animals such as God has made them; for they have an infinite number of parts which he would have had to know before looking for the causes of their formation. But apparently he did not think of that. For we would not be wise if we wished to explain exactly how a clockmaker makes a clock without knowing beforehand of what parts the work is composed.

To specify what Descartes did not wish to do also implies stressing that which he did not do but which he should have done, at least more clearly. Descartes is content to assert that the seeds of the male and the female are mixed together in the uterus, and then begin a vortical motion because of the heat due to the fermentation of the mixture. Then the heart forms, the brain and the other parts, until there is a complete living being. But Descartes does not explain either the first principle of mechanism (on this point, Plempius and Harvey are correct), nor the extraordinary complexity of the living being, or what La Forge, in his defence of the 'machine' described, still without using the term, as a true 'organism'.

For reasons at once scientific (using now the work of Nehemiah Grew (1641–1712), Nicolas Hartsoecker (1656–1725), Antoni van Leeuwenhoek (1632–1723), Marcelo Malpighi (1628–1694), Francesco Redi (1626–1697), and Jan Swammerdam (1637–1680) and metaphysico-theological, Malebranche comes to offer his solution: each animal, each plant, each human body, must be preformed, entirely created by God at creation in miniature, then waiting for an opportune moment to take its adult size. All preformed organisms are stored at creation in the first female of each species. What we observe and designate as 'creation' is thus only the visible development of what was previously invisible, and is not an advent or a new formation.[59]

[59] There is to date no comprehensive reconstruction of the context and challenges of the Malebranche's "solution". For the most recent and enlightening works on this point, see Karen Detlefsen, "Biology and Theology in Malebranche's Theory of Organic Generation", in *The Life Science in Early Modern Philosophy*, Ohad Nachtomy and Justin Smith eds., OUP, 2014; Keith Hutchison, "Supernaturalism and the Mechanical Philosophy", *History of Science* 21 (1983), 297–333 and Andrew Pyle, "Malebranche on Animal Generation: Preexistence and the Microscope", in

Cartesian mechanism thus finds itself at the same time preserved by Malebranche and innocent of the reproaches of incompleteness that had been made against him. But what intrinsically falls outside the scope of mechanism must nevertheless be justified by principles of another order: 'supernatural' principles that adopt the point of view of 'divine providence'. Otherwise, the vacant space will be quickly reseized by defenders of forms. The particularity of the Malebranchean solution is that it occupies this terrain by using the most recent scientific research on embryology and by instituting a new intersection between domains of knowledge considered as separate, by Descartes and by the 'anatomists': that of human reason or natural philosophy and that of divine or supernatural reason. In this way, the wisdom and omnipotence of God themselves become visible to the smallest gnat, the littlest tulip bulb, the smallest part of the human body.

1.6 *L'Homme* Embodied

To tell the story of *L'Homme*, we have started from a double rejection: in the name of the progress of anatomy and in the name of metaphysics, the ghost in the machine no longer making an appearance to haunt the contemporary positivist scene. Between Descartes and us there are interlocking receptions that enhance one or another of the reasons, and even the articulation of this rejection (for the eighteenth century, see *Stephen Gaukroger** and *Claude Gautier**).

However, there are at least two ways of questioning this rejection and locating, in allowing it to grow in a very different context, the hermeneutic project of the first readers of Descartes.

The first consists in returning to the Cartesian meaning of physiology, as distinct from anatomy, in order to reinsert Descartes in the history of modern anthropology. It is a matter of reallocating him a place in the movement of secularisation characteristic of the radical Enlightenment, without restricting the project to this, but by the same token moving the conventional boundaries of a historiography strongly constructed by Jonathan Israel and strengthened by Antonio Damasio, putting it simply, in relation to the reduction of Spinoza himself to a Spinozism constructed by certain receptions of it (on Spinoza and *L'Homme*, see *Julie Henry**). Here we try to hold together the scientific approach which excludes metaphysics from its domain of application, and the requirement that we account for the complexity of human actions which scientific explanation does not allow us to completely reduce, even today (*Gary Hatfield**). It is in this space which still remains vacant for human reason, which can allow meta-physical considerations to resurface, and which Descartes' project fully manifests.

The second way of relocating *L'Homme* precisely complements the first. It consists in the idea of an embodied Descartes (*Barnaby R. Hutchins, Christoffer Basse*

The Problem of Animal Generation in Early Modern Philosophy, ed. by Justin E.H. Smith, Cambridge: Cambridge University Press, 2006, 194–214.

*Eriksen and Charles T. Wolfe**), i.e. a unitary Descartes, allowing us, for this reason, to explain cognition, memory, sensation and human health.

These two current research perspectives open up paths to a new story of *L'Homme*. One could reproach them for being just a new stage in this story. Nevertheless, what makes them distinctive is that they are motivated by the same questioning as Descartes and his first readers: the search for the 'true' man, or complete man, which human reason can never pretend to have exhausted.

The current state of research on *L'Homme* thus opens up at one and the same time the path of progress and the path of humility in the vast philosophical programme of seeking to know oneself.

Part I
Editions and Translations of *L'Homme*

Chapter 2
The Primacy of *L'Homme* in the 1664 Parisian Edition by Clerselier

Annie Bitbol-Hespériès

Abstract How can we understand the priority ascribed to *L'Homme* in the Préfaces by Clerselier and Schuyl and in the *Remarques* by La Forge? What are the reasons of this imbalance? What does the editor's decision to change the title of the "second treatise" mean? Why, also, the silent attitude of La Forge on important sections of *L'Homme*? In other words, we wish to understand how Descartes's specific new way of dealing with important biomedical questions, linked with philosophical implications, in *L'Homme*, and even more in *La Description du corps humain*, was concealed. Such is the relevant issue of the content of the Préfaces by Schuyl and Clerselier, and of the *Remarques* of Louis de La Forge for this posthumous edition of *L'Homme* and the "second traité". There is a huge gap between the Préfaces and the content of *L'Homme* and *La Description du corps humain*. For instance, the words "heart" and "body" are given less importance in the Préfaces than the word "soul", which sounds somewhat strange, not to say highly paradoxical, because this is in complete contradiction with Descartes' writings.

How can we understand the priority ascribed to *L'Homme* in the 1664 Parisian edition, from both the long title and long *Préface* by Clerselier to the copious *Remarques* by La Forge[1]? This question raises other important issues. Why did Clerselier focus on promoting *L'Homme* in the title given to the volume and in his *Préface* and why did La Forge strive to provide a better understanding of *L'Homme* in his *Remarques* and figures? Should Clerselier's *Préface* only be judged in relation to, or even in reaction to, Schuyl's *Foreword* to the Latin translation of *L'Homme*, the *De Homine*, published in Leyden in 1662[2]? Why is such importance granted to annotating *L'Homme* at the expense of the 'second Treatise' *De la formation du fœtus (On the*

[1] *L'Homme de René Descartes et un traitté de la formation du fœtus, du même auteur, avec les Remarques de Louis de la Forge, Docteur en médecine demeurant à La Flèche, sur le Traité de l'Homme de René Descartes, et sur les figures par lui inventées*, Paris, Ch. Angot, 1664.

[2] Schuyl's *Foreword* is included at the end of the 1664 edition in the French translation by Clerselier's son.

A. Bitbol-Hespériès (✉)
Centre d'Etudes Cartésiennes, Paris, France
e-mail: annie.bitbol@wanadoo.fr

D. Antoine-Mahut, S. Gaukroger (eds.), Descartes' *Treatise on Man* and its Reception, Studies in History and Philosophy of Science 43,
DOI 10.1007/978-3-319-46989-8_2

formation of the foetus)? What are the reasons dealing with such an emphasis on *L'Homme*? Are they related with the question of the illustrations highlighted by Schuyl, Clerselier and La Forge? And what about the absence of any comments by La Forge on important moments of *L'Homme*? For what reasons is the exact content of the 'second Treatise' left aside in the *Préface* by Clerselier and so little commented by La Forge?

In other words, there is a huge gap between the content of both *L'Homme* and of the 'second Treatise' and their presentation in the *Préface* and the comments in the *Remarques*.

To understand this complex situation, it is thus appropriate to compare the two posthumous published texts and to bear in mind the background in which these *Préface* and *Remarques* were written.

(I) The two texts: titles and contents.

L'Homme was written from the end of 1629 to 1633. It is a part of the ambitious treatise called *Le Monde* that remained unpublished during Descartes's lifetime because of the condemnation of Galileo's *Dialogue on the Two Chief World Systems* by the Congregation of Cardinals established to censor books in Rome. The unpublished *Le Monde* including *L'Homme* is alluded to in *Le Discours de la méthode (The Discourse on Method)*, the first book published by Descartes in 1637.[3] But there is a significant difference between *L'Homme* and the *Discourse*. For instance, the explanation of the movement of the heart and of the blood was extended and deepened between 1633 and 1637 and was given a major role in the fifth part of the *Discourse on Method*.[4] Parts dealing with the explanation of sensations and especially of the sight in *L'Homme* found their definitive place in *La Dioptrique*, the first of the *Essays* of the *Method*.[5] *L'Homme* is indeed a starting point in Descartes' medical thoughts as can be seen from the explanation of pain, tackled in *L'Homme* and developed in depth in the *Sixth Meditation* and in *Principles* IV, article 196. The same is true as regards the topic of passions, analysed in depth in *Les Passions de l'âme (Passions of the Soul)*, published in 1649. Twelve years after Descartes's death, and thanks to one of the copies of Descartes's manuscript of *L'Homme* that had been circulating, a Latin translation was published in Leyden in 1662, under the faithful title *De Homine*. Using another manuscript and claiming, at the beginning of his *Préface*, that it was the 'original' one, Clerselier published *L'Homme* in Paris two years later. If there was no mention of *L'Homme* in the inventories made after Descartes's death in 1650, it was not only because of the controversy with Regius, but also because the text of *L'Homme* had become obsolete, *passé*, compared to Descartes' published books.

[3] AT VI, 45–46, CSM, I, 134.

[4] See my Introduction to the edition of Descartes, *Le Monde, L'Homme*, Paris, Le Seuil, 1996, XL-XLV.

[5] See *Dioptrics*, Discourses IV and V, and my Introduction to *Le Monde, L'Homme*, op. cit., XXVII–XXIX, XXXV–XXXVI.

While allusions to *L'Homme* were contained in some of Descartes's Letters dealing with medicine and published by Clerselier in the two volumes of his edition of Descartes's Correspondence,[6] by contrast, there was no mention of the title of the second text, *Un traité de la formation du fœtus (A Treatise on the Formation of the Foetus)*. Moreover, there was no mention of such a title in the inventories made after Descartes's death but the Stockholm Inventory indicated a text called *La Description du corps humain (The Description of the Human Body)* registered under the letter G.[7] This specific title matches with the *Treatise on the Formation of the Foetus* and is to be found in its entirety at the very beginning of this *Treatise*: *Description of the Human Body and of all its functions both those which do not in anyway depend on the soul, and those which do, and including the chief cause of the formation of the parts of the body* (*La description du corps humain et de toutes ses fonctions, tant de celles qui ne dépendent point de l'âme, que de celles qui en dépendent, et aussi la principale cause de la formation de ses membres*), on p. 109 in the 1664 edition, just after the text of *L'Homme*. What is the aim of such a change in the title? What are the explicit and implicit elements of this modification?

We first have to mention that Clerselier was fully aware of the situation as an editor of Descartes's Correspondence which contains letters dealing with medicine. There are indications about Descartes's interest and experiments in embryology, especially in one letter written at the end of January 1648, to Elisabeth, Princess of Bohemia, published in volume I in 1657. This letter highlighted this important topic, claiming also that the new text was a rewriting of *L'Homme*.[8] Indeed, compared to *L'Homme, La Description* contains a new topic: the study of the formation of the foetus, or, in seventeenth century terms, of the generation of animals (animals meaning living creatures). This theme was explicitly emphasized by Clerselier in his edition. He pinpointed the part dealing with generation in the Cartesian treatise, so much so that he modified the title *The Description of the Human Body* into *On the formation of the foetus*, and this new title also stands as a running head in the original edition.

Descartes had given up trying to explain reproduction and embryogenesis in the *Treatise on Man*, 'because it' would have taken him 'too long'.[9] He was to return many times to this question, which required 'free time and the convenience of practising some experiments'[10] and the most complete statement of his thoughts on this issue are to be found in the *Description of the Human Body*.

The difficult question about how animals are generated was still viewed as a 'secret' or a mystery of Nature, and a method was needed to disclose this secret. In the seventeenth century, the importance of human dissections and of independent observation on animals began to be felt more and more strongly, especially thanks

[6] Before the publication of Descartes' s unpublished texts in 1664, Clerselier had issued two volumes of Descartes' Letters, in 1657 and 1659. The third volume was published after 1664.

[7] AT X, 9–10.

[8] 31 January 1648, AT V, 112. Letter XXV, Clerselier, vol. I.

[9] To Mersenne, June 1632, AT I, 254, CSMK, 39.

[10] See also, *La Description*, AT XI, 252–253.

to Vesalius and Bauhin in anatomy and to Fabricius of Aquapendente in embryology. These anatomists are Descartes's acknowledged sources in medicine.[11] But in medical treatises, as well as in the teaching of medicine, the authority of Aristotle and Galen remained strong. Aristotle still played an important part in the history of medicine because of his views about the link between the soul and the body and his teleological outlook, and also, more specifically, in the history of embryology through his thesis on the nature of the semen. Yet, according to Galen, followed by all the anatomists, it is difficult to explain the ingenuity Nature employs in forming a foetus. This arouses the wonder and admiration of whoever observes them in dissections.[12]

In the *Description of the Human Body* Descartes was obviously trying to avoid the passive attitude of wonder of many physicians, the obscure controversies and all the prolix discussions in medical treatises. Descartes is very clear on an important debate about whether or not the female had semen (*'semence'*, the secretions of the organs devoted to generation). For Descartes, both the male and female semen are necessary for producing an animal.[13] What is also important to notice is the Cartesian method in medicine of eradicating admiration towards Nature, eradicating finalism in medicine, eradicating the Galenic 'attractive faculty'[14] of the semen (supposed to attract blood), as well as of the 'formatrix' faculty of the male semen. These are the implicit aspect of Descartes' method to address the issue of generation in the *Description*. As for the explicit aspect of the Cartesian method, it is represented by the following principles: the existence of the two semen, male and female, as mentioned previously and also the idea that heat and fermentation are the driving force that explains generation. These principles are expressed by Descartes in the *Description of the Human* Body.[15] Due to the emphasis given to *L'Homme*, Clerselier and La Forge were more interested in the importance of the power of the imagination of the mother on the foetus, a theme also tackled in the *Dioptrics*.[16]

What is also clearly expressed in the *Description of the Human Body* is the question of the order of the formation of the parts during the development in the womb of the mother. The controversies were widespread between the Galenists and the Aristotelians. The Galenists argued that the heart, liver and brain, the three 'principal' organs in the body, become defined at the same time, as explained in Galen's *De usu partium, (On the usefulness of parts)*. As for the Aristotelians, they were influenced by the *De generatione animalium, (On the Generation of Animals)*,

[11] Cf. Annie Bitbol-Hespériès, « Cartesian Physiology », in *Descartes'Natural Philosophy*, S. Gaukroger, J. Schuster and J. Sutton Eds, Routledge, 2000, 349–382. See also my Foreword to the edition of *Le Monde, L'Homme*, Paris, Le Seuil, 1996.

[12] See in particular *De usu partium*, ed. Kühn, vol. 4, 224 ff, French edition by Daremberg, vol. 2, 137 ff.

[13] *Description*, beginning of the fourth part, AT XI, 253.

[14] On the importance of the 'faculties', see Galen, *De naturalibus facultatibus, (On the natural faculties)*, Kühn, vol. 2, p. 1–214, Daremberg, vol. 2, 212–320.

[15] *Description*, beginning of the fourth part, AT XI, 253.

[16] *L'Homme*, AT XI, 177, *Préface* np (p. 33), *Remarques*, 334–345. See *Dioptrique*, AT VI, 129.

where Aristotle explained that the heart was the first organ that appears in the embryo.

In the *Discourse on Method*, the heart was to become the first organ to appear, and the movement of the heart was to become the 'first and most widespread movement that we observe in living beings'.[17] This thesis was formulated more under the influence of Harvey's Treatise, *On the Movement of the Heart and Blood in living creatures, Exercitatio anatomica de motu cordis et sanguinis in animalibus*, published in 1628 and read by Descartes at the end of 1632,[18] than of Aristotle. Once the priority of the rule of the heart had been admitted and checked with dissections of calves at different stages of their development and observations on incubated eggs,[19] Descartes promoted his conception of the heat in the heart which is a kind of fire, as already mentioned in *L'Homme* and in the *Discourse on Method.* This 'principle of movement and life', as mentioned at the end of *L'Homme* is a 'corporeal principle' as stated in the *Passions of the Soul*,[20] written in the same years as the *Description* and published in 1649, and as quoted in some letters.[21] This 'principle of life'[22] is fully detailed in the last text dealing with medicine written by Descartes, *La Description du corps humain*.

Indeed, the heart, its detailed structure, its movement, its cause, and its formation, and also the circulation of the blood, hence the formation of the veins and arteries, are the most important topics in *La Description.* They are more crucial than the question of generation, although the latter was emphasized by Clerselier when publishing *La Description*. The editor's decision to change the title focusing on the question of embryology entails the consequence of leaving aside other important topics and affecting the composition of the *Description*. It has to be mentioned that after the short Preface called *part One*, *part Two* is entitled 'On the Movement of the Heart and the Blood', while *part Three* is dealing with Nutrition. *Parts Four and Five* deal with the problem of embryology, but it has to be noted that part Four is called a 'digression' and that the presence of this specific word, in keeping with its etymology, is not consistent with the modification of the title by Clerselier. So, it becomes obvious that the new title given by Clerselier does not reflect the exact content of the book.

(II) The medical context and the controversies about Harvey's discoveries:

What is implicit with the change in the title is that Clerselier avoided the medical subject of the circulation of the blood, which was still controversial in France. The Medicine Faculty in Paris was still rejecting the circulation of the blood in 1664, when Clerselier published *L'Homme* together with *La Description*, more than 36 years after the publication of Harvey's book and more than 27 years after the publi-

---

17 AT VI, 46, CSM, I, 134.

18 AT I, 263, CSMK, 40.

19 Cf. *Excerpta anatomica.*

20 Cf. *Passions*, Art. 8.

21 See Annie Bitbol-Hespériès, *Le principe de vie chez Descartes*, Paris, Vrin, 1990.

22 Cf. *Passions*, Art. 107.

cation of the *Discourse on Method*, which publicized in French the discovery of the circulation of the blood. The circulation of the blood was admitted in *L'Homme* but no details were given about this new discovery. The expression 'perpetual circulation', first mentioned in *L'Homme* in order to define the movement of the blood in living organisms, is also found in the *Discourse on Method*, where details are given in support of this brilliant discovery: experimental proofs with the name of his discoverer in Latin, *Hervaeus*, and the title of his book, *De motu cordis* in the margin of the *Discourse*.[23] This expression also appears in *La Description*, with a refined explanation.[24] The laudatory approval given by Descartes to the genuine novelty of Harvey's 1628 demonstration of the circulation of the blood also appears in his correspondence[25] and the *Passions of the Soul*.[26] This explicit quotation is interesting to note since Descartes is usually reluctant to mention his sources. It is also worth noting because the recognition of the Harveian thesis of the circulation of the blood took a long time and provoked many objections, especially in France, see of course, the play *Le Malade Imaginaire* by Molière, written in 1673, the year after King Louis XIV decided to have the circulation of the blood taught in Paris, in his garden (in those days *Le jardin du Roi*, now *Le jardin des plantes*), by Dionis, a surgeon.

But Schuyl, Clerselier and also La Forge to a lesser extent tend to ignore the fundamental relation of Descartes' writings about the body, -from the *Discourse* to the *Passions* and *The Description*-, to Harvey's treatise about the movement of the heart and the blood. It is thus important to read *La Description* to realize how challenging it was to study the human body at this turning point in the history of medicine. Descartes' *'second Treatise'* offers a remarkable insight into this significant moment.

La Description is therefore closely linked with Harvey's two brilliant demonstrations on the movement of the heart and on the movement of the blood. This is the major interest of the *Description*, which is also a more accurate text related with the two published books where Descartes had claimed his approval of the circulation of the blood: the *Discourse* and the *Passions*, in particular the 'Brief account of the parts of the body and of some of their functions', article 7. As in the *Discourse* and in the *Passions*, in *La Description*, Descartes quotes Harvey's name in Latin. Concerning the true interest of the *Description*, it is worth noting that one can find in this treatise a re-writing of the 'proofs' of the circulation of the blood, in a more precise manner than in the other Cartesian texts. What is striking is the careful study of the experimental evidence of the circulation of the blood given by William Harvey. This other major interest of the text is linked with the various dissections Descartes had performed on animals.

As in the *Passions*, article 7, Descartes condemns those who still have a very prejudiced view and are unable to accept the circular movement of the blood and to

---

[23] AT XI, 127, AT VI, 50–51.

[24] Cf. AT XI, 239, C.S.M.I, 316.

[25] Cf. letters to Beverwick, 5 July 1643, AT IV, 4, to Newcastle, April 1645, AT IV, 189.

[26] Cf. *Passions*, Art. 7, C.S.M.I, 330.

distinguish 'the true and certain reasons' from the false and probable ones.[27] What is striking is that Schuyl in his Latin *Foreword* to his edition of the *De Homine*, and Clerselier in his French *Préface* ignored Descartes's approval of the circulation of the blood. What is also striking about La Forge's *Remarques*, is that he commented neither the expression 'perpetual circulation' in the *Treatise on Man* nor the detailed experimental proofs given in the 'second Treatise'. Yet La Forge, in his vast set of *Comments* on *L'Homme* that included some references to the 'second Treatise', referred to some 'experiments' that confirm his own agreement with the 'truth of the circular movement of the blood',[28] but this is far from being an important part of this Commentary.

The *Description* is also linked with the fifth part of the *Discourse on Method* and the *Passions of the Soul*, articles 8 and 9, concerning the explanations about the movement of the heart and of its cause, different from Harvey's explanation. As an indication about the importance of Descartes' relation to Harvey in the *Description*, it has to be noted that the words 'diastole' and 'systole' are present neither in *L'Homme* nor in the *Discourse on Method* and the *Passions of the Soul*. By contrast, they appear in *La Description*, not in the second part dealing with the motion of the heart and of the blood, but in the fourth part concerning 'the parts of the body which are formed in the seminal material',[29] and in the fifth part 'on the formation of the solid parts',[30] especially the heart.[31] However, the words 'diastole' and 'systole' are not defined by Descartes in the same way as Harvey did in chapter two of his *De motu cordis*. It is because Descartes had not agreed with Harvey's new definitions of the diastole and systole. If (as Harvey asserts[32]), the expulsion of blood occurs during systole, which is the phase of contraction of the heart and thereby the phase of its diminution in volume, there must be something in the heart that is the cause of its contraction. In *La Description*, Descartes explained clearly : 'Now if we suppose that the heart moves in the way Harvey describes, we must imagine some faculty which causes this movement; yet the nature of this faculty is much harder to conceive of than whatever Harvey purports to explain by invoking it'.[33]

Descartes then details his alternative explanation, quoting the experiments given by Harvey and providing other examples, some of them already mentioned in the letter written to Plempius on the 15 February 1638, and published in the first volume of Clerselier's edition of Descartes' Letters.[34] According to Descartes, the crucial phenomenon is the rarefaction and expansion of the blood connected with the production of heat in the heart. This expansion distends the heart, thereby provoking diastole which is the phase of blood expulsion according to Descartes (and unlike

[27] AT XI, 240–241.

[28] *Remarques*, p. 190.

[29] AT XI, 252, CSM, I, 321.

[30] AT XI, 273, not in CSM.

[31] AT XI, 267 and 280–282.

[32] Cf. *De motu cordis*, cap. 2.

[33] AT XI, 243–244, CSM, I, 318.

[34] AT XI, 241–242. See to Plempius, AT I, 527. Clerselier, vol. I, letter 78.

Harvey). It has to be noted that it is in his *Comment* about the valves of the heart in *L'Homme* that La Forge commented the words diastole and systole and referred with approbation to the 'second Treatise' and therefore disagrees with Harvey's explanation of the movement of the heart and more accurately with the French translation of Bartholin's *Anatomie Réformée*.[35]

The Cartesian explanation of the movement of the heart in *La Description* included the discussion dealing with Aristotle, based on the letter written by Plempius, one of the first readers of the *Discourse on Method.* In his objections against Descartes, mentioned in a letter written in January 1638, Plempius had discussed the fifth part of the *Discourse*, and referred to Aristotle, *De Respiratione*, cap. 20.[36] But this reference to Aristotle in the *Description* is the only explicit one to be found in Descartes' medical writings and is a mere clarification, especially for physicians. Yet in his *Remarques* about the cause of the movement of the heart, La Forge referred the readers to the letters to 'a physician from Louvain' without quoting Plempius' name and making any reference to Aristotle.[37] In *La Description* Descartes treated once again with the greatest respect Harvey's discovery of the circulation of the blood, but dissociated it from its Aristotelian context. Yet Descartes tended to obliterate Harvey's own Aristotelian background, which meant both a vitalistic approach and a teleological context. Descartes had already eliminated this background, in an implicit manner, in the *Discourse on Method* when explaining the structure of the heart and of its valves, as well as the circulation of the blood.[38]

Indeed 'anatomy and mechanics' are of paramount importance in Descartes' Works and this connection between anatomy and mechanics is of the highest significance since it rules out the alternative connection between anatomy and finality. At the beginning of *La Description*, Descartes thus explicitly condemns 'our ignorance of anatomy and mechanics'.[39] Descartes invoked 'the rules (or laws) of mechanics' in order to explain the formation of the valves in the heart, that were a traditional subject of wonder.[40] In the *Discourse on Method*, Descartes had already stated that 'the laws of mechanics [...] are identical with the laws of nature'.[41] The 'production' of these valves is taken as an example of the action of matter on itself and of its mechanical forces. Descartes' detailed descriptions of the valves in the heart in *La Description*, after their mention in *L'Homme*,[42] the *Discourse on Method* and the *Passions*, are also based on personal observation of hearts of calves and oxen, as

[35] *Remarques*, p. 191–192. Bartholin written Bartolin.

[36] AT XI, 244–245. See Plempius to Descartes, January 1638, AT I, 497. See also, Descartes to Plempius, 15 February 1638, AT I, 522.

[37] *Remarques*, p. 186, 188.

[38] AT VI, 47–51, CSM, I, 134–136.

[39] AT XI, 224, CSM, I, 314.

[40] AT XI, 279–280, not in CSM. On the valves of the heart, see also, AT XI, 229–230, *L'Homme* (AT XI, 124), The *Discourse on Method*, V, AT VI, 47.

[41] AT VI, 54, CSM, I, 139.

[42] AT XI, 124–125 ('les petites peaux').

shown in the *Exerpta anatomica.*They are not related only to Harvey's demonstration of their function.[43]

La Description is also noteworthy when Descartes used the comparisons with fermentation both for the generation of heat in the heart and for the mixing of the male and female semen.[44] This was a major breakthrough in the history of medicine, because the heat in the heart was a deep mystery. It was often qualified of 'divine', with a supposed origin in the heavens, and a mysterious origin was also assigned to the male semen, as can be seen from the *Works* of Fernel in his *Physiologia* and the *Comments* by Riolan the Elder.[45] In his *Remarques* about the heat in the heart in *L'Homme*, La Forge implicitly acknowledged the novelty of Descartes's comparison with fermentation when he opposed the fermentation to the 'faculty of the soul called pulsific', and added: 'for me, I owe so much respect to my Teachers that I can maintain this word of 'faculty' not to displease them'.[46] Then La Forge admitted that 'claiming this is a quality of the body or a property of the soul, is not helpful to explain what it is'. He thus subscribed to the Cartesian explanation of the fermentation in the heart saying that 'this has to be called the pulsific faculty'.[47] In his *Préface*, Clerselier had already granted that 'the countless number of faculties' coined by physicians, was a way of 'getting rid of all the greatest difficulties'.[48]

Descartes began to face these 'greatest difficulties' when he was writing *L'Homme.* Among these complex questions, La Forge asked the one of the movement of the muscles, 'the most difficult to understand', an echo of Clerselier's *Preface* where the movement is qualified of 'the most important action for the author to describe and explain'.[49]

However this insistance on this 'most important' question according to Clerselier entailed an omission that indicated a great reluctance towards the Cartesian explanation. This 'most important' question dealt with the 'principle of these movements', as explained in *L'Homme* and *La Description*, and as already mentioned in some published texts : the *Discourse on Method* and the *Passions of the Soul.* In other words, the fundamental question of the 'principle' of all the functions performed by a living body, explained by Descartes in a new way, as confirmed in the *Passions*, article 8, is avoided and nearly ignored by Schuyl, Clerselier and La Forge. Clerselier adressed this issue once and significantly weakened the scope of Descartes' texts. Clerselier praised La Forge (not Descartes), and added 'for what regards the discipline of animals it is explained without admitting in them any soul that thinks, knows or reasons, nor any other principle of life or movement except the blood and the (animal) spirits moved by the heat in the heart'.[50] Such a statement

---

[43] *De motu cordis et sanguinis*, 1628, Introduction, cap. VII, cap. XVII.

[44] AT XI, 231 and 253, and the comparison with yeast.

[45] See *Le principe de vie chez Descartes*.

[46] *Remarques*, 183.

[47] *Remarques* about « fort chaud ».

[48] *Préface* np, (61).

[49] *Préface* np, (22).

[50] *Préface*, np. (32).

altered not only the meaning of Descartes' new explanation on a fundamental topic but also its context, because with Clerselier, as well as with La Forge and before them with Schuyl, it became associated to the question of the difference between humans and animals and linked to the only relevant issue for the three of them, namely the question of the soul of beasts.

It is crucial to note that the prolix annotation by La Forge and the long Forewords by Schuyl and Clerselier tended to hide a key point in Descartes' writings, namely the radical innovation introduced by Descartes in medicine: the way he imposed on medical questions a new conceptual framework. This new framework consisted in the systematization of mechanism with the heat in the heart becoming 'the principle of movement and life' as stated at the end of *L'Homme* and explained in the first part of *La Description*. This was a break with the medical and surgical books, deeply rooted in the Hippocratic and Galenic tradition, and also in the Aristotelian philosophy.

(III) The philosophical context and medicine:

It has to be noted that all the physicians and surgeons in Europe agreed that the human body draws its life, that is to say its animation, its motion, from the soul or 'vital principle' or 'principle of life'. Death meant the 'departure' of that soul, which is life. The human body was the 'dwelling of the soul'.[51] Riolan the Younger clearly stated: 'Man is composed of two really different natures, the soul and the body. The former, which is linked with the body, is the principle of life and of all the actions, and therefore, the form and perfection of the body'.[52] Over the course of the different chapters describing the different parts of the body in medical treatises, the soul is particular in that it is divided into a 'vegetative soul' linked with nutrition and generation, a 'sensitive soul' linked with the senses and movement, and a 'rational soul', which belongs only to rational animals, that is to human beings. This usual assertion of the 'triple soul' was put forth in the medical theses written by the physician Henricus Regius (Hendrik De Roy or Le Roy), who taught medicine in the University of Utrecht and who sent his medical theses to Descartes before they were defended by Regius' students in medicine.[53] This 'controversy concerning the threefold nature of the soul',[54] led Descartes to assert firmly in the following letter written in May 1641 to Regius: 'a man's soul is unique, that is to say rational'.[55] Clerselier was fully aware of these letters, since he referred to them at the beginning of his *Préface*.[56] The eradication of the non-conscious or non-cogitative functions of the soul is the major philosophical reason for the Cartesian study of the 'Nature of

[51] Du Laurens, *Historia anatomica*, Paris, 1600, French translation, 1610, by Sizé.

[52] Riolan (The Younger), *Anthropographia*, Paris, 1618, 1626, French translation 1629.

[53] Cf. Regius' medical theses called *Physiologia sive cognitio sanitatis*, 1641. See my paper "Descartes et Regius: leur pensée médicale", in *Descartes et Regius, Autour de l'Explication de l'esprit humain*, Theo Verbeek (ed.), Rodopi, Amsterdam-Atlanta, GA, 1993, p. 47–68.

[54] AT III, 369, CSMK, 181.

[55] AT III, 371, CSMK, 182.

[56] *Préface*, np (14–15)

Man'. This eradication is found at the end of *L'Homme*, in the *Dicourse on Method*, in the *Passions* and in *La Description*.

But in the 1664 Parisian edition of *L'Homme* and *La Description*, this major aspect of Descartes' thought was also concealed by the lack of *Remarques* by La Forge on the last paragraph of *L'Homme*, that had become article 151 according to Clerselier who gave it a simplistic title: 'that all the functions assigned to the 'machine of the body' followed the disposition of its organs'. The Cartesian rejection of these vegetative and sensitive souls and the affirmation of the unity of the rational soul, without any link to the vital principle had already been delivered to the public in the fifth part of the *Discourse on Method*. They had been confirmed in the *Passions of the Soul* (articles 5, 6, 8, 47), as well as in the posthumous *La Description* where Descartes clearly denounced the 'error' of believing 'that the soul is the principle responsible for all bodily movements'.[57]

The fundamental methodological challenge of dualism and the clear-cut distinction between the principle of life and the soul which breaks with the medical tradition legitimates the comparison of the body with a 'machine', and changed the definition of death, not linked with the fault of the soul.[58] And Descartes asserted that 'our soul, in so far as it is a substance which is distinct from the body, is known to us merely through the fact that it thinks, that is to say, understands, wills, imagines, remembers, and has sensory perceptions ; for all these functions are kinds of thought'.[59] It should be noted that 'remembers' is an addition in *La Description*, linked with the 'organ or seat of the 'common' sense, the imagination and the memory', namely the conarium or pineal gland.[60]

It becomes obvious that Clerselier's *Préface* and La Forge's lack of comment on the last paragraph of *L'Homme* pursued the same goal, namely to expunge the strong Cartesian opposition to the traditional conception of the soul as the principle of life.

Interestingly, Clerselier and La Forge were not the first to do so. Indeed, Schuyl had been clear in his Latin Foreword about his reluctance to admit that the principle of life can be reduced to the heat in the heart.

Schuyl wrote that for Descartes 'it is not difficult to grant life to animals, in agreement with the Holy Scripture ; at the same time he [Descartes] explains that for him, life consists in heat and in the convenient arrangement of the parts; he also grants them sensation, in so far as it depends on a bodily organ, without any thought'. This sentence involves an implicit reference to the letter from Descartes to Henry More, -Morus, also a pious man and an avid reader, concerned as Schuyl by impiety-, letter of February 5th, 1649, published by Clerselier in the first volume of Descartes' Correspondence.[61] However, Schuyl changed once more the initial context of the

[57] AT XI, 224–226, CSMI, 314.

[58] AT XI, 224–225, CSMI, 314–315, and *Passions*, art. 5.

[59] AT XI, 224, CSM, I, 314.

[60] AT XI, 227. See the *Meditations*, II and beginning of III, *the Principles*, I, article 9. See also in *La Description*, the 'little gland called *conarium*', AT XI, 270.

[61] Cf. to Morus, 5 February 1649, AT V, 278–279, CSMK, 366. See also, to Newcastle, 23 November 1646, AT IV, 573–575 and CSMK 301, *Passions*, art. 50.

original text, since Descartes avoided mixing theology to medicine and did not invoke the Holy Scripture in his explanations on the principle of life. In addition, Schuyl modified the scope of the Cartesian text, since the implicit quotation is cut short. Indeed, Schuyl substituted the Cartesian definition of life, reduced to 'the heat in the heart', '*calore qui est in corde*', with a vague reference to 'heat','*calore*'. Clearly, nothing in the medical tradition is similar to the Cartesian 'fire without light', or 'heat in the heart' which is reduced to fermentation.

It becomes clearer why Froidmont, (Fromondus), one of the first readers of the *Discourse on Method*, was deeply surprised with the importance given by Descartes to 'some heat of the same nature as the heat that warms hay' in the explanation of the functions of the body. It also becomes clearer why Froidmont voiced this objection: 'noble actions cannot result from so ignoble and brutish a cause as heat'.[62] And, further on, Froidmont refers to Descartes' philosophy as being 'too blunt and too mechanical', '*nimis crassa et mecanica*'.[63] Descartes replied to this fundamental criticism by asserting that accusing his philosophy of being like mechanics amounted to rejecting everything that should be praised in his opinion. Descartes added that mechanics, a part of philosophy has remained 'more true and less corrupted' than the others, '*verior et minus corrupta*'.[64] There is no doubt that this exchange of letters influenced both writers of the Prefaces of *L'Homme*, Schuyl and Clerselier, since Descartes' reply to Plemp was published in French in the second volume of the Letters edited by Clerselier.

In a new mechanistic context, Descartes granted to the heart and the blood a primacy that will never stop growing in all his published works, see e.g. *Passions*, article 71.

And it is exactly this systematization of mechanism that is successively targeted by Schuyl, La Forge and Clerselier, for instance by 'adding' comments to the original text of *L'Homme* and providing it with a different conceptual framework. They did not feel allowed to do this simply because of the 'unclarities' ('obscurités') of the treatise, that La Forge often mentioned in his *Remarques*, but also because of its incompleteness, which was stressed, with an emphasis on the issue of figures.

The text of *L'Homme* was intended to be illustrated but Descartes was not good at drawing, as can be seen from the schematic figures on the folding accompanying the *Excerpta anatomica*.[65] I agree with Clerselier's thought that the physician Regius was the one who could have illustrated *L'Homme* in the best way. In the copy of the manuscript supplied to Schuyl, Clerselier also mentioned a draft copy ('brouillon') by Descartes. But can we trust Clerselier? Remember his similar attitude concerning what he wrote about the existence of this 'brouillon' and about the 'original' of *L'Homme*.[66] And when Clerselier re-drew the figure, he was influenced by the pub-

[62] Froidmont (Fromondus) to Plemp (Plempius), 13 September 1637, AT I, 403.

[63] Froidmont to Plemp, 13 September 1637, AT I, 406, and Descartes, 3 October 1637, I, 413. See *Le principe de vie chez Descartes*.

[64] Descartes to Plemp for Froimont, 3 October 1637, AT I, 420–421.

[65] Inserted just after AT XI, 634.

[66] *Préface*, np (11 and 26).

lished one in the *De Homine* and by Gutschoven' figures for *L'Homme*. The emphasis given to the question of the figures is all the most paradoxical that Clerselier noted that the published drawings by Gusthoven and La Forge 'do not look like' the parts in the body and claimed that their aim was 'not to publish an anatomical book'.[67] It is worth pointing out, however, that Descartes explicitly referred to 'a learned anatomist' at the beginning of *L'Homme* and quoted many times 'anatomists', and 'anatomy'.[68] Indeed anatomical knowlege is explicitly implied by Descartes, and not only in *L'Homme*,[69] but also in the *Discourse, the Dioptrics* and the *Passions*. In *La Description*, the 'ignorance of anatomy' is condemned at the beginning of the text,[70] and anatomical evidence plays an important part, not only with Harvey's demonstrations, but also with a new discovery of the lacteal veins by Asellius and Descartes insists on the conditions of this discovery.[71] But Descartes could not have known the discovery of Pequet, though mentioned by La Forge, after Asellius.[72] Descartes could not have known either Harvey's treatise on the *Generation of Animals*, also quoted by La Forge.[73] Pecquet's discovery and Harvey's *De generatione animalium* were published after Descartes' death.

The Description of the Human Body is indeed a major text in the Cartesian *summa medicinae*. Firstly, because this text has strong links with the other published books by Descartes and not only with the *Dioptrics*, the *Principles of Philosophy* and the *Meteors*, explicitly quoted in the *Description*.[74] And secondly, because the *Description* is dealing with the most important and most contemporaneous medical problems in the middle of the seventeenth Century, after the explanation of sensations and especially of sight already given in the *Dioptrics*. These most important and most contemporaneous medical problems were: Harvey's demonstration of the movement of the heart and of the blood, Asellius' demonstration of the lacteal veins, the question of generation, and the importance of dissections and vivisections in medicine, not to mention the authority of Aristotle and Galen.

Schuyl and Clerselier in their *Préfaces*, as well as La Forge in his *Remarques*, impose a different context to *L'Homme* and to *La Description*. First, Schuyl is concerned with atheism. Second, the references quoted in his *Foreword* reveal his fascination for '*magia naturalis*' (Della Porta is cited), but also his interest for collections of 'rarities' that are supposed to be admired, such as Van Hoorn's collec-

---

[67] *Préface*, np (33).

[68] Cf. AT XI, 120,123 124,125, 138,147, 152, 166, 194, 200.

[69] In my edition of *L'Homme*, published with *Le Monde*, I showed some anatomical plates from the *Theatrum anatomicum* by C. Bauhin, a book of paramount importance for Descartes, as confirmed by my translation and annotation of the *Primae cogitationes circa generationem animalium* and *Excerpta anatomica*, with *L'Homme* and *La Description*, for the forthcoming vol. II of Descartes *Œuvres complètes*, Gallimard-Tel.

[70] Cf. AT XI, 224, 226, 228, 231.

[71] Cf. AT XI, 243.

[72] *Remarques*, 180–181.

[73] *Remarques*, 175.

[74] Cf. AT XI, 248, 255, 275.

tion, which includes a calcified pineal gland (Van Hoorn's *Admirandorum Thesaurus* is cited in the Latin Preface).[75] The latter reference is even more notable given those to Liceti (misnamed Licette by Clerselier), Aldrovandi and Gessner (note that Clerselier suppressed these references). In his *Remarques*, La Forge mostly relied on *L'Anatomie réformée* by Bartholin (written Bartolin), including his anatomical mistakes, for instance on the so-called nervous filaments (*nerveus funiculus*) attached to the pineal gland,[76] that Descartes never wrote about and that the dissections of brains he had performed did not show. La Forge was also very attentive to his forthcoming book, the *Treatise On the Human Mind, traité de l'Esprit de l'homme*, and referred to it on several occasions. Clerselier who vowed not to 'make the book thicker' -but who did so with his *Préface*-, and who claimed to be faithful to Descartes, in fact tended to erase the deep originality of the two texts he edited by often mentioning the soul and by making long references to Augustine, not quoted by Descartes in *L'Homme*, the *Discourse*, the *Meditations*, the *Principles*, the *Passions* and the *Description*.

Schuyl himself already quoted Augustine several times, first referring to *De libero arbitrio*, chap. VIII, 18, then to *De quantitate animae*, on the bodily nature of the so-called soul of animals. The latter topic was further illustrated by Schuyl through the example of the scolopendra, which was then further developed, as it enabled Schuyl to allegedly demonstrate that the animal's soul can be divided, and is therefore material.[77] Scolopendras resemble centipedes, with their bodies made of numerous segments with two pairs of legs each. Schuyl mentioned observing several times in experiments that scolopendras can 'live for more than a month after cutting off the head from the rest of the body'. However, this assertion, motivated by the anatomical structure involving numerous segments of scolopendras, which allegedly enabled it to stay alive after being cut in two parts, is erroneous. The parts of a cut scolopendra can move, as Schuyl pointed out, after Aristotle, since the action of the legs –if not that of the soul-, goes on in each segment, but only for a short time. This is not the only inaccuracy in this example, since the two verses by Ovid are wrongly attributed to Lucretius by Clerselier and/or by his son.[78]

More importantly, Schuyl's arguments contradict the Cartesian explanation, as evidenced by Descartes' writings. Descartes' November 1641 letter to Regius is particularly illuminating, and Schuyl and Clerselier were aware of this letter since it appeared in the first volume of the Letters edited by Clerselier. Descartes did not refer to scolopendras in this letters, but he mentioned the parts of eels that move after being cut off, which amounted to the same idea. He gave an explanation linked with the arrangement of the solid parts and the movement of the animal spirits or of the fluid parts entering into the solid ones.[79]

---

[75] *Préface*, 445–446. *Hoorn= the anatomist Johannes Van Horne, Hornius (1621–1670).*

[76] *Remarques* on art. 63.

[77] *Foreword*, 434–440.

[78] *Foreword*, 437. No mention of Lucretius nor Ovid in the Latin Foreword.

[79] Letter to Regius, November 1641, AT III, 445.

It does seem very difficult to establish a continuity of thought between Augustine and Descartes. And this task becomes even more difficult at the point when Clerselier too referred to Augustine. Clerselier actually chose to quote a long extract of the *De Trinitate* (book X, end of chapter IX and chapter X), which spans almost 7 full pages in the original edition of his *Préface*: 'Let us now listen to Augustine, who, at the end of the ninth and in the whole tenth chapter of book 10 of the *Trinity*, says these very wise words: "know yourself"'.[80] This is put in relation to the soul, which is linked to life according to Augustine, as stressed by Clerselier: 'a living being and its life are precisely attributed to the soul', which, conversely, is not the case in Descartes' thought. Isn't this a strange *Préface* to *L'Homme*, and an even stranger one to *La Description du corps humain*? With this reference to Augustine, Clerselier's Préface to *L'Homme* and to *La Description* stands in contradiction to three fundamental aspects of Descartes' medical thought.

- First, Descartes has rejected the view of soul as the principle of life since the writing of *L'Homme*, and has strongly reasserted this rejection in his later works, the *Passions of the Soul* and *La Description*, as well as in his correspondence. He advocated the systematisation of mechanism, as discussed above, and the reduction of the principle of life and motion to the heat in the heart, a heat which is 'ordinary'.
- Second, Descartes applied the 'know yourself' ('know thyself'),[81] principle to the body and not to the soul, as stated at the beginning of the first part (Préface) of *La Description du corps humain.* While this point was not entirely new, his use of the Socratic precept in the context of medicine was original, compared to the two backgrounds linked with this precept: the first one, widespread and associated with praises to the Glory of God. Acquiring knowledge about the human body meant leading to admire Nature or/and God for the perfection of the 'fabric' of the human body. The second one linked with a moralizing context.[82] Indeed, from the beginning, *La Description* is a true manifest for the new medicine, built on Cartesian principles.
- Third, Descartes took care to separate in his works medicine and praises to God, i.e. medicine and theology, and also medicine and teleology, which was original for his time.

In conclusion, Clerselier's Foreword does not simply provide some words of introduction. Rather, it is a long guide that not only promotes another medical and metaphysical context than the one conveyed by these texts, but also aims to alter the goals Descartes had in mind when he was writing these texts. For instance, the words 'heart' and 'body' are less present in Clerselier's *Foreword* and in La Forge's *Comments* than the word 'soul', which sounds somewhat strange, if not paradoxical, because this is at odds with the content of Descartes's writings.

---

[80] *Préface*, np (p. 52–59). This reference to Augustine influenced La Forge who quoted the same text in his Foreword to his *Traité de l'esprit de l'homme.*

[81] AT XI, 223 (first page, first sentence of the *Description*).

[82] See my paper 'Connaissance de l'homme, connaissance de Dieu', in *Les Études Philosophiques*, 1996, n° 4, 507–533.

Chapter 3
New Indications for Critical Edition of *L'Homme*

Franco A. Meschini

Abstract The chapter reflects on some variants from the Latin edition of 1662 edited by Fl. Schuyl: the partitioj, the cross-references, the omission of the Latin passage on the French edition (*Hic notari...*), an omission in the fifth part (*Et notez en passant...*). Finally, starting from the compliance (all but one case) of the text of the two editions as far as it concerns the pineal gland, the authors asks himself about the "mysterious" presence/absence of the pineal gland in *L'Homme*.

I have already explained the contents of this text once at the conference held in Lyon (January 2014); for an enlarged version of that conference and of the text published here, see Franco A. Meschini, *Per un'edizione critica de* L'Homme *di Descartes. Nuovi materiali ed altre suggestioni*, in *L'utopia: alla ricerca del senso della storia*, scritti in onore di Cosimo Quarta, a cura di G. Schiavone, Milano, Mimesis, 2015, pp. 527–550.

Abbreviations: AT, followed by the roman number of the volume and the indication of the page = R. Descartes, *Oeuvres*, éd. par Ch. Adam et P. Tannery, nouv. présent. par J. Beaude, P. Costabel, A. Gabbey et B. Rochot, 11 vols, Paris, Vrin, 1964–1974; BL, followed by the number of the letter and the indication of the page = R. Descartes, *Tutte le lettere*, [trad. it. con testi originali a fronte], a cura di G. Belgioioso, con la collaborazione di I. Agostini et *alii*, Milano, Bompiani, 2009²; BO, followed by the indication of the page = R. Descartes, *Opere 1637–1649*, [trad. it. con testi originali a fronte], a cura di G. Belgioioso, con la collaborazione di I. Agostini, F. Marrone e M. Savini, Milano, Bompiani, 2009; BOP, followed by the indication of the page = R. Descartes, *Opere Postume 1650–2009*, [trad. it. con testi originali a fronte], a cura di G. Belgioioso, con la collaborazione di I. Agostini, F. Marrone e M. Savini, Milano, Bompiani, 2009; *De homine* = R. Des Cartes, *De homine*, figuris et latinitate donatus a Fl. Schuyl, Lugduni Batavorum apud P. Lefen et F. Moyardum, 1662; *L'Homme* = R. Descartes, *L'Homme* [...] et un *Traitté de la formation du foetus* [...], à Paris chez Charles Angot, 1664; *Remarques* = L. De La Forge, *Remarques* [...] sur le *Traitté de L'Homme*, de René Descartes et sur les figures par luy inventées, in R. Descartes, *L'Homme* [...] et un *Traitté de la formation du foetus* du mesme autheur [...], à Paris chez Charles Angot, 1664.

F.A. Meschini (✉)
Dipartimento di Studi Umanistici, Università del Salento,
Piazzetta Angelo Rizzo, 1, 73100 Lecce, Italy
e-mail: francoaurelio.meschini@gmail.com

D. Antoine-Mahut, S. Gaukroger (eds.), Descartes' *Treatise on Man* and its Reception, Studies in History and Philosophy of Science 43,
DOI 10.1007/978-3-319-46989-8_3

A philological matter about Descartes' *L'Homme* already arises in the seventeenth century. The correspondence from Descartes itself contains annotations on the text of *L'Homme* that we need to consider for a critical edition.[1] The same attention should be paid to two letters, one *ad lectorem* and one *ad amicum*, two Latin editions edited by F. Schuyl (1662, 1664) and the *Remarques* [...] *sur L'Homme* by Louis de la Forge. Clerselier in particular clearly poses a real philological problem and yet for him, this question is does not arise, because, from the first French edition (1664), the possibility of editing a critical text was in fact precluded by Clerselier's negative judgement of Schuyl's edition.[2]

Because of this, we can say that, paradoxically, the best edition of *L'Homme* is still the one from AT, albeit this edition is flawed by prejudice inherited from the first French publisher. This edition represents the best example because it gives a list of all the places (at least this is the purpose) in which there are differences between the two editions. At the same time we can say that it is flawed because it doesn't give proper consideration to the "innovations" of Schuyl, which while they are not true variations because they are in translation, there are nevertheless traces of variants which may be author's variants, so we need to keep them in consideration regarding the apparatus of a critical edition. After all, the same Latin translation (1677) based on the Clerselier's edition[3] is in some cases, one in particular, helpful in establishing the text.

That said, Clerselier was the first to raise the question of the text, at two different times (in the preface to the second volume of the letters, in 1659, and in the preface to *L'Homme*, in 1664) regarding three or four different aspects. The issues posed by Clerselier are still the same today. Briefly, they are:

1. figures;
2. text partitioning;
3. the text;
4. the relationship between *L'Homme* and *Monde ou Traité de la Lumière*.

Clerselier, however, considers that a history of the text is not needed, because, apart from other possible considerations, he possessed (or believed he possessed) the autograph (the only autograph).

Discussion of this last point, which starts with the excellent reconstruction of van Otegem,[4] is something which I have discussed elsewhere,[5] and I will not discuss it further here. Points 1, 3, 4 are important points of comparison with the edition of

[1] See F.A. Meschini, *Note per un'edizione critica de* L'Homme *di Descartes*, in F.A. Meschini (a cura di), *Le opere dei filosofi e degli scienziati. Filosofia e scienza tra testo, libro e biblioteche*, from the panel held in Lecce 7–8 February 2007, Firenze, Olschki, 2011, pp. 165–204.

[2] See Clerselier, *Préface*, in BOP 598–600.

[3] R. Descartes, *Tractatus de homine et de formatione fœtus*, quorum prior notis perpetuis Ludovici De La Forge, M.D. illustratur, Amstelodami, apud Danielem Elsevirium, 1677.

[4] M. van Otegem, *A bibliography of the works of Descartes (1637–1704)*, 2 vols., («Quaestiones infinitae», n. 38), s.l., Publications of the Zeno Insitute of Philosophy, 2002, vol. II, pp. 485–536.

[5] See Meschini, *Note... cit., passim.*

Schuyl; with regard to issue 2, Clerselier doesn't makes any references to the Latin edition, while, as we briefly shall show, the division in the edition by Schuyl raises a significant question about the tradition of the text of *L'Homme* and, therefore, for the text itself. I will not say anything about the figures, which I have discussed in my previous work.[6] I will focus, instead, in what follows, on certain aspects of points 2 and 3, and this will also raise some considerations concerning the relationship between the Clerselier's text and the one from Schuyl.

3.1 Text Partitioning

L'Homme, as is known, is divided into five parts and one hundred and six articles with a continuous numbering. Clerselier explicitly declares that he himself had introduced the partitioning and the further articulation and also that he had assigned the titles of each article and of each part. He justifies this triple intervention as an adaptation from the first treatise (*L'Homme*) to the second, the *Description du corps humain*, which already had a partial partitioning.[7] Clerselier, once he had completed the division of the second treatise, considered it useful to proceed in the same way also for the first treatise. After all, the intervention of the editor doesn't prejudice the text at all, since, as stated by Clerselier (and the statement is quite acceptable), someone who wants to read it in continuous way can still do this without taking into account the partitioning. In some way, Clerselier presents to the reader the same concern that Descartes had already shown in the preface of *Discours*.[8]

[6] I already mentioned the question in G. Cimino, F.A. Meschini, *Le origini della neurofisiologia. Le ricerche sul sistema nervoso nella seconda metà del Seicento*, in *Storia della Scienza*, 10 voll., Roma, Istituto della Enciclopedia Italiana, 2001–2004, vol. V (2002): *La rivoluzione scientifica e l'Età dei Lumi*, pp. 658–676: pp. 666–667.The preface by Clerselier (BOP 602–630) is essential for the history of the text's composition de *L'Homme* and for a comparison with the ones from *De homine*. About the illustrations on *De homine*, see the opinion gave by Nicolas Steensen in the letter to Th. Bartholinus (August, 26th 1662): «Prodiit hisce diebus Cartesii tractatus de homine una cum figuris a Florentio Schylio inclytae Urbis Sylvae Ducis Senatore, et ibidem Philosophiae Professore editus, in quo figurae conspiciuntur non inelegantes quas ex ingenioso cerebro prodisse certum est., an vero tales in ullo cerebro conspiciendae valde dubitarem» (in Th. Bartholinus, *Epistolarum medicinalium. Centuria IV* [...], Hafniae, typis Matthiae Godicchenii, sumptibus Petri Haurold, Acad. Bibl., 1667, p.113). About the studies on the illustrations, see Claus Zittel, *Conflicting Pictures: Illustrating Descartes'* Traité de l'homme, in Sven Dupré, Christoph Lüthy (eds.), *Silent Messengers. The Circulation of Material Objects of Knowledge in the Early Modern Low Contries*, Berlin, Lit Verlag, 2011, pp. 217–260. See the introduction note curated by G. Belgioioso and J.-R. Armogathe in BOP 207–209.

[7] See Clerselier, *Preface*, in BOP 630.

[8] «Si ce discours semble trop long pour être tout lu en une fois, on le pourra distinguer en six parties» (AT VI 1/BO 24).

Among modern publishers, AT doesn't take into account either the partitioning or the division of the articles, followed by André Bridoux (1953),[9] Ferdinand Alquié (1963),[10] Gianni Micheli (1966)[11] for the Italian translation, Karl E. Rothschuh (1969)[12] for the German translation, Thomas Steele Hall (1972)[13] for the American edition and Annie Bitbol-Hespériès (1996)[14] in her annotated edition; Gianfranco Cantelli (1960)[15] for the first Italian translation maintained the division into five parts, but not into articles, followed by an important edition for Cartesian studies in Italy, that of Eugenio Garin published by Laterza (1969),[16] while the most recent edition by Bompiani (2009), which has as a reference text the one from Clerselier's edition (the second edition: 1677) repeats the divisions exactly.[17]

The Schuyl's edition is not divided into parts, but the text is divided into 36 articles (without titles). Is it an editorial intervention as in the case of the first French edition? Schuyl, who in two editions published by himself, gives, in a wealth of detail, indications referring to the copies which he used to lead (1662) or revise (1664) the translation. He, doesn't say, however, anything about the division in articles, neither in the letter *ad lectorem* (1662) nor in the letter *ad amicum* (1664).

Based on the statement of Clerselier, who affirmed that the autograph manuscript in his possession consists of a continuous text, without any partitioning, the modern editor will have to answer the question about the authorship of this division in the Latin translation. Who is the reponsible for this division? It seems, according to the *usus scribendi* of Schuyl, that the translation should be traced back to a stage which precedes the translation itself.

---

[9] [R.] Descartes, *Œuvres et lettres*, textes présentés par A. Bridoux, Bibliothèque de la Pléiade, Paris, Garnier, 1953, pp. 83–873.

[10] R. Descartes, *Œuvres philosophiques*, éd. par F. Alquié, 3 tomes, Paris, Garnier, 1963–1973, tom. I, pp. 379–480.

[11] R. Descartes, *Opere Scientifiche*, I vol.: *La Biologia*, a cura di G. Micheli, Torino, UTET, 1966. This text was the first one, at least in Italy, which gave value to the *Remarques* of La Forge. About La Forge's *Remarques*, see E. Scribano, *Macchine con la mente. Fisiologia e metafisica tra Cartesio e Spinoza*, Roma, Carocci, 2015, pp. 77–106.

[12] R. Descartes, *Über den Menschen* (1632) sowie *Bescheibung des Menschlichen Körpers* (1648), nach der ersten französischen Ausgabe von 1664 übersetzt und mit einer historischen Einleitung und Anmerkungen versehen von Karl E. Rothschuh, Heidelberg, Verlag Lambert Schneider, 1969.

[13] R. Descartes, *Treatise of Man*, French Text with Translation and Commentary by Th. S. Hall, Cambridge (Massachusetts), Harvard University Press, 1972.

[14] R. Descartes, *Le Monde, L'Homme*, introd. de A. Bitbol-Hespériès, textes établis et annotés par A. Bitbol-Hespériès et J.-P. Verdet, Paris, Éditions du Seuil, 1996.

[15] R. Descartes, *L'uomo*, introduzione e traduzione di Gianfranco Cantelli, Torino, Boringhieri, 1960. For the reason of this choice, see introduction, pp. 30–31.

[16] R. Descartes, *Il mondo Trattato della luce. L'uomo*, trad. it. di M. Garin, introduzione di Eugenio Garin, Bari, Laterza, 1969. Also Stephen Gaukroger for the English edition maintained just the division into five parts: R. Descartes, The World and Other Writings, transleted and edited by Stephen Gaugroker, Cambridge, Cambridge University Press, 1998, pp. 99–169.

[17] For the reason of this choice, see introduction note by G. Belgioioso, J.-R. Armogathe, in BOP 209–211.

An intervention of this kind, even if doesn't change the text directly, remains an important intervention that affects the presentation of the text, and only through negligence could it be silent and not motivated by the publisher. This doesn't seem the case for a diligent publisher, as Schuyl seems to be. It is not unlikely, therefore, that Schuyl, unlike Clerselier (responsible for his own admission of the division), has already found the division in its sources and has not felt the need to talk about it or to justify it. As studies stand, it is difficult to add more, but it is certain that the division that we find in the Schuyl edition raises new questions about the transmission of the text of *L'Homme*, even raising doubts about to its drafting.

On the question of the division into parts and articles, the future editor of *L'Homme* has to answer another question: When did Clerselier introduce the division? At which stage of preparing the text for printing did he do it? Before or after sending the copy to La Forge? In the *Remarques*, in fact, there is no mention of the division, and this tells us that the text on which La Forge worked, —the annotated text from La Forge—is a continuous text. And yet, as seems plausible, Clerselier wrote (or revised) his preface, keeping in mind, and being in possession of La Forge's notes. This means that we need also to investigate the relationship (and debt) between Clerselier's *Prefatio* and La Forge's *Remarques*.

We give below a synopsis of the division in Clerselier articles and Schuyl. Remember that the articles in the French edition are distributed as follows: Part I: Articles 1–14; Part Two: 15–26; Part Three: 27–51; Part Four: 52–57; Part Five: 58–106.

Schuyl [1662]			Clerselier [1664]	
p.	1	art. 1	art.	1
	1–3	art. 2		2
	3–5	art. 3		3–5
	5–7	art. 4		6
	7	art. 5		6
	7–9	art. 6		7
	9–11	art. 7		8
	11–13	art. 8		9–11
	13	art. 9		12
	13–26	art. 10		13–23
	26–29	art. 11		24
	29–31	art. 12		25
	31–37	art. 13		26–31
	37–39	art. 14		32–33
	39–41	art. 15		34
	41–43	art. 16		35–36
	43	art. 17		36
	43–46	art. 18		37–39
	46–54	art. 19		40–46
	55–65	art. 20		47–51
	65	art. 21		51

Schuyl [1662]			Clerselier [1664]	
	65–66	art. 22		52
	66–67	art. 23		52–54
	67	art. 24		55
	67–68	art. 25		55
	69–70	art. 26		56
	70–72	art. 27		57–62
	72	art. 28		63
	73–78	art. 29		63–65
	78–80	art. 30		66–67
	80–103	art. 31		68–90
	103–104	art. 32		91
	104–113	art. 33		92–100
	114–115	art. 34		101
	115–120	art. 35		102–105
	120–121	art. 36		106

We can observe a discrepancy in Clerselier's partitioning, between the text and the partition (which reveals a conservative attitude on the part of the publisher, not so different from that demonstrated in the *incipit* of the work),[18] in particular in the passage from the fourth part to the fifth. I refer to the beginning of the fifth part, in which it is seems unlikely that Clerselier didn't notice that beginning a new part with the adverb *secondement* was forced.

According to the text we have, Clerselier decides to begin a part (the fifth) with an numeral adverb which refers to the word *before* that should be elsewhere in the text: «Secondement pour ce qui est des pores ...».[19] This flawed correspondence between the text (Descartes) and Clerselier's partitioning, gives us a clue, I think, to the text itself, more precisely a sign of the text's authenticity. This is above all because of the adverb *secondement*, which can only be the wording of Descartes. In fact, for what we have called the the forcing of the partitioning, Clerselier had no reason to introduce the numeral adverb, but rather could have had an interest in removing it. This is a matter on which it would be worthwhile to dwell further, were it not for the fact that Schuyl himself presents a different text here. Here we read: «Quod *vero* ad cerebri poros attinet, eos nobis aliter imaginari non debemus, quam per modum quorundam intervallorum, quae inter fila lintei observare licet».[20] «Quod *vero* ad cerebri poros attinet»: in Schuyl's text the adverb (*secondement*) has been replaced by the word *vero*; we find the same thing a few pages later: «Ut *vero* omnia explicem, quae in illo plexu occurrunt notatu digna, operae pretium erit agere de spirituum distributione».[21]

[18] See Meschini, *Note* … cit., p. 171.

[19] AT XI 170/BOP 446.

[20] *De homine*, p. 72. Quoted words not italicized in the original.

[21] *De homine*, p. 74. Quoted words not italicized in the original.

In this case, it is the third point that Descartes believes requires an explanation (the distribution of animal spirits) and also here the adverb (*troisièmement*) is replaced, as in the previous case, by the word *vero*. This word is certainly used in enumerations, but it doesn't say anything about the place occupied in the enumeration itself. In the next Latin translation, in the Clerselier edition, by the way, we find the numerals adverbs preserved.

What is the relationship between the Schuyl and the Clerselier edition? If, as it seems, it is Schuyl himself who introduced the innovation, not Clerselier (though we have to consider that the Schuyl sources are, probably, later than manuscript possessed by Clerselier) and if, as I have said elsewhere, we cannot exclude the possibility that Descartes himself had revised the text, it remains to determine whether, in this case, the changes are due to an intervention of the author (author variant = **A**) or from a source used by Schuyl (tradition variant = **T1**) or from Schuyl himself (translation variant = **T2**). The question is not so easy, in fact. If we were sure that the only autograph was the one owned by Clerselier (as Clerselier himself believed), there would be no doubt about the correct (and last) reading. But once again, allowing the possibility of author variants of the edition that Schuyl seems to present traces of, the Clerselier autograph loses its absolute authority and is up to the editor evaluate the text and give an account in the apparatus.

In this case the editor will be faced with a textual problem that is not so common. In short, there seems to be no reason why this variant is attributed to **A** rather than in **T1** or, on the other hand, to **T2**. It is a situation that reflects that in which there is a publisher confronted with two *adiaphore* variants. In this case, we have only a text (the one published by Schuyl) and three possible authors, between whom there is no reason to one rather than another. But, paradoxically, this indifference shows in favour of an intervention from **A**, the only one not to be bound by any text, and therefore the one allowed to innovate without reasons (apparently) and explanations.[22]

3.2 A Passage Latin in the French Edition

Another point on which the Latin translation (or rather, in this case, the Latin translations) allow us with greater confidence to tell something about the text and also something about the autograph possessed by Clerselier is related to the Latin passage, the only Latin passage which we can find in the edition by Clerselier. Here we can compare both the passage in the text of Clerselier, and the one in text of Schuyl:

[22] See Meschini, *Per un'edizione critica…*, cit., (pp. 534–540) for the differences between Schuyl's edition and Clerselier's edition about the cross reference (*ci-desssus*: 2 *alibi*, 11 *supra*, 1 *superius;* 3 not translated); and see Meschini, *Note…*, cit., (pp. 173–174) for the inner differences in Schuyl's editions of 1662 and 1664.

52. Lorsque les liqueurs, que j'ai dit ci-dessus servir comme d'eau-forte dans son estomac, et y entrer sans cesse de toute la masse du sang par les extrémités des artères, n'y trouvent pas assez de viandes à dissoudre pour occuper toute leur force, elles la tournent contre l'estomac même, et agitant les petits filets de ses nerfs plus fort que de coutume, font mouvoir les parties du cerveau d'où ils viennent: ce qui sera cause que l'Ame étant unie à cette machine concevra l'idée générale *de la faim*. Et si ces liqueurs sont disposées à employer plutôt leur action contre certaines viandes particulières que contre d'autres, ainsi que l'eau-forte commune dissout plus aisément les métaux que la cire, elles agiront aussi d'une façon | particulière contre les nerfs de l'estomac, laquelle sera cause que l'Ame concevra pour lors l'appétit de manger de certaines viandes, plutôt que d'autres. (*Hic notari potest mira hujus machinæ conformatio, quod fames oriatur ex jeiunio; sanguis enim circulatione acrior fit; et ita liquor ex eo in stomachum veniens nervos magis vellicat; idque modo | peculiari, si peculiaris sit constitutio sanguinis; unde pica mulierum.* L'on peut ici remarquer la structure admirable de cette machine, qui est. telle que la faim lui vient d'avoir été trop longtemps sans manger; dont la raison est. que le sang se subtilise et devient plus acre par la circulation; d'où il arrive que la liqueur qui va des artères dans son estomac agite et picote plus fort que de coutume les nerfs qui y sont, et même qu'elle les agite d'une certaine façon particulière, si la constitution du sang se trouve aussi avoir quelque chose de particulier: et c'est. de là que viennent ces appétits désordonnés, ou ces envies des femmes grosses). Or ces liqueurs s'assemblent principalement au fond de l'estomac; et c'est là qu'elles causent le sentiment de la faim.	22. Cum igitur liquor, quem supra dixi in stomacho velut aquae fortis officio defungi, et ex sanguinis massa per arteriarum extremitates in ipsum continuo illabi sunt, |sufficiens alimentum istic non invenit, quod consumat; aut in quod omnes suas vires exerceat, in ipsum stomachum illas convertens, filamenta nervorum ejus solito vehementius agitat, commoventur cerebri partes, ex quibus illa oriuntur. Adeoque in causa sunt, quod anima, illi machinae unita, generalem famis ideam concipiat. Si vero liquor ille ejus ejus sit temperamentum, ut vires suas magis exercere possit in certos quosdam cibos, quam in alios; quemadmodum communis aqua fortis citius metalla solvit, quam ceram: peculiari quoque modo in nervos stomachi: unde anima tum concipiet appetitum hos potius cibos, quam alios comedendi. *Hic notari potest mira hujus machinae conformatio, quod fames oriatur ex jejunio. Sanguis enim circulatione acrior fit, et ita liquor ex eo ad stomachum veniens, nervos magis vellicat; idque modo peculiari, si peculiaris sit constitutio sanguinis: unde [1664: inde] pica mulierum.* .... 23. Liquor autem iste praecipue in stomachi fundo congregatur, ibique sensum famis proritat

We know that La Forge in *Remarques* judges this Latin passage to be an interpolation. Clerselier was the last to review the text, including the text of the *Remarques*. He reaffirms the validity of the editorial choice to keep the text as it is because it is authentic. He affirms this in *Remarques*, following the observation of La Forge:

> *Hic notari potest. p. 56. l. 8.* Ie croy que ce passage n'appartient pas au texte, mais que c'est la remarque de quelqu'un à qui cet écrit est tombé entre les mains, laquelle il avoit mise à la marge, mais le copiste peu intelligente ou trop fidele, l'a inseré dans le texte, qui ne laisse pas d'estre entier sans cela; outre qu'il n'y a point de raison pourquoy L'Atheur auroit

changé de langage. *L'observation est iudicieuse, et veritable, sinon que la remarque mesme est de Monsieur Descartes; c'est pourquy on ne l'a pas voulu obmettre.*[23]

How does this debate proceed? First, let's see how modern editors have solved the issue:

AT XI 163–164 The edition adopts the Latin text and expunges, taking into footnotes, the eleven lines which, in the Clerselier edition, follow the Latin passage, and forms a translation which is considered to have been made by Clerselier and not by Descartes. The editor mentions that in Schuyl's edition there are the five Latin lines, but in italics, as if they were unrelated to the text. The Micheli, Garin, Cantelli, Hall editions all followed on this point AT.

Alquié (I 434–435) The editor expunges, taking into footnotes, the Latin passage and the translation. In so doing, that he accepts the observations of La Forge (who, by the way, doesn't refer to the French text) and reinforces them with an observation that is based upon the *usus scribendi* from Descartes: Descartes switches, it is true, from one language to another, but only in the letters, never in the treatises.[24] In this case the switch of language seems contrary to his own habits. The edition of Bitbol-Hespériès/Verdet explicitly follows the choice from the Alquié edition (pp. 146, 193, n. 131). Also Rothschuh and Gaukroger expunges the Latin passage and the French translation.

BOP 2009 Bompiani edition publishes the entire text of Clerselier (both Latin and French) considering the tradition of the text of *L'Homme* to be much too uncertain to depart from that of the first publisher (pp. 210–211).

So, three different solutions. The one from AT agrees with Clerselier for the Latin text, but doesn't accept the *lectio* of the first French publisher for the French text. His decision to expunge the French text is based on a plausible conjecture, but without any real historical philological support. Alquié expunges both Latin and French text agreeing with La Forge, in the Latin text, but, in my opinion, misinterpreting the French one. La Forge, in fact, makes no reference to the French text, probably because in his copy the translation has not been inserted. The translation appears in the printed text and Clerselier may have added it at a later date, probably after the remarks of La Forge about the Latin text in *Remarques*.

The solution from the Bompiani edition is the most conservative because it doesn't propose to establish the text, but to give, as an hallmark, Clerselier's French edition. For this reason the Bompiani edition published the edition of the text by Clerselier both in Latin and French.

Personally, I agree with the solution of AT but with different motivations and evidence.

[23] *Remarques*, p. 286. The two final lines, italics in the text, represent an addition to La Forge's remark. They are an intervention by Clerselier in La Forge's text.

[24] The Latin passage in *Meteores* is a quote. I give thanks to the young fellow Grigore Vida who drew my attention to this.

1. Clerselier himself, owning the autograph, tells us that the Latin text is not an interpolation; the most compelling evidence it supplies, in fact, is in the text of Schuyl which shows the same Latin passage. The fact that the Latin text is also in Schuyl (without significant differences) can be taken as an argument against the hypothesis of interpolation formulated by La Forge. In fact, as van Zurck, the main witness of Schuyl, copied directly from the autograph of Descartes[25] (in which we can find the Latin passage), this would mean that the interpolation was already in the manuscript even when it was in the hands of Descartes and this is impossible (by definition).

And, more than that, we cannot assume just because of the italics that the mentioned passage was already in the antigraph of Schuyl. AT noted the italics, but he didn't draw all the consequences. The italics in this case are not used to underline the idea that the text is in some way extraneous to the work, but rather to indicate that it was already in Latin. In Clerselier the passage was in italics to mark the transition from one language to another, but in Schuyl italics are intended to alert the reader to the fact that in the manuscript that passage was already in Latin and therefore is not the result of translation.

2. AT expunges the French text that translates the Latin text passage because it is a translation of the Latin passage, so probably not from Descartes, but probably from the editor itself. The French text is expunged by Alquié because he extends the observation of La Forge to the passage enclosed in round parenthesis, and then also to the translation.

Nothing tells us that La Forge intends to extend his observation also to the French text. In favour of his interpolative conjecture he takes into consideration the switch of language (from Latin to French, so that it means it is referring only to the Latin lines), which he considers totally unjustified. Concerning the language switching, Alquié takes gives more weighting to the *usus scribendi* of Descartes, from French to Latin, that was used only in private texts and letters but never in the treatises (and even in *L'Homme*). This is good point, but not in this case because we are dealing with an incomplete text, unpublished, revised only in part, so we cannot rely on this habit of Descartes in this case. After all, as hypothesized by AT, it is quite likely that the French text is an addition of Clerselier, and there are two reasons for this: (1) it is not present in the Schuyl edition; (2) it is not present—something I consider that quite important—even in the Latin translation of the Clerselier edition, which, however, reports the three Latin lines as they are. The French text is not translated, because the translator or Clerselier himself doesn't consider it (apparently) useful. The passage seems to have only had a translation function and was not by Descartes, but by the editor.

At the end, the Latin passage will be kept in the text (based on the affirmation of authenticity by the French publisher and based on a claim of Schuyl that reports the

[25] Schuyl, *Ad lectorem*, in *De homine*, p. 32 n.n. See. van Otegem, *op. cit.*, pp. 488, 495; Meschini, *Note…*, cit. pp. 171–187.

passage in italics being taken from a different tradition than the one from Clerselier); the French passage will be removed from the text and reported only in apparatus, based on the Schuyl's edition and on the Latin translation from Clerselier .

3.3 An Omission in the Latin Text of Schuyl

Another matter I would like to draw attention to an *omission* in the Latin text of Schuyl. This is article LXIV of Clerselier, the second of the fifth part, entitled by Clerselier: *Comment se fait la distribution des esprits et d'où vient l'eternüement et l'éblouissement, ou vertige*. Here Descartes describes the distribution of animal spirits, focusing on some phenomena originating in the brain, sneezing and dizziness or vertigo.

Both phenomena are explained on the basis of the movement of the animal spirits which are considered in terms of direction and strength.[26] The interesting part, however, is not the explanation of the phenomena that Descartes gave, an explanation that is integrated by La Forge in introducing a reference to fermentation,[27] but rather the consideration that comes immediately after it, about the weakest parts of the spirits. These are the last lines of the mentioned passage:

> Et notez *en passant* que *ces plus foibles parties* des Esprits, ne viennent pas tant des arteres qui s'inserent dans la glande H, comme de celles qui se divisant en mille branches fort déliées tapissent le fond des concavités du cerveau[28]

These lines are missing in the Latin text. The text describes *scotomia* (dizziness) and *vertigo*, caused by the reflux of those weaker spirits: as they come out from the gland H they don't have the required strength to enter either into small tubes marked *aa*, or to go through I (representing the pores through which the animal spirits pass into the nostrils) or through K and L (pores that lead the spirits to the palate), so they

[26] «Iamais ils [scil.: les esprits animaux] ne s'arrestent un seul moment en une place, mais à mesure qu'ils entrent dans les concavitez du cerveau EE, par les trous de la petite glande marquée H, ils tendent d'abord vers ceux des petits tuyaux a, a, qui leur sont le plus directement opposez; et si ces tuyaux a, a, ne sont pas assez ouverts pour les recevoir tous, ils reçoivent au moins les plus fortes et les plus vives de leurs parties, pendant que les plus foibles et superflües sont repoussées vers les conduits I, K, L, qui regardent les narines, et le palais; à sçavoir les plus agitées vers I, par où, quand elles ont encore beaucoup de force, et qu'elles n'y trouvent pas le passage assez libre, elles sortent quelquefois avec tant de violences, qu'elles chatoüillement les parties intérieures du nez, ce qui cause l'*Eternüement*; puis les autres vers K et vers L, par où elles peuvent facilement sortir, pource que les passages y sont fort larges; où si elles y manquent, estant contraintes de retourner vers les petits tuyaux a, a, qui sont en la superficie interieure du cerveau, elles causent aussi-tost un *éblоüissement*, ou *vertige*, qui trouble les fonctions de l'*imagination*./Et notez en passant que ces plus foibles parties des Esprits, ne viennent pas tant des arteres qui s'inserent dans la glande H, comme de celles qui se divisant en mille branches fort déliées tapissent le fond des concavités du cerveau» (AT XI 171–172; BOP 450–452).

[27] See *Remarques*, p. 327.

[28] See here *supra*, n. 26.

return back to *aa.* In Schuyl's text, this is followed by: «*In transitu* vero notare licet, *particulas spirituum debiliores* in pituitam facile condensari…», which corresponds to the passage in Clerselier that follows the omitted text by Schuyl. The text matches, but doesn't translate it. It doesn't translate it, not only because Schuyl possessed a manuscript different (as we documented) from the autograph used by Clerselier for his edition (and, therefore, also a text partially different), but also because, compared to Clerselier, the translation of Schuyl seems to be more like an adjustment. In Clerselier, in fact, at this point we read: «Notez *aussi* qu'*elles* se peuvent aisement épaissir en pituite…». By contrast, Schuyl recovers the *en passant* (*in transitu*) from the previous sentence (where the French text present the word *aussi*) and replaces the pronoun (*elles*), which refers to *plus foibles parties des Esprits* from the previous sentence, with *particulas spirituum debiliores.* This expression in the Latin text becomes necessary just because the previous sentence drops it (the equivalent to: *Et notez ... des concavités du cerveau*). We can now quote the entire Latin passage comparing it with the French text and with the Clerselier Latin text edition from 1677.[29]

Clerselier	Schuyl
… où si elles y manquent, estant contraintes de retourner vers les petits tuyaux *a, a*, qui sont en la superficie interieure du cerveau, elles causent aussi-tost un ébloüissement, ou vertige, qui trouble les fonctions de l'imagination.	… vel si illis ibidem introitus negetur, ita ut coacti sint reverti ad tubulos *a a*, qui interiori cerebri superficiei insunt, confestim scotomiam, vertiginemve cuasantur, quibus functiones imaginationis turbantur.
Et notez *en passant* que *ces plus foibles parties des Esprits*, ne viennent pas tant des arteres qui s'inserent dans la glande H, comme de celles qui se divisant en mille branches fort déliées tapissent le fond des concavités du cerveau.	……………………….. ………………………
Notez aussi qu'*elles* se peuvent aisément épaissir en pituite, non pas jamais étant dans le cerveau, si ce n'est. par quelque grande maladie, mais en ces larges espaces qui sont au-dessous de sa base, entre les narines et le gosier;	*In transitu* vero notare licet, *particulas spirituum debiliores* in pituitam facile condensari: non quidem quamdiu in cerebro continentur, nisi vehemens quidam morbus adsit: sed in istis amplioribus spatiis, quae infra basin illius inter nares et guttur deprehenduntur.

How can we explain this omission? Firstly, we shouldn't consider the omission to be an accidental act, just because of these 'adjustments'. In fact, if it was unintentional (and this is true whether the omission is attributed to the translator or directly from the copy or the author), there would be no explanation for the use of *in transitu* and, above all, the 'repetition' of the sentence *particulas spirituum debiliores.* But, if this is not an intentional omission then, paradoxically, it is not question

[29] Italics and underlining are not in the text.

of omission, but a real intervention. Those lines were not simply omitted but deleted. Who was responsible for it? The possibilities are, as in previous cases, at least three: Schuyl, the copy or copies he used to conduct the translation, and Descartes himself.

La Forge, commenting this text, doesn't notice anything abnormal, but is quick to point out that:

> Elles [*scil.*: les plus faibles parties des Esprits] ne viennent pas tant des artères qui composent le lassis choroïde que des autres branches de l'artère carotide, que nous avons dit se répandre le long de la supeficie des ventricules, et la tapisser presque toute pendant que l'Animal est en vie; car apres sa morte ces artères venant à se plier et à se retirer, cette tapisserie ne paroistre presque plus.[30]

A clarification just to exclude the possibility that the weakest parts of the animal spirits may come from the choroid plexus, which we know is particularly rich in arterial blood and animal spirits. Nothing else. Still, La Forge should have recognized some aspect of a not entirely orthodox Cartesian doctrine about the pineal gland. In fact, the pineal gland is supported by the choroid plexus, it is almost suspended, but here we can read *artères s'inserent dans la glans H*. It is possible that it is because of this verb, because of this insertion of the arteries into gland, because of this inaccuracy, if we want to consider it as an unfortunate choice of words, or an anatomical inaccuracy, that the innovation has been introduced into the text of Schuyl.

This comparison allows us a further consideration: except this case, I beleive that we can't find any other variants between the Clerselier and Schuyl edition related in some way to the pineal gland. The two texts, in respect of that topic, are entirely the same. Still, this passage about the pineal gland should have caused some surprise in the context of Descartes' *work*: at some stage of drafting or transmitting the text, or subsequently in commenting and the studying the text. Because, from this point of view the work, both in the French and in the Latin text, is still surprising.

What is surprising, I mean absolutely surprising, is the transition (taking the French edition, but we could apply the same method to the Schuyl edition) from the first to the second part. A few lines apart, and on two consecutive occasions, we see in fact a 'forgetfulness', which is nevertheless inexplicable as an omission on the part of the author.

First time:

> Or à mesure que ces Esprits entrent ainsi dans les concavités du cerveau, *ils passent de là dans les pores de sa substance, et de ces pores dans les nerfs*; où selon qu'ils entrent, ou même seulement qu'ils tendent à entrer plus ou moins dans les uns que dans les autres, ils ont la force de changer la figure des muscles en qui ces nerfs sont insérés, et par ce moyen de faire mouvoir tous les membres.[31]

The first part, as we mentioned before, ended up talking about animal spirits (*elles cessent d'avoir la forme du sang, et se nomment les Esprits Animaux*) and the

[30] *Remarques*, pp. 328–329.

[31] AT XI 130 /BOP 378.

beginning of the second part is entirely coherent, especially in the Latin text, whereas in the French text at the beginning of the second part, there is seamlessly, without even a new paragraph, on the same line, the closing words of the first part. The transition from concave or cerebral ventricles to the nerves through the pores of the same brain substance takes place as if the pineal gland is « uninformed », where a few lines earlier its said that in this gland enters all the thinner parts and is agitated from the blood and that it must be imagined « comme une source fort abondante, d'où ells [*scil.*: the thinner parts and agitated from the blood] coulent en meme temps de tous côtés dans les concavitez du cerveau».[32]

The second forgetfulness is immediately below and is even more striking: it is, in fact, the most famous Cartesian metaphor, the one about the fountains of the royal gardens. In that metaphor, so apparently meticulous, the pineal gland has no place. Furthermore the pineal gland is never in question in any of the metaphors or similarities occurring in *L'Homme*.

Actually, if we take the partitioning in the French edition as our reference point, in the three central parts of the work the pineal gland is missing. This is even more surprising because in the parts where the pineal gland is not mentioned we can nevertheless find feelings and movement, that is to say, we can read about functions, and in the explanation of these—once they are attributed the role that Descartes attributed in the first part (or the fifth!) of *L'Homme*—the pineal gland becomes essential.

How could we explain this absence—or, we can say, this presence—that suggests a underground path? The pineal gland emerges in the first part, then it disappears in the following three parts, then again it appears and, this time, consistently, in the fifth and final part. What has happened? There is no question that every step in this direction is treacherous. The comparison between the French text and the Latin one can not be of any help because the two texts are, as we have already explained, similar about this aspect. However we cannot consider it as a forgetfulness. If anything, this similarity allows us to fix the time when the van Zurk copy autographed by Descartes was a drafting phase of *L'Homme*. And, at the same time, it allows us to imagine an earlier stage without pineal gland.

In my opinion, at this phase of the studies, we can't affirm much more. Neither can we expect more from a future and desirable critical edition, because of the reasons we have given and because of the similarity between the witnesses. The underground presence of the pineal gland in *L'Homme* is in fact a challenge for historians. A study of the pineal gland in Descartes will also have to face, at least, another apparent anomaly. This time it is in the *Passions de l'âme*, especially the fact that in articles VII–XVI, even if we read about animal spirits, nerves, and muscles, there is no mention of the pineal gland— the *petite glande* —which will be the central starting point from articles XXX, and for all the rest of work.[33]

---

[32] AT XI 129/BOP 376.

[33] For further considerations, see F.A.Meschini, *Materiali per una storia della medicina cartesiana. Dottrine, testi, contesti e lessico*, («Filosofie», n. 299), Milano-Udine, Mimesis Edizioni, 2013, pp. 104–128.

Chapter 4
L'Homme in English

Stephen Gaukroger

Abstract The earlier lack of interest in *L'Homme* and how this has changed over the last few decades, with attention to the physiological ingredients in Cartesian epistemology and account of the passions. The translations of Hall and Gaukroger are compared.

There have been two English translations of *l'Homme*. The first was a copiously annotated edition by Thomas Hall, which appeared as *René Descartes: Treatise on Man,* published by Cambridge University Press, in 1972. In 1998 my own translation appeared along with translations of *Le Monde* and *La Description du corps humain*, also published by Cambridge University Press. Besides these, there is a selection of short extracts from *L'Homme* that appeared in the first volume of the most comprehensive English translation of Descartes' works, the three-volume *The Philosophical Writings of Descartes*, translated by John Cottingham, Robert Stoothoff, and Dugald Murdoch, published by Cambridge University Press in 1985.

The interesting questions arise not from translation difficulties—the language of the text itself is straightforward—but primarily from how one places the work in a proper context. It is striking that there were no English translations before Hall's. This is in spite of the fact that the treatise offers Descartes' most extensive treatment of perceptual cognition, a theme that lies at the core of the Cartesian epistemology to which philosophers have devoted a great deal of attention. In this respect, it would seem that the work might have been included in the collections of 'philosophical writings' that have appeared since the two-volume English translation published by the Haldanes in 1911. But it was effectively quarantined from what was considered to be Descartes' philosophical work, and until recently I doubt if the vast majority of Anglophone writers on Descartes' epistemology were even aware of it, or at least were aware of its contents: and Anglophone philosophers were certainly not unique in this regard.

S. Gaukroger (✉)
Unit for the History and Philosophy of Science, University of Sydney,
Level 3, Carslaw Building (F07), Eastern Avenue, Sydney, NSW 2006, Australia
e-mail: Stephen.gaukroger@sydney.edu.au

D. Antoine-Mahut, S. Gaukroger (eds.), Descartes' *Treatise on Man* and its Reception, Studies in History and Philosophy of Science 43,
DOI 10.1007/978-3-319-46989-8_4

In what follows, I look at some of the reasons for the absence of *L'Homme* from the philosophical canon, and then consider the question of its contextualization.

4.1 The History of Epistemology

The modern understanding of Descartes' philosophy begins in earnest with the Kantian/Hegelian historiography of philosophy of Kuno Fischer. The Descartes volume of his *Geschichte der neueren Philosophie* (6 vols., Berlin, 1852–77) was translated as *Descartes and his School* in 1887, and it effectively formalized an interpretation that remained unquestioned over the next 100 years, and which is still the dominant interpretation. Fischer's account did two things. First, it described what it took to be a major shift in the nature of philosophy, effectively starting with Descartes, from metaphysics, which had occupied ancient and medieval philosophy, to epistemology. Second, with the shift from metaphysics to epistemology, the basic fault-line underlying philosophical disputes shifted from the dichotomy between Platonism and Aristotelianism to that between rationalism and empiricism. Rationalism and empiricism were construed as competing and mutually exclusive epistemologies, the former basing everything on truths of reason, the latter basing everything on sensation.

Descartes' successors, whether self-styled Cartesians, whether critical but sympathetic, or hostile, did not think of the Cartesian project in these terms. For them, Descartes was a natural philosopher, and his work in mechanics, cosmology, and physiology lay at the core of his enterprise. This is particularly the case with his Cartesian successors such as Rohault and Régis, whose understanding of Cartesianism led them to do away with epistemological and metaphysical foundations and to go straight to natural-philosophical questions in their textbooks.[1] Malebranche, Spinoza, and Leibniz each engaged Descartes at the level of natural philosophy. Spinoza and Leibniz abandoned the idea of a sceptically-driven epistemology and approached metaphysics more directly, whereas Malebranche's treatment of epistemology was resolutely physiological. It is of importance here that Malebranche was the most influential representative of Cartesianism, and indeed the conduit to Descartes' thought to the middle decades of the eighteenth century.

Other than scepticism, there was one ingredient in Descartes' thinking about knowledge and cognition that was largely independent of questions in physiology however: the doctrine of innate ideas. It was Locke who developed the most sustained criticism of this doctrine, and it was the French Lockeans—particularly Condillac and Diderot—who pursued the question in greatest detail. It is in this context that epistemology becomes transformed into a struggle between Descartes and Locke. In the Plan of the *Encyclopédie*, for example, d'Alembert writes:

---

[1] See, for example, Theo Verbeek, 'The Invention of Nature: Regius and Descartes', in S. Gaukroger, J. Schuster, and J. Sutton, eds., *Descartes' Natural Philosophy* (London: Routledge, 2000),149–67.

> The Multiplicity of these Sensations, the concurring Agreement of their Evidence, the Degrees we observe them in, the involuntary Affections they excite in us, compar'd with the voluntary Controul we have over our Ideas of Reflection, which operate only upon our sensations; all this, we find, produces in us an irresistible Impulse to ascertain the real Existence of external Objects; and to regard them as the Cause of our Sensations. Many philosophers have held this Impulse to be the Effect of a Supreme being, and the most convincing Argument of the real Existence of an external World. But as there is no relation, that we know of, betwixt any single Sensation, and the Object thus suppos'd to occasion it, we cannot reason from the one to the other: and nothing but a kind of Instinct, more certain than reasoning itself, could oblige us to draw so remote a Conclusion.[2]

What has happened here is that a profound and complex epistemological question about sense certainty has been translated into a simple choice between divine guarantee versus what might be termed psychological certainty. One element in Descartes' elaborate epistemological argument is held up as if it were the issue on which everything else hinged (which it certainly is not for Descartes), so that what is now at stake is a choice between religion and reason. This is a choice made easier by the fact that whatever epistemological rationale the divine guarantee may have had in the original Cartesian argument, its removal from the context of this argument robs it of any epistemological rationale, so that it now appears as devoid of epistemological function, and hence wholly gratuitous as a means of securing the veridicality of sense perception. D'Alembert does not so much deny that there may be other epistemological questions at stake, as deny that consideration of them could be of any value, suggesting that for fear of 'obscuring a Truth acknowledg'd even by the Sceptics, when not heated in Dispute, we leave the capable Metaphysicians to discover the transcendental Cause in this Case.'[3] Similarly with the question of God's existence. He does not deny this, telling that it follows from reflection[4]: it is just that it can play no fundamental role in our cognitive or moral thinking, for the ideas that underlie these derive exclusively from natural sources, notwithstanding that revelation may occasionally 'serve as a Supplement to Natural Knowledge'.[5]

On this reading, Descartes and Locke are treated as being concerned with essentially the same kinds of questions, one offering divinely-guaranteed innate ideas as the solution, and the other offering sensation. Such a reading is that which Kant will take up and develop, offering his own solution to questions that, on his account, Descartes and Locke were manifestly unable to answer. For Kant, epistemology is a strictly a priori exercise, and this is the view that has been taken by the philosophical tradition since, at least up to the last decades of the twentieth century, when empirical issues started to be introduced into the consideration of cognition. If one confines one's attention to the *Meditations*, then it does indeed look as if Descartes treats cognition as a matter of conceptual analysis. But once one not only looks at some of his other writings—notably the *Treatise on Man*, the *Description of the*

[2] Denis Diderot and Jean le Rond d'Alembert, *Encyclopédie* (2nd. edn., 40 vols., Geneva, 1777–9), i. vii. Contemporary translation from *The Plan of the French Encyclopedia* (London, 1752), 5–6.

[3] *Plan*, 6; *Encyclopédie*, i. vii.

[4] *Plan*, 10–11; *Encyclopédie*, i. ix.

[5] *Plan*, 23; *Encyclopédie*, i. xv.

Human Body, and the *Passions of the Soul*—but also at how he spent his time in the 1620s and 1630s, with regular visits to the butchers for eyes and brains of cows and sheep for dissection, not to mention his extensive lens-grinding activity in the mid-1620s, then it soon becomes clear that his approach to cognition, particularly perceptual cognition, is thoroughly naturalized. This is in sharp contrast to the conceptual analysis approach, whereby perceptual cognition almost becomes a form of intellectual cognition, a form of non-sensory contemplation.[6] Perceptual cognition is treated as if the mind is essentially disembodied—Ryle's 'ghost in the machine'—but this disembodied mind somehow interacts with a body to which it is mysteriously connected.

In this way, the *Treatise on Man* is an essential corrective to the *Meditations*. As Descartes himself remarks in the Conversation with Burman: 'A point to note is that you should not devote so much attention to the *Meditations* and to metaphysical questions, or give them elaborate treatment in commentaries and the like ... They draw the mind too far away from physical and observable things, and make it unfit to study them. Yet it is precisely these physical studies that it is most desirable for men to pursue.'[7]

4.2 The Treatise on Man and the History of Medicine and Physiology

Hall's translation of *L'Homme* is an exercise in the history of medicine. Hall was author of a two-volume history of physiology—*Ideas of Life and Matter: Studies in the History of General Physiology 600BC-1900AD* (Chicago, 1969)—and the copious annotations to his translation of *L'Homme* make it an invaluable source in the history of physiology and medicine. Although what is offered is very much a stand-alone text, which does not relate it to Descartes' wider concerns, it offers a very extensive explanatory apparatus. As a result, the volume is invaluable resource, and provided the only detailed reliable commentary on *L'Homme* until the French edition of Annie Bitbol-Hespériès appeared in 1996 (*Le Monde, L'Homme*, Éditions du Seuil). It is still an important resource, even for Francophone readers.

My own translation, *Descartes: The World and Other Writings* (Cambridge University Press, 1998) provided annotated translations of *Le Monde* and *L'Homme*, and (unannotated) translations of cognate texts: in the case of *Le Monde*, extracts from *La Dioptrique* (Chap. 2) and *Les Météores* (Chap. 8); and, in the case of *L'Homme*, the first full English translation of *La Description du corps humain*. The primary aim of the translation was to place the work in context, and in this respect it was designed in some respects to complement my *Descartes: An Intellectual*

[6] See, for example, S. Gaukroger, 'Descartes' Theory of Perceptual Cognition and the Question of Moral Sensibility', in J. Cottingham and P. Hacker, eds., *Mind, Method, and Morality* (Oxford: Oxford University Press, 2100), 230–51.

[7] AT v. 165.

Biography (Oxford University Press, 1995). In this book I had set out to show how Descartes' work was driven, throughout his career, by natural-philosophical considerations. This is not too difficult to show for the 1620s and early 1630s, despite the fact that the *Regulae*, the only text from this period that philosophers have studied, has been assimilated to an epistemologically-driven account of scientific method. But, I argued, the subsequent work, particularly the *Meditations*, was also part of a natural-philosophical project, in this case one that offered a metaphysical legitimation of a mechanical conception of the cosmos which Descartes had already established on the appropriate natural-philosophical grounds. The translation was designed to bring to the attention of those studying Descartes a number of texts in which he uses natural-philosophical resources to establish positions which they may otherwise encounter only in epistemological or metaphysical versions (e.g. in the *Principia*), versions which often do not make much sense because their motivation appears mysterious.

Part II
The Early Reception of *L'Homme*

Chapter 5
The Early Dutch Reception of *L'Homme*

Tad M. Schmaltz

Abstract This is a consideration of the connection of *L'Homme* to two very different forms of early modern Dutch Cartesianism. On the one hand, this work was central to a dispute between Descartes and his former disciple, Henricus Regius. In particular, Descartes charged that Regius had plagiarized *L'Homme* in order to distance himself from a form of Cartesian physiology in Regius that is not founded on a proof of the spirituality of the human soul. Despite this repudiation, Regius remained a prominent proponent of Cartesian medicine. On the other hand, Florentius Schuyl published a Latin translation of *L'Homme* that included a preface in which he defends Descartes's doctrine of the "beast-machine" by invoking the authority of Augustine. This preface set the stage for the emphasis in the work of Clerselier and other French Cartesians on the presence in Descartes of a kind of Augustinian spiritualism.

Descartes's *Traité de l'homme* (hereafter, simply *L'Homme*) was published posthumously first in the United Provinces, in Latin translation, in 1662, and subsequently in France, in the original French, in 1664. However, this text made an impact on early modern Cartesianism even prior to the publication of these editions. This is clear from its role in the bitter dispute between Descartes and Henricus Regius (Hendrik de Roy),[1] a Dutch medical professor who was once a trusted follower. Toward the end of his life, Descartes charged that Regius had plagiarized *L'Homme*, intending thereby to distance himself from a form of Cartesian medicine in Regius that does not require a proof of the spirituality of the human soul. Though Descartes's renunciation of Regius split the Dutch Cartesian community, Regius remained a prominent proponent of the sort of mechanistic physiology that we find in *L'Homme*.

In contrast, the Dutch translator of *L'Homme*, Florentius (Florent or Florens) Schuyl,[2] prepared the way for a form of Cartesianism in which acceptance of the

[1] 1598–1679.

[2] 1619–1669.

T.M. Schmaltz (✉)
Department of Philosophy, University of Michigan,
2215 Angell Hall, 435 South State Street, Ann Arbor, MI 48109-1003, USA
e-mail: tschmalt@umich.edu

D. Antoine-Mahut, S. Gaukroger (eds.), Descartes' *Treatise on Man* and its Reception, Studies in History and Philosophy of Science 43,
DOI 10.1007/978-3-319-46989-8_5

spirituality of the soul is central. In particular, Schuyl included in his translation a lengthy preface in which he invokes Augustine in support of Descartes's doctrine of the "beast machine," according to which non-human animals are mere mechanisms. There is a kind of "Augustino-Cartesian" medicine here that contrasts not only with the Regius's anti-metaphysical Cartesian medicine, but also with the "Arisotelico-Cartesian" physics of Regius and Schuyl's Dutch contemporary Johannes de Raey.[3] Though De Raey is on the side of Schuyl, and opposed to Regius, in emphasizing the foundational importance of the claim that mind is really distinct from body, he nonetheless came to be opposed to Schuyl and Regius alike when he held in his later work that medicine is outside the bounds of Cartesian natural philosophy. A consideration of the initial Dutch reception of *L'Homme* therefore serves to draw attention to the presence in the early modern period of fundamentally incompatible forms of Dutch Cartesianism.

5.1 Descartes v. Regius[4]

Regius was educated in medicine during the 1620s in Padua by Sanctorius Sanctorius,[5] a famous Italian physicist who applied quantitative methods to medical issues.[6] Regius's Paduan degree gave him the right to teach private lessons in physics, and the lessons that he taught in Utrecht toward the end of the 1630s were well attended. His success was such that he was appointed extraordinary (outside of ordinary, perhaps most similar to an adjunct) professor of medicine and botany in 1638 at the newly established university at Utrecht, and the following year he was promoted to ordinary professor.

As we will discover, there is some controversy over whether Regius had access to the manuscript version of *L'Homme*. However, he certainly did have access to the summary versions of Descartes's physiology published in the fifth part of the *Discours* and in the accompanying *Dioptrique* and *Météores* (1637). Regius's medical training predisposed him toward the sort of mechanistic physiology that he found in these texts. He also served as something of a lightening rod in controversies over Cartesianism in Utrecht. Indeed, a set of medical disputations that he sponsored in 1641 led to the first official condemnation of the teaching of Cartesianism. Among the most controversial claims in these disputations was the thesis that a human being is merely an *ens per accidens*, a human soul only accidentally related to the human body. For theological critics, this was tantamount to a denial of the

[3] 1622–1702.

[4] This section draws on the discussion of the relation between Descartes and Regius in §5.2 of Schmaltz 2017.

[5] Aka Sanctorio Sanctorio; 1561–1636.

[6] On Sanctorius's approach in medicine and its influence on Regius, see Farina 1975.

claim—essential for the doctrine of the resurrection of the body—that the human soul bears an essential relation to the body to which it is united.[7]

Descartes was initially quite taken with Regius, and indeed wrote concerning Regius in 1643 that he is "so confident of his intelligence" that there is "hardly anything in his writings that I could not freely acknowledge as my own."[8] That was then. The situation changed in 1645 when Regius sent Descartes a draft of a "complete physiologia" with the title, *Fundamenta physices*. Descartes reports that he reacted with "astonishment and grief" to claims in "De Homine," the final chapter of Regius's text, concerning the human mind.[9] In the draft that he sent to Descartes, Regius disputes Descartes's demonstration of the real distinction of mind from body when he insists that natural reason alone cannot preclude the possibility that the human mind is a mode of body. Descartes notes that whereas Regius had previously claimed that a human being is an *ens per accidens*, he now falls into the opposite error of denying that the human mind must be an immaterial substance, an error which is "far worse" than his original one.[10]

In his final letter to Descartes, Regius protests that he in fact emphasized in his text that even though natural reason alone cannot prove the real distinction, Scripture places beyond doubt the claim that our soul is immaterial and immortal.[11] There is the charge in the literature that Regius appealed to Scripture merely in order to placate his theological critics.[12] However this strategy would hardly have been effective against such critics, who were as insistent as Descartes that natural reason can establish both the existence of God and the immateriality of our soul.[13] In taking a fideist stand on these issues, Regius was staking out his own distinctive—and controversial—position.

This sort of position was obviously unacceptable to Descartes.[14] But what must have been more galling to him was Regius's suggestion that Descartes cannot have been serious in making the sort of metaphysical claims concerning the soul that he did in the *Meditationes*. In his last contribution to his correspondence with Descartes, Regius reports, concerning Descartes's text, that

> many honorable gentleman have often told me that they have too good an opinion of the excellence of your mind to believe that you have, in the bottom of your soul, no sentiments contrary to those which have appeared in public under your name. And to conceal nothing

[7] Descartes told Regius that in making this claim "you could scarcely have said anything more offensive and provocative" (*Descartes to Regius*, second half of Dec. 1641, Descartes and Regius 2002, 91). Nonetheless, Descartes also held that Regius's heart was in the right place when he said that a human being is an *ens per accidens*, since he does not "understand otherwise than what everyone admits, namely that is is composed of two things that are really distinct" (*Descartes to Regius*, late Jan. 1642, Descartes and Regius 2002, 98).

[8] AT 8-2:163. AT = Descartes 1964–1976.

[9] *Descartes to Regius*, July 1645, Descartes and Regius 2002, 187–88.

[10] Descartes and Regius 2002, 188.

[11] *Regius to Descartes*, 23 July 1645, Descartes and Regius 2002, 189.

[12] See, for instance, Hallyn 2006, 173–201.

[13] Here following Bos 2013, 59.

[14] But cf. the claim in Wilson 2000 that Regius's position is in fact latent in Descartes's work.

> from you, several here are persuaded that you have greatly discredited your philosophy in publishing your metaphysics. You promise nothing but what is clear, certain and evident; but, to judge by this beginning, they claim that there is nothing but what is obscure and uncertain, and the disputes that you have had with skilled people prompted by this beginning serve only to multiply the doubts and darkness.[15]

There is in fact an example of this sort of reaction to Descartes in a letter from the Arminian theologian Caspar Barlaeus,[16] who offered the following in response to a request for comment on the *Meditationes*:

> Where [Descartes] rebukes and condemns the trite, he offers nothing better. He promises proofs such that "none more solid can be proffered by the human mind," much to my surprise, since I do not expect from the human mind more than probable reasons. He promises us geometrical evidence, and leads us into Cimmerian darkness and Egyptian obscurity.[17]

Though I have not found evidence that he knew Barlaeus's reaction to the *Meditationes*, Regius's remarks to Descartes indicate that this sort of reaction was not unusual in the Dutch intellectual community.

Regius prefaces his report to Descartes of the negative views of the "honorable gentlemen" with the warning to his correspondent that "you would perhaps do yourself more injury if you declare that … you have sentiments remote from mine touching on metaphysics." Regius's suggestion is that he was doing Descartes a favor by providing a version of Cartesian natural philosophy in the *Fundamenta physices* that jettisons Descartes's controversial metaphysical prolegomena. Nonetheless, Regius also closes his letter by referring to his text as "my book, or to speak better, your book, since it truly comes from you."[18] Clearly he did not want at this point to cut his ties to Descartes. Indeed, there is evidence that he altered the *Fundamenta* to make it more acceptable to Descartes, for in the first edition of this text, published in 1646, there is no hint of the suggestion that the human soul may be a mode of body.

Matters had gone too far by this point for such gestures to have any effect. When Descartes saw the published version of the *Fundamenta*, he could take it to be only a dangerous repudiation of the metaphysical foundations of his natural philosophy. Descartes had threatened to disown Regius if he published this text, and Claude Picot presented him with an opportunity to do just that when he asked Descartes to write a preface for Picot's French translation of the *Principia*. Descartes took advantage of this opportunity, writing in his preface:

> Last year [Regius] published a book entitled *Fundamenta physicœ* [*sic*] in which, concerning physics and medicine, it seems he has taken everything from my writings, those I have published as well as a still imperfect work on the nature of animals that fell into his hands; nevertheless, because he transcribed it poorly and changed the order, and denied certain truths of metaphysics on which all physics must be founded, I am obliged to disown the work entirely.[19]

---

[15] *Regius to Descartes*, 23 July 1645, Descartes and Regius 2002, 190.

[16] 1584–1648.

[17] *Barlaeus to Constantijn Huyghens*, 7 Aug. 1642, quoted in McGahagan 1976, 127.

[18] Descartes and Regius 2002, 190.

[19] AT 9-2:19.

Descartes's emphasis on the importance of certain metaphysical views for his physics can be seen as a response to the view of the "honorable gentleman" Regius mentioned, namely, that Descartes could not have been serious in embracing these questionable views. What is new here, though, is Descartes's implicit charge that Regius illicitly and incompetently plagiarized his "work on the nature of animals," that is to say, the unpublished and unfinished *L'Homme*.[20]

We find more information about this additional charge of plagiarism in Descartes's correspondence dating from this time. Thus, in a 1647 letter to Elisabeth he indicates that what Regius mistranscribed and misunderstood was

> the whole section where, dealing with the motions of the muscles, I take as an example two of those that move the eye. So fond was he of this passage that twice in his book he repeats, word for word, two or three pages from this section. And yet he has not understood what he wrote, for he has omitted the main point, namely that the animal spirits that flow from the brain to the muscles cannot return by the same passage through which they came. Without this observation, everything he writes is worthless, and because he did not have my diagram, he produced one that clearly shows his ignorance.[21]

The section from *L'Homme* Descartes has in mind here concerns what he calls elsewhere his "nice piece on the motion of the muscles,"[22] that is, his account there of the operation of the antagonistic muscles of the eye. The version of this section in the Latin translation of *L'Homme* (1662) includes a crude diagram from Descartes's own hand (see Fig. 5.1). However, the diagram that Clerselier used in his edition of *L'Homme* is considerably more elaborate (see Fig. 5.2). Clerselier tells us in his preface that this diagram was based on a copy of a "broüillon" included among the material he inherited from Descartes, which Clerselier then attempted "to draw as best I could" in order to correct some mistakes he found in it.[23]

In this diagram (Fig. 5.2), the tubes *ce* and *bd* are hollow nerves, *E* and *D* are muscles, and *g* and *f* are valves. If the "animal spirits" (that is, the subtle bodily fluid carried through the nerves) flow with more force into *bd* than into *ce*, they close *g* and open *f*. The spirits in *E* are enabled to flow into *D* through the diagonal connecting tube *ef*, but since *g* is closed, the spirits in *D* are prevented from flowing into *E*. In this way *D* is filled as *E* is depleted, and the eye is turned toward *D*. If, on the

[20] That this is the reference is clear from Descartes remark in a 1646 letter that "it is now twelve or thirteen years since I described all the functions of the human or animal body" in the work in question (*Descartes to Mersenne*, 23 Nov. 1646, AT 4:566–67).

[21] *Descartes to Elisabeth*, Mar. 1647, AT 4:626. Cf. *Descartes to [Huygens]*, 5 Oct. 1646, AT 4:517–18; *Descartes to Mersenne*, 23 Nov. 1646, AT 4:566.

[22] *Descartes to Mersenne*, 23 Nov. 1646, AT 4:567. This account of the muscles also receives special attention both in Clerselier's preface to his edition of *L'Homme* and in La Forge's *Remarques* included in this edition.

[23] AT 11:xii–xiii, xix–xx. Clerselier described this figure as "petit, déchiré, et défiguré," and noted that he saved the original so that it could be reviewed by "ceux qui en auront curiosité." Clerselier noted that he needed to improve the design because Descartes's drawing suggested that there were three folds on the valves regulating the flow of animal spirits into the eye muscles, rather than the two folds mentioned in the text of *L'Homme*.

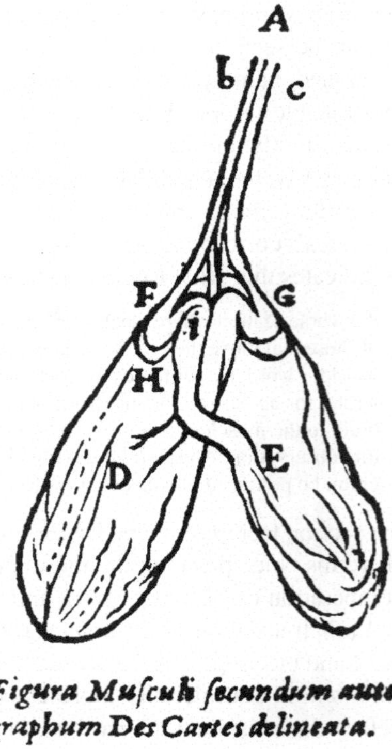

Fig. 5.1 Descartes's Diagram of Eye Muscles in *De Homine* (1662)

other hand, the spirits flow with more force into *ce*, the mechanisms ensure that the eye is turned toward *E*.[24]

In the section of the *Fundamenta physices* concerning the antagonistic muscles, Regius provides his own diagram (see Fig. 5.3). This diagram is obviously more similar to the simple diagram in the Latin edition than the more elaborate diagram in Clerselier's French edition. Nonetheless, we have good reason to reject the claim of one commentator that the resemblance between the diagrams of Descartes and Regius provides "the most damning evidence" of Regius's plagiarism.[25] After all, in his letter to Elizabeth Descartes himself explicitly attributed the errors in Regius's diagram to the fact that he had not seen Descartes's own diagram.

Appealing to his diagram, Regius asks us to imagine a flow of spirits in muscle *B*, which closes valve *E*.[26] In his letter to Elisabeth, Descartes emphasizes that Regius lacks any means of preventing the spirits from leaving the muscle by the same tube from which they entered. In fact, however, Regius claims in his text that

---

[24] AT 11:133–37. During the 1660s, Descartes's hydro-mechanical theory of the eye muscles was disproved by Jan. Swannerdam (1637–1680) in the United Provinces and Jonathan Goddard (1617–1675) in London, who showed that muscle volume did not increase during contraction. Nonetheless, this theory remained influential in the early modern period. See Donaldson 2009.

[25] Gariepy 1991, 179.

[26] *FP* X, Regius 1646, 233–35.

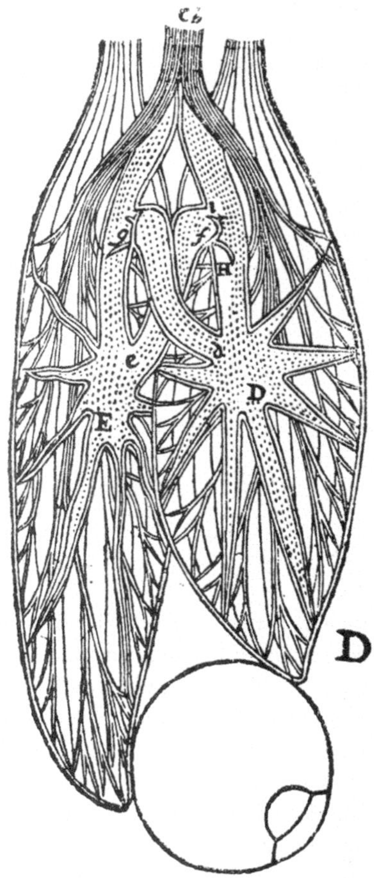

Fig. 5.2 Clerselier's Diagram of Eye Muscles in *L'Homme* (1664)

valves *D* and *G* prevent the return of animal spirits to the brain.[27] The difficulty for Regius seems to be rather that he does not have a way of accounting for reciprocal antagonistic action. For his diagram suggests that the stronger flow into *B* leaves *G* open, and so allows for a flow of spirits into muscle *C*, thus preventing the disparity that causes the eye to turn toward *B*. It is the lack of Descartes's connecting diagonal tubes with different directional flows that seems to be the real source of the problem for Regius's account.[28]

Even if Descartes's diagnosis of Regius's error is not correct, he seems to be justified in thinking that Regius's discussion of the mechanism involved in antagonistic muscles bears a remarkable similarity to that provided in *L'Homme*. Regius

[27] Regius 1646, 234.

[28] There is a similar account of Regius's error in Mouy 1934, 87–89.

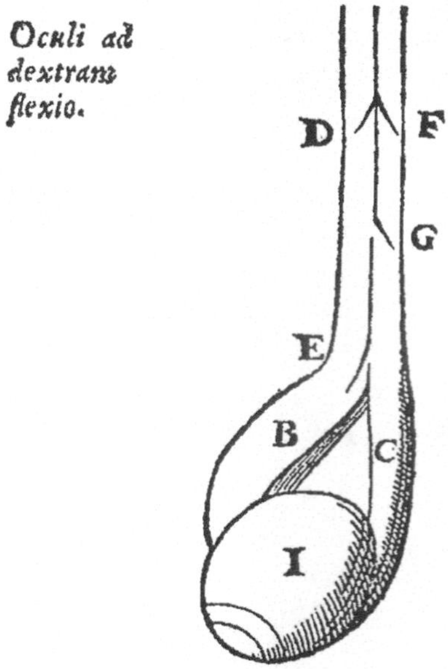

Fig. 5.3 Regius's Diagram of Eye Muscles in *Fundamenta physices* (1646)

himself insisted to Clerselier that he had never seen Descartes's unpublished text,[29] and it is possible that he learned the details of the account there through some means other than his own reading of *L'Homme*. However, it must also be noted in Regius's defense that the charge of plagiarism concerns only a relatively limited section of Descartes's work on what is after all a matter of detail. There certainly is no support for Descartes's suggestion in his 1647 preface that Regius has "taken everything" from his writings. For instance, immediately preceding Regius's account of the anagonistic muscles in the *Fundamenta* is his appeal—absent from Descartes's writings—to a kind of circulation of the animal spirits that is similar to the circulation of the blood. In support of this claim, Regius cites his own experiments with slugs.[30] There is also Descartes's own earlier admission in his 1642 *Epistola ad Patrem Dinet* that Regius had constructed a complete physiologia on his own, having seen at that point only Descartes's *Discours* and *Essais*.[31] Moreover, in a letter to Regius concerning his work in medicine Descartes notes that

> there are many other things in your theses that I have ignored, and also much, so far as I have knowledge of it, that I have explained in detail otherwise than you have explained it here. This however does not surprise me; for it is much more difficult to give one's opinion

---

[29] In correspondence with Clerselier, which I discuss presently. It should be noted, however, that there is some evidence that Regius did see a draft of Descartes's *Le Monde*; see the editorial comments on Descartes's letter to Regius of May 1641, in Descartes and Regius 2002, 67n.18. The question is whether this draft included *L'Homme*.

[30] Regius 1646, 231–32. Regius takes his experimental work to show that some portion of the animal spirits sent to the muscles reach the heart by means of venules and then are returned to the brain through arteries.

[31] AT 7:582–83.

> on all things which concern medical matters, which is the job of the teacher, than to choose the things that are easiest to know, and precisely to leave aside the rest, as I myself do in the other sciences.[32]

The indication here is that it is Regius, as the teacher of medicine, who has covered more territory than Descartes, the dabbler in the sciences, an indication that seems to be confirmed not only by the proposal concerning the circulation of the spirits, but also by other details of Regius's physiology.[33] Regius was no mere plagiarizer, but had a mind of his own in developing the details of Descartes's sketchy physiology.

In any event, Descartes's remarks in the 1647 preface precipitated a final break from Regius. That same year, as part of a medical disputation on the inflammation of the feet and legs, one of Regius's students, Petrus Wassenaer,[34] composed a short corollary that summarizes the metaphysical issues on which Regius departs from Descartes.[35] Though the disputation itself was suppressed due to the official prohibition at Utrecht of discussions of Descartes, Wassenaer circulated copies of this summary under the title *Explicatio mentis humanae*.[36] Included in the summary are the points regarding the difficulties with the demonstration of the real distinction that Regius had suppressed in the initial edition of his *Fundamenta*.[37] Also, this summary includes an emphatic rejection of a pure human intellect, stating that "as long as it is in the body, [the human mind] is organic in all its actions" and that "all common notions that are engraved in the mind have their origin in observation of things or in verbal instruction."[38]

Descartes responded immediately by composing his *Notæ in programma quoddam*, in which he insists on the soundness of his own proof of the real distinction, as well as on the need to attribute to mind a pure intellect that is not dependent on the body. There was a rejoinder to Descartes's response in Regius's *Brevis explicatio mentis humanae* (1648). However, Descartes failed to engage further, and thus left it to his Dutch supporter, Tobias Andreae, to provide a response to this work. This Andreae did in his *Brevis replicatio reposita Brevi explicationi mentis humanæ Henrici Regii* (1653), in which he argues that it can be demonstrated that the human

---

[32] *Descartes to Regius*, Nov. 1641, AT 3:443, Descartes and Regius 2002, 87.

[33] For instance, Regius recognized sooner than Descartes the importance of Gaspare Aselli's discovery in 1627 of lacteal vessels that carry chyle from the intestines. Moreover, Regius offered an account of the motion of the heart in terms of the flow of animal spirits that cannot be found in Descartes. I discuss these points further in §5.2 of Schmaltz 2017.

[34] d. 1680.

[35] Included as part of Regius and Wassenaer1647, and re-published in Regius 1648.

[36] I am drawing here on the discussion in the editorial introduction to Verbeek 1993a, 1–3.

[37] See Descartes summary of *Explicatio* at AT 8-2:342–43. Regius included the suppressed claims in the second and third editions (1654 and 1661b, respectively) of his *Fundamenta*, retitled *Philosophia naturalis*.

[38] AT 8-2:344 and 345. The lack of pure intellect and dependence of all human thought on bodily sense organs is emphasized in Regius 1654, 404, and Regius 1661b, 477, passages that have no counterpart in Regius 1646. On the organic constitution of mind, see Regius 1654, 343, and Regius 1661b, 407.

mind is a substance really distinct from body, and thus that it cannot be a mode of body; that the human mind has "inorganic faculties," such as pure intellect and will, that do not depend on body; and that universal and purely intellectual ideas are innate to the mind, and thus not drawn from sense experience.[39]

On the French front, Clerselier took up the defense of Descartes against Regius. In the preface to his first volume of his edition of Descartes's *Lettres*, published in 1657, Clerselier alludes to the charge of plagiarism when he enjoins Regius to admit publicly that there is nothing good in his *Fundamenta physices* that he has not taken from Descartes, including from works that "have fallen into [Regius's] hands, which I hope soon to make public," namely, *L'Homme*. Clerselier also complains that Regius confirms his unfaithfulness in excising the acknowledgment of his debt to Descartes in the first edition (1646) of the *Fundamenta* from the second edition (1654) of this work, retitled *Philosophia naturalis*. He finally exhorts Regius to return to orthodox Cartesianism by embracing Descartes's own metaphysical conclusions regarding the human soul.[40] In order to highlight those conclusions, Clerselier included not only several of Descartes letters to Regius, but also, at the end of the volume, a French translation of Descartes's *Notæ*.

Regius reponded to the criticisms of him in Clerselier's preface by publishing, also in 1657, a second edition of his *Brevis explicatio* that adds the preface of a certain Carolus Fabricius, the identity of whom is uncertain.[41] Fabricius angily denounces Cerselier for publishing Regius's personal correspondence with Descartes without Regius's consent, and thus for being "a nefarious and wicked violator of sacred and intimate familiarity and friendship." Fabricius further indicates that the published letters "do not themselves merit trust" since they are "merely fictitious and made up after the fact."[42] In response to Clerselier's charge that Regius showed ingratitude by removing his praise of Descartes, Fabricius insists that Regius was forced to do so by Descartes's mistreatment of him in the *Notæ*. Moreover, he responds to the charge of plagiarism by noting that it is not Regius who was the plagiarist, but Descartes, since Descartes "has appropriated the entire collection of Regius's complete Physiologia, which having seen a number of years beforehand, he retained, plundered and adapted for his use."[43] In closing, Fabricius claims that Descartes actually came to Regius's position in the end, appealing to

[39] In Andreae 1653, 21–81, 98–113 and 149–63, respectively.

[40] AT 5:754.

[41] There is perhaps some temptation to identify Fabricius with Regius himself. However, in the preface to Fabricius 1648, the author indicates that he is writing from Roermond, a town in the province of Limburg. Unless this is a mere invention, it would seem that Fabricius is not Regius. Thanks to Erik-Jan. Bos for drawing my attention to this evidence.

[42] Regius 1657b, 6. Fabricius is alluding here to the fact that Regius and not Clerselier had the originals of the letters that Descartes sent him, and that Clerselier's version of these letters relied on incomplete drafts that he needed to augment. For discussion of the history of Clerselier's publication of the Descartes-Regius correspondence, see the editorial introduction to Descartes and Regius 2002.

[43] Regius 1657b, 9. Fabricius also notes—correctly—that Regius has addressed the question of how the nerves prevent "the return of the spirits to the brain."

reports that Descartes expressed his approval of Regius in 1649, after the move of the former to Sweden.[44]

Undeterred by the clear rebuke of him in Fabricius's preface, Clerselier wrote to Regius in 1659 to ask for his assistance in producing figures for the editions of *Le Monde* and *L'Homme* that Clerselier was preparing. As Clerselier indicates in his letter to Regius, his thought was that Regius's participation in this project would provide a means of restoring his former connection to Descartes.[45] Still smarting from the rebuke of him in Clerselier's preface, Regius declined the request, noting that he did not want to be seen as supporting the slanderous view that he had plagiarized this work.[46] With this response there was an end to any hope of a grand Cartesian alliance between Regius, as leader of an established Dutch front of Cartesianism, and Clerselier, as leader of a newly emerging French front.

Regius's rebuff of Clerselier, along with Descartes's earlier rebuff of Regius, would seem to indicate that Regius was an outsider to early modern Cartesianism. Indeed, this was the official Clerseliean position on this matter. Clerselier passed his own negative view of Regius on to Adrien Baillet. In his *Vie de Descartes*, which serves as something of the culmination of Clerselier's Cartesian campaign, Baillet notes that though he was "the first Disciple of Descartes," Regius later "resolved to sacrifice the honor of his Master for his own" in publishing his *Fundamenta physices*.[47] Baillet adds that "Aristotle perhaps never carried his ingratitude so far toward his master Plato," and that "Maximus the Cynic has never treated his master Gregory of Nazianzus with more insolence."[48] Regius's insolence and ingratitude are reflected in the fact that he was "the first plagiarist of Descartes" who also was "the first rebel among his disciples, or the first schismatic among his supporters."[49]

Nonetheless, Regius did not conceive of himself as a schismatic and rebel against the Cartesian cause. Though he certainly had no desire to join Clerselier's Cartesian campaign, we have seen that Regius did include in the volume he published in response to Clerselier the report of Descartes's late affection for him. Regius further included in the third edition (1668) of *Fundamenta medica* a prefatory letter—not found in the previous editions—in which he emphasizes the influence of Descartes's

---

[44] Regius 1657b, 10. As evidence, Fabricius cites two letters from Robert Creighton (1593–1672), sent from Sweden to his friend Regius, that were appended to the second edition of *Brevis explicatio*.

[45] See Clerselier's report of his request in his preface to *L'Homme*, AT 11:xiv-xv. In this preface Clerselier also presents a more moderate discussion of the plagiarism charge, admitting that it is "not impossible" that the resemblances are coincidental, and leaving it to the reader to decide "who between Monsieur Descartes and Monsieur le Roy is the master or the disciple, and which of the two is the first inventor of things where they agree, or if they are both invented" (AT 11:xv).

[46] Regius's last letter to Clerselier is published in Regius 1661a, which is prefaced by remarks from none other than Carolus Fabricius. Fabricius's reference in these remarks to the "slanderous preface" of Clerselier is repeated in Regius's letter.

[47] Baillet [1691] 1970, 2: 21 and 269.

[48] Baillet [1691] 1970, 2:271.

[49] Baillet [1691] 1970, 2:171.

views in the *Discours* and appended *Essais*, and also reproduces the passage from the *Epistola ad Patrem Dinet* in which Descartes endorses his views.[50] At least with respect to issues in physics and medicine, Regius continued to see himself as developing a genuinely Cartesian line.

In fact, after Descartes's death Regius was recognized by his contemporaries as an important proponent of Cartesianism. This is indicated by a "Discourse on the Heart" that the Rotterdam physician James de Back[51] included in his 1653 English translation of Harvey's *De motu cordis*. In this discourse, De Back refers not only to Descartes's position that the motion of the heart consists in the diastole, but also to the elaboration of that position in the work of "the most learned *H. Regius*, Professor of Physick in the University of *Utrecht*, and a notable follower of *de Cartes*."[52] This reference to Regius as Descartes's "notable follower" may seem to be surprising given that it occurs only a few years after Descartes's public repudiation of his former disciple, a repudiation of which De Back, as a Dutch physician, would undoubtedly have been aware. However, in this medical context views on issues such as circulation and the motion of the heart are more to the point in determining an ideological connection to Cartesianism than the sort of metaphysical issues that separated Descartes from Regius. With respect to the former issues, Regius did indeed adhere to a recognizably Cartesian line.

Through his long medical career in Utrecht, Regius served as a primary source for a new generation of physicians. In fact, he set the agenda for the Cartesian branch of Dutch medicine. We find evidence of this in the titles of medical texts. There is, for instance, the *Lumen rationale medicum, hoc est. praxis medica reformata, sive annotationes in praxim Henrii Regii* (1686) of Theodore Craanen, a student of Regius, as well as the *Animadversiones medicae … in Henrici Regii Praxim medicam* (1695) of Johann Broen,[53] a student of Craanen. There was a network of Dutch Cartesian physicians and medical professors who owed their training in mechanistic physiology either directly or indirectly to Regius. Regius the Cartesian outcast therefore became something of a Cartesian authority with respect to issues in medicine.

5.2 Schuyl and De Raey

I have referred to the Latin translation of *L'Homme*. The man primarily responsible for this translation was the Dutch Cartesian Florentius Schuyl. As a student at Utrecht, Schuyl was a partisan of the traditional Aristotelian philosophy, and at his graduation in 1639 he defended the thesis that magnetism is an irreducible "occult quality" that causes magnetic effects. An interesting side note is that during the

[50] This edition was retitled *Praxis medica*; the preferatory letter there is unpaginated.

[51] ca. 1594–1658.

[52] Harvey 1653, 114.

[53] 1663–1703.

course of Schuyl's defense, the traditionalist Arnauld Senguard felt compelled to come to the defense of his student, whereupon none other than Regius rose to question the supervising professor's performance.[54] Despite this interruption, Schuyl went on to provide an accepted defense of his broadly Aristotelian position.

However, the fact that Schuyl produced a Latin translation of *L'Homme* in 1662 indicates that he had converted to the new Cartesian mechanistic philosophy by that time. This change is reflected in a 1667 speech at Leiden in which he insists that magnetism is in fact a "manifest quality" that can be understood in terms of the mechanical features of bodies.[55] At this point Schuyl was stepping down from his position of *Rector Magnificus* in Leiden, but 5 years earlier he held the less magnificent position of philosophy teacher at a "Latin school" in 's-Hertogenbosch, something roughly equivalent to the current prep school. The success of his translation was such that within a short time he was able to exchange this position for a prestigious university chair in medicine in Leiden. Clearly, *De Homine* made Schuyl's career.

In his preface to *De Homine*, Schuyl notes that a principal result in Descartes has been "to ruin this dangerous opinion, which dishonors and profanes the image of God," according to which there is "little difference between the human Soul (this incorruptible and immaterial Mind) and the soul of beasts."[56] Most of the preface is in fact devoted to a defense of Descartes's doctrine that non-human animals are, in contrast to human beings, mere mechanisms devoid of thought and feeling; this is now widely known as the doctrine of the "beast-machine." This doctrine is not defended explicitly in *L'Homme*, though it is so defended in Descartes's summary of his mechanistic physiology in the fifth part of his *Discours*.[57] As Paul Dibon has documented with respect to the situation in the United Provinces, however, the doctrine of the beast-machine "appears only as a minor theme of discussion as much in the first Cartesian polemics as in philosophical teaching in the university."[58] Schuyl's preface served to raise awareness of this issue not only in the United Provinces but also in France. Moreover, it is significant in light of the later development of Cartesianism that Schuyl's initial discussion appeals explicitly to passages in Augustine that he takes to provide an endorsement of the position that the operations of an animal body do not involve any spiritual element.[59] Though Descartes himself

[54] As indicated in the 1642 *Testimonium Academiae Ultrajectinae, et Narratio Historica quà defensae, quà exterminatae novae Philosophiae*, reproduced in French translation in Descartes and Schoock 1988, 86–87.

[55] *De Veritate Scientiarum et Artium Academicarum*, reprinted in Lindeboom 1974, 125 ff. I am drawing here on the discussion of Schuyl in Ruler 2008, 159–68.

[56] Descartes 1664, 412 (from the French translation of Schuyl's preface).

[57] See AT 6:56–59.

[58] Dibon 1990, 683.

[59] Schuyl cites passages from *De libero arbitrio* VIII.18 and *De quantitate animae* XIV and XXXIII, in Descartes 1664, 428–29 and 435, respectively.

arguably had little interest in linking his views to those of Augustine,[60] Schuyl indicated the importance of promoting a grand Augustino-Cartesian alliance.

To be sure, Schuyl's Cartesian appeal to Augustine was not unprecedented among Dutch partisans of Descartes. Indeed, prior to the publication of Schuyl's translation of *L'Homme* we find such an appeal in the work of the Leiden theologian Abraham Heidanus.[61] As early as 1645, and thus during Descartes's own lifetime, Heidanus had attempted to link Cartesian philosophy to Augustinian theology to defend the former against the charge of heterodoxy.[62] Though the invocation of Augustine is not as prevalent in earlier Dutch Cartesianism as it was to become in later French Cartesianism, one can find examples of it in Dutch-trained Cartesians other than Heidanus. There is, for instance, the proclamation in Clauberg's *De cognitione Dei et nostri* (1656) that "Augustine approves of Cartesian metaphysics" (*Augustinus cartesianae Metaphysicae favet*).[63]

However, it is Schuyl's invocation of Augustine in his preface to *De Homine* that had the most dramatic effect. Prior to the publication of Schuyl's edition, the appeal to Augustine was largely absent from the initial post-Descartes campaign in France to promote Cartesianism. Thus, Augustine is not mentioned in Jacques du Roure's early textbook summary of Cartesianism, *La philosophie divisée* (1654), and the Church Father is mentioned only incidentally in Clerselier's preface to the first volume of his edition of Descartes's *Lettres* (1657) and not at all in his preface to the second volume (1659). However, before the publication of Clerselier's French edition of *L'Homme* in 1664, Descartes's views were condemned as theologically dangerous in Louvain (1662) and were placed on the *Index librorum prohibitorum* in Rome (1663). Thus there was a need to emphasize the religious orthodoxy of Descartes's philosophy. It is no accident that Clerselier's edition of *L'Homme* includes a French translation of Schuyl's preface, and that in his own preface Clerselier provides an extended commentary on Augustine's views in *De Trinitate* X.10. One main goal here is to draw attention to the fact that such views reinforce the theologically orthodox result in Descartes that "the human soul is of a spiritual nature, and is really distinct from that of body."[64] Clerselier's preface marks the attempt to emphasize Descartes's embrace of a kind of Augustinian spiritualism that, as Schuyl's preface indicates, has as its flip side the doctrine of the beast-machine.

Clerselier's edition of *L'Homme* includes a set of *remarques* of the Saumur physician Louis de la Forge. In *L'Homme* Descartes had promised an account of the

---

[60] For instance, when a Dutch contemporary Andrea Colvius noted the Augustinian roots of his *cogito* argument, Descartes responded that he made a very different use of this argument than did Augustine (AT 3:247). For a further defense of the notion of an essential connection between the views of Augustine and Descartes is one that Descartes himself did not embrace, see §3.1 of Schmaltz 2017.

[61] 1597–1678.

[62] On this point, see Ruler 2008, 164.

[63] Clauberg [1691] 1968, index, n.p. In the corresponding §26 of IV, Clauberg cites several texts from Augustine.

[64] Descartes 1664, préf., n.p., citing *De Trinitate* IX and X. On this emphasis in Clerselier, see Kolesnik-Antoine 2012.

human mind that he never delivered. La Forge attempted to provide this missing account in his *Trait[t]é de l'esprit de l'homme ... suivant les principes de René Descartes* (1666). In this text La Forge includes a lengthy preface, the main thrust of which is indicated by its subtitle, "In which the author displays the conformity of the doctrine of Augustine with the opinions of Descartes, concerning the nature of the soul" (*Dans laquelle l'Auteur fait voir la conformité de la Doctrine de Saint Augustin, avec les sentimens de Monsieur Descartes, touchant la Nature de l'Ame*).[65] On one count La Forge cites Augustine in his preface no less than 51 times,[66] and in the course of his discussion of Augustine he quotes or summarizes passages from eleven different Augustinian texts. La Forge may well have consulted the collection of Augustinian passages that the Oratorian André Martin published as Ambrosius Victor in 1653 (*Sanctus Augustinus, De existentia et veritate Dei*) and 1656 (*Sanctus Augustinus, De anima*). But in contrast to the case of Martin, La Forge—in line with Schuyl and Clerselier—is citing Augustine in order to defend Descartes's identification of soul with the immaterial *res cogitans* and his complementary denial that there is any other kind of soul that animates the beast-machine.[67]

Prior to Schuyl's attempt to "Augustinize" Descartes, however, there was the attempt to "Aristotelianize" Cartesian natural philosophy in the work of Schuyl's Dutch contemporary Johannes De Raey. De Raey was trained in medicine by Regius at Utrecht, and he taught medical students in philosophy in Leiden during the 1650s. At the same time, he was engaged in the project of reconciling Aristotle with Descartes so as to make the new Cartesian philosophy suitable for use in the Dutch universities, where Aristotelian scholasticism still reigned. This project is reflected in the title of his *Clavis philosophiae naturalis seu Introductio ad contemplationem naturae Aristotelico-Cartesiana* (1654), which indicates that the "key" (*clavis*) to constructing a philosophy of nature lies in an "Aristotelico-Cartesian" contemplation of nature.[68] In this text, De Raey begins by decrying the "perversions and corruptions" of Aristotle's philosophy introduced first by Averroes and Arabic commentators, and then by Catholic scholastics.[69] Once the later corrupting elements are eliminated, De Raey is confident that we will find in Aristotle an account of the material world similar to what one finds in Descartes. Thus he claims that Aristotle himself endorsed the view that whereas God and the human mind are conceived to be "incorporeal and spiritual," body is conceived to be merely extended,

[65] La Forge 1974, 75.

[66] Manning 2012, 150.

[67] Henri Gouhier has famously distinguished those, such as Martin, who offer a form of "cartesianized" Augustinianism—*augustinisme cartésianisé*—from those, such as Schuyl, Clerselier and La Forge, who offer a form of "augustinized" Cartesianism—*cartésianisme augustinisé*; see Gouhier 1978.

[68] This work was based on a series of disputations on the (pseudo-) Aristotelian *Problemata* on which the Leiden curators allowed De Raey to lecture starting in 1651.

[69] Raey 1654, Epistola, n.p.

"the object of pure mathematics."[70] Cartesian dualism is therefore central to De Raey's Aristotelico-Cartesian physics.

De Raey's attempt to offer this sort of physics anticipated similar attempts later in France. Perhaps the most enthusiastic French proponent of the effort to reconcile the physics of Descartes and Aristotle was the Génévofain René le Bossu.[71] Le Bossu's *Parallèle des principes de la Physique d'Aristote det de celles de René Descartes* (1674)[72] was devoted to the argument that "there is no contradiction between the Principles of Aristotle and those of Descartes, and the truth being on one side, can equally be on the other."[73] More subtle, however, and certainly more influential, was the reconciliationist line in the *Traité de physique* (1671) of Jacques Rohault.[74] In the preface to this text, Rohault announces that.

> I have taken from Aristotle all general notions, either for the establishment of the principles of natural things, or also for what concerns their principal properties. And I am concerned to reject the Void, and Atoms or Indivisibles of Epicurus, which were things contrary to what I believed to be solidly established by Aristotle.[75]

In the *Traité* itself, Rohault embraces an Aristotelian hylomophism that distinguishes the matter common to all bodies from the forms that distinguish them into particular bodies.[76] With De Raey, moreover, Rohault insists Aristotle allowed for the identification of the matter with extension.[77]

However, there is an important difference between the approaches to Cartesian physics in De Raey and Rohault. In the preface to his *Traité*, Rohault claims that one of the principal impediments to progress in physics is the fact that "one treats it too metaphysically, and that one often stops at so abstract and so general questions."[78] Rohault in fact is particularly concerned in this text to provide empirical explanations of particular phenomena in nature. In contrast, De Raey insists that physics must rest on *praecognita* that derive not from sensory experience, but rather from "a purely mental intuition" (*solius mentis intuitu*).[79] According to De Raey, the "vulgar

---

[70] Raey 1654, 40.

[71] 1631–1680.

[72] An earlier version of this work was published in Pierre le Gallois's *Conversations de l'Academie de Monsieur l'Abbé Bourdelot* (1672). Cf. the discussion of Le Bossu's *Parallèle* in Grene 1993, 81–85.

[73] Le Bossu [1674] 1981, 293.

[74] 1620–1672. The contrast between the texts of Le Bossu and Rohault is indicated by the fact that whereas Le Bossu's *Parallèle* had only one modern edition, between 1671 and 1739 there were more than 25 reprintings of Rohault's *Traité*, including several Latin and English translations.

[75] *TP*, Préf., Rohault 1671, n.p.

[76] On Rohault's endorsement of hylomorphism, see Manning 2012, 28–32. As Manning notes, Rohault does distinguish himself from Aristotle in confining his hylomorphism to matter and form and in excluding the third Aristotelian principle of privation. One can find a similar exclusion in Le Bossu [1674] 1981, 283.

[77] Rohault 1671, 1:39.

[78] *TP*, Préf., Rohault 1671, n.p.

[79] Raey 1654, 36; cf. Descartes's claim in the Second Meditation that we know the nature of the wax *solius mentis inspectio* at AT 7:31.

notions" deriving from the sensory faculties allow us to determine "what conduces or not to the conservation of life," but can in no way replace the purely intellectual *praecognita* that provide the basis for a philosophical contemplation of nature.[80] The inspiration here, of course, is Descartes's distinction in the Sixth Meditation between the "teachings of nature," which have the purpose of informing us of what is helpful or harmful to the preservation of the mind-body union, and the "natural light," which informs us of the essential natures of things.[81]

We have seen that De Raey takes our *praecognita* to reveal the real distinction of mind from body, and in this respect he is on the side of Schuyl against Regius. However, there is an important difference from both Schuyl and Regius that emerges in De Raey's later thought. Though in his earlier years De Raey seems to have treated medicine as part of Aristotelico-Cartesian natural philosophy, there is a dramatic shift in his view of the relation between medicine and physics that we can date to around 1665.[82] This shift is indicated in a 1680 letter, in which De Raey considers the claim of the Dutch physician Franciscus Sylvius (Franz de le Boë)[83] that "in medicine as in physics what is known truly is known only by experience."[84] De Raey responds to this claim by noting

> first, that one should not philosophize outside of philosophy, and, second, what follows from this, that neither medicine nor the mechanical arts have ever been or can ever be a part of philosophy.[85]

De Raey's conclusion here is that medicine has its own "foundation and subject" that it does "not owe to philosophy, of which physics is a part, which we distinguish from medicine."[86]

Sylvius's empiricist conception of physics and medicine is grounded in the general principle that

> whatever is in the external senses, from which the beginning of all of our cognition comes about, is so certain to me, as certain as possible; whence it is even confirmed as true by the ancient philosophers: Nothing is in the intellect, which was not first in the senses.[87]

Nor was Sylvius alone in adhering to such a principle. Indeed, in the sixth article of his *Brevis explicatio* Regius asserts explicitly that "so long as it is in the body, [the human mind] is organic in all its actions. Thus as the disposition of the body varies, so the mind has different thoughts."[88]

---

[80] Raey 1654, 16.

[81] AT 7:83.

[82] As indicated in Verbeek 1995, 81.

[83] 1614–1672. Sylvuis was a proponent of an "iatrochemical" approach to medicine that sometimes contrasted with the more "iatromechanical" approach of Regius. For this point, see Ragland 2012.

[84] The claim from Sylvius's "De hominis cognitione" (1658), in Sylvius 1679, 896.

[85] Raey 1692, 654.

[86] Raey 1692, 259.

[87] Sylvius 1679, 896.

[88] AT 8-2:344.

The objection in De Raey's 1680 letter does not concern the claim in Sylvius and Regius that the senses play a foundational role in medicine. Indeed, his comparison of medicine to the mechanical arts suggests that De Raey considers the former to be an art rather than a philosophical science. As such, medicine could be rooted in the teachings of nature deriving from sense experience, as opposed to the intellectual *precognita* that issue from the natural light. De Raey's objection is rather to the generalization of claims about cognition in medicine to claims about human cognition in general, including the sort of cognition involved in physics. De Raey's conception of philosophy led him to insist that physics, as a part of philosophy, has a purely intellectual foundation.

Schuyl's Augustinianism also led him to reject the sort of empiricist view of human cognition that we find in Sylvius and Regius. For Schuyl, there is a sharp divide between the clear and distinct ideas of pure intellect and the confused ideas of sense and imagination, just as there is for De Raey.[89] However, it is clear that Schuyl takes medicine to be a theoretical enterprise with an essential connection to Augustino-Cartesian spiritualism. In this respect Schuyl is arguably closer than De Raey to Descartes himself. After all, in the same preface to the *Principia* in which he attacks Regius, Descartes famously compares philosophy to a tree with metaphysics as the roots, physics as the trunk and medicine as one of the branches.[90] This image—which is perfectly in line with Schuyl's own conception of Cartesian medicine—is in considerable tension with De Raey's claim that medicine has its own "foundation and subject" that it does "not owe to philosophy, of which physics is a part."

Of course, I have noted that De Raey's conception of the sharp distinction between medicine and physics is linked to Descartes's distinction between the teachings of nature and the natural light. However, in emphasizing this distinction De Raey is interested less in being faithful to Descartes than in addressing difficulties in accommodating Cartesian philosophy to the Dutch university curriculum. We have seen that Regius confronted the charge that his version of Cartesianism conflicts with theological doctrines such as the resurrection of the body. More generally, the objection is that Cartesian philosophy is not suitable for training students for work in the "higher faculties" of law, medicine and—most important here—theology. De Raey's response to this objection is that since philosophy and theology are radically distinct disciplines, the two cannot conflict. Whereas philosophy is based on abstract concepts and is directed to purely theoretical ends, theology is based on revelation and is directed to the practical goal of salvation. In De Raey's later thought this line of argument is extended to the case of medicine, with this discipline being presented as a purely practical enterprise that differs in kind from a philosophical contemplation of nature.[91]

---

[89] For documentation of this point, see Ruler 2008, 165–66.

[90] AT 9-2:14.

[91] Theo Verbeek has defended this reading of De Raey in several works. See, for instance, Verbeek 1993b and 1995.

However, this was not the only route to making Cartesian philosophy academically acceptable in a Dutch context. For the case of Schuyl reveals that there also was the option of linking Cartesian mechanistic physiology to an unimpeachable Augustinian metaphysics. Moreover, Regius showed how Cartesian physics as well as Cartesian medicine could be practiced as thoroughly empirical disciplines, detached from the sort of dogmatic dualistic metaphysics on which De Raey and Schuyl alike placed so much emphasis. The initial legacy of *L'Homme* is thus not a single system that perfectly reflects Descartes's intentions, but rather a variety of early modern Cartesianisms that absorbed the lessons of this text in very different ways.

Work Cited

Andreae T. Brevis replicatio reposita Brevi explicationi mentis humanae Henrii Regii. Amsterdam: L. Elzevier; 1653.
Baillet A. La vie de monsieur Descartes. 2 vols-in-1. Geneva: Slatkin; [1691] 1970.
Bos E-J. Henricus Regius et les limites de la philosophie cartésienne. In: Kolesnik-Antoine D, editor. Qu'est-ce qu'être cartésien? Lyon: ENS Éditions; 2013. p. 53–68.
Clauberg J. Opera Omnia Philosophicum. 2 vols. Hildesheim-New York: Georg Olms Verlag; [1691] 1968.
de Raey J. Clavis philosophiae naturalis sive Introductio ad contemplationem naturae Aristotelico-Cartesiana. Leiden: J. & D. Elzevier; 1654.
de Raey J. Cogita de interpretatione. Amsterdam: Wetstein; 1692.
Descartes R. In: C. Clerselier, editor. Lettres de Mr Descartes. 3 vols. Paris: Angot; 1657–67.
Descartes R. De homine figuis et latinitate donatus a Florentio Schuyl. Leiden: Moyardum & Leffen; 1662.
Descartes R. L'Homme de René Descartes, et un Traitté de la formation du foetus, du mesme auteur. Avec les Remarques de Louis de la Forge, docteur en medecine. Paris: Angot; 1664.
Descartes R. Œuvres de Descartes. Nouvelle présentation. Edited by Adam C, Tannery P, 11 vols. Paris: J. Vrin; 1964–1974.
Descartes R, Regius H. In: Bos E-J, editor. The correspondence between Descartes and Henricus Regius. Utrecht: Utrecht Department of Philosophy; 2002.
Descartes R, Schoock M. La Querelle d'Utrecht. (trans. and editor Verbeek T). Paris: Les impressions nouvelles, 1988.
Dibon P. Regards sur la Hollande du Siècle d'Or. Vivarium: Naples; 1990.
Donaldson IML. The Treatise of man (De Homine) by René Descartes. J R Coll Physicians Edinb. 2009;39:375–6.
Fabricius C. De Antidoto Primirosiano Epistola, Secunda editio, Priori auctior et emendatior. Rotterdam: Isaci; 1648.
Farina P. Sulla formazione scientifica di H. Regius: Santorio e il 'De statica medicina'. Rivista Critica de Storia della Filosofia. 1975;33:363–99.
Gariepy TP. Mechanism without metaphysics: Henricus Regius and the establishment of Cartesian medicine. Ph.D. dissertation, Yale University; 1991.
Gouhier H. Cartésianisme et Augustinisme au XVIIe siècle. Paris: Vrin; 1978.
Grene M. Aristotelico-Cartesian themes in natural philosophy: some seventeenth-century cases. Perspect Sci. 1993;1:66–87.
Hallyn F. La Philosophia naturalis de Regius et l'écriture athée. In: McKenna A et al., editors. Les libertines et la science. Saint-Étienne: University of Saint-Étienne; 2006. p. 37–46.

Harvey W. The anatomical exercises of Dr William Harvey professor of physick, and physician to the Kings Majesty, concerning the motion of the heart and blood … To which is added Dr James De Back his Discourse of the heart, physician in ordinary to the town of Roterdam. London: Leach; 1653.
Kolesnik-Antoine D. Les voies du corps. Schuyl, Clerselier et La Forge lecteurs du traité de L'Homme de Descartes. Conscutio temporum: Rivista critica della postmodernità. 2012;2:118–28.
La Forge L d. In: Clair P, editor. Œuvres philosophique, avec une etude bio-bibliographique. Paris: Presses Universitaires de France; 1974.
Le Bossu, R. Parallele des principes de la physique d'Aristote & de celle de René Des Cartes. Paris: Vrin; [1674] 1981.
Lindeboom GA. Florentius Schuyl (1619–1669) en zijn betekenis voor het Cartesianisme in de geneeskunde. The Hague: Nijhoff; 1974.
Manning G. Three biased reminders about hylomorphism in early modern science and philosophy. In: Manning G, editor. Matter and form in early modern science and philosophy. Leiden: Brill; 2012. p. 1–32.
McGahagan TA. Cartesianism in the Netherlands, 1639–1676: the new science and the calvinist counter-reformation. Ph.D dissertation, University of Pennsylvania; 1976.
Mouy P. Le Développement de la physique cartésienne, 1646–1712. Paris: Vrin; 1934.
Ragland E. Chymistry and taste in the seventeenth century: Franciscus dele Boë Sylvius as a chemical physician between Galenism and Cartesianism. Ambix. 2012;59:1–21.
Regius H. Fundamenta physices. Amsterdam: A. Elzevier; 1646.
Regius H. Brevis explicatio mentis humanae, sive, Animae rationis … a Notis Nobil. Cartesii vindicata. Utrecht: Trajecti ad Rhenum; 1648.
Regius H. Philosophia naturalis. 2nd ed [of Regius 1646]. Amsterdam: L. and D. Elzevier; 1654.
Regius H. Praxis medica medicationum exemplis demonstratur. 2nd ed. Utrecht: Trajecti ad Rhenum; 1657a.
Regius H. Brevis explicatio mentis humanae. 2nd ed. Utrecht: Trajecti ad Rhenum; 1657b.
Regius H. Epistola Henrici Regii ad V. Cl. Clerselierum JCtum Parisiensem. Utrecht: Trajecti ad Rhenum; 1661a.
Regius H. Philosophia naturalis. 3rd ed [of Regius 1646]. Amsterdam: L. and D. Elzevier; 1661b.
Regius H, Wassenaer P. Medicatio viri cachexia leucophlegmatica affecti. Utrecht: Trajecti ad Rhenum; 1647.
Rohault J. Traité de physique. 2 vols. Paris: Savieux; 1671.
Schmaltz TM. Early modern cartesianisms: Dutch and French constructions. New York: Oxford University Press; 2017.
Sylvius F (Franz de le Boë). Opera medica. Amsterdam: Elzevier & Wolfgang; 1679.
van Ruler JA. Substituting Aristotle: platonic themes in Dutch cartesianism. In: Hedley D, Hutton S, editors. Platonism at the origins of modernity: studies on platonism and early modern philosophy. Dordrecht: Springer; 2008. p. 159–75.
Verbeek T, editor. Descartes et Regius: Autour de l'Explication de l'esprit humain. Amsterdam: Rodolpi; 1993a.
Verbeek T. Tradition and novelty: Descartes and some cartesians. In: Sorell T, editor. The rise of modern philosophy: the tension between the new and traditional philosophies from Machiavelli to Leibniz. Oxford: Clarendon Press; 1993b. p. 167–96.
Verbeek T. Les Cartésiens face à Spinoza: l'exemple de Johannes de Raey. In: Cristofolini P, editor. L'Hérésie spinoziste: la discussion sur le Tractatus Theologico Politicus, 1670–1677, et la reception immediate du spinozisme. Amsterdam-Maarssen: Holland University Press; 1995. p. 77–88.
Wilson C. Descartes and the corporeal mind: some implications of the Regius affair. In: Gaukroger S et al., editors. Descartes' system of natural philosophy. Cambridge: Cambridge University Press; 2000. p. 659–79.

Chapter 6
The Critical Reception of Cartesian Physiology in Tommaso Cornelio's *Progymnasmata Physica*

Raffaele Carbone

Abstract This article highlights certain key moments in the dissemination of Cartesianism in Naples in the 17th century. It focuses, in particular, on the *Progymnasmata physica* (1663), written by Tommaso Cornelio (1614–1684), who derived a great deal of his conceptions of physics and physiology from Descartes. Although precise references to Descartes' texts are thin on the ground, we hypothesize that Cornelio was familiar with *L'Homme*, probably also on the basis of the fifth part of the *Discours de la méthode*, in which, as is well known, there is a summary and a completion of the treatise that Descartes declined to publish. Finally, we stress the critical aspects of Cornelio's reception of Cartesianism and the fact that he introduces the novelty of Cartesian teachings and positions to a wider context.

6.1 Descartes' Reception in Naples and the Work of Tommaso Cornelio

It is known that, from the fourth decade of the seventeenth century, Descartes' works and thinking slowly began to catch on in Italy, even in the southern regions of the peninsula.[1] A letter from Father Mersenne to Galileo Galilei reveals that, the

[1] Cf. E. Garin, "Cartesio e l'Italia", in *Giornale critico della filosofia italiana*, XXIX, 1950, p. 395–405; G. Belgioioso, *Cultura a Napoli e Cartesianesimo. Scritti su G. Gimma, P. M. Doria, C. Cominale*, Lecce, Congedo Editore, 1992; M. Torrini, "Cartesio e l'Italia: un tentativo di bilancio", in *Giornale critico della filosofia italiana*, a. LXXX, fasc. II, 2001, p. 214–230. On the Neapolitan philosophical context in the seventeenth century, cf. B. De Giovanni, *Filosofia e diritto in Francesco d'Andrea. Contributi alla storia del previchismo*, Milan, Giuffrè, 1958; P. Piovani, "Il pensiero filosofico meridionale tra la nuova scienza e la Scienza Nuova", in *Atti dell'Accademia di scienze morali e politiche della Società nazionale di scienze, lettere e arti in Napoli*, LXX, 1959, p. 77–109, now in Id., *La filosofia nuova di Vico*, edited by F. Tessitore, Naples, Morano, 1990, p. 11–53.

R. Carbone (✉)
IHRIM (UMR 5317) – ENS de Lyon,
15 parvis René Descartes BP 7000 69342, Lyon Cedex 07, France
e-mail: raffaele.carbone@ens-lyon.fr

D. Antoine-Mahut, S. Gaukroger (eds.), Descartes' *Treatise on Man* and its Reception, Studies in History and Philosophy of Science 43,
DOI 10.1007/978-3-319-46989-8_6

very year in which the *Discours de la méthode* and the *Essais* were published, news and information regarding the essays – if not actually the texts themselves – began to circulate in Italy[2] and in 1641, the Genovese mathematician and physicist Giovanbattista Baliani (1582–1666), in a letter to the Jesuit Luigi Confalonieri, alludes to the *Discours* and discloses the impressions that his reading of the *Dioptrique* and the *Géométrie*[3] made upon him. In Naples in particular, it was in scientific circles that the author of the *Discours de la méthode* became an object of interest. The people who operated within this context, such as Tommaso Cornelio, Leonardo Di Capua and Giovanni Alfonso Borelli, favoured Cartesian physics, but were not completely ignorant of the metaphysical principles at the basis thereof. However, it was not long before even thinkers and scholars of an essentially philosophical background – from Michelangelo Fardella to Giacinto Sigismondo Gerdil – discovered and embraced the Cartesian orientation. Descartes has been quoted in Italian philosophical and scientific treatises since the day he died, some time in 1650, and the first to sing his praises and disseminate his thinking were Tommaso Cornelio (1614–1684), professor of mathematics, medicine and astronomy at the University of Naples, and Francesco D'Andrea (1625–1698), jurist and philosopher, and member of various associations, including the *Accademia degli Investiganti.*[4] It was this association that recognized his pioneering role in introducing Cartesian philosophy to Naples:

> It didn't take me long to dedicate myself to the study of humane letters and other sciences; thus, upon the arrival in Naples, in 1649, of Mr. Tommaso Cornelio, to whom our city owes everything we know that is worth knowing about philosophy and medicine, I was the first to embrace the manner of philosophizing that he promoted by bringing to Naples the works of Renato delle Carte, whom we had never even heard of until that time [...].[5]

---

[2] From this letter, it is clear that Mersenne sent Galileo Descartes' texts: "Tandem, Vir Illustrissime, perfectam habes de dioptricis theoriam et praxim, quibus dioptrica instrumenta deinceps construantur, quae siderum phaenomena nobis perfectius detegant, et geometriam, quae hactenus incognita detegat. Accipe igitur librum Domini de Cartesii, quem ad te Batavia mittit, nosque fac certiores de huius libri receptione, ut ad illum confestim scribam te illum accepisse. Quod si difficultates quascunque proponendas habueris circa quaedam quae forte satis explicata non fuerint aut quae videbuntur negotium aliquod facessere, pollicetur se statim ea. soluturum atque responsurum. Tuas autem litteras fidelissime sum ad illum missurus, si quas scripseris. Utinam nostra te Gallia, quemadmodum R. P. Campanellam, teneret, ut duobus summis viris eodem saeculo eodemque loco frueremur, et ea. esses libertate qua Gallos esse contingit. Verumtamen sit summa laus Deo Optimo Maxime, eiusque voluntas in nobis, sicut in coelis, ex omni parte perficiatur. Quem veneror tibi firmissimam salutem atque valetudinem impertiatur, quamdiu tui fuero" (M. Mersenne a Galileo, 27 novembre 1637, in *Opere di Galileo*, Florence, Barbera, 1968, vol. XVII, p. 226).

[3] Cf. Baliani's letter of May 1642 to Confalonieri, in C. Costantini, *Baliani e i Gesuiti. Annotazioni in margine alla corrispondenza del Baliani con Gio. Luigi Confalonieri e Orazio Grassi*, Florence, Giunti-Barbera, 1969, p. 26. On the dissemination of these Cartesian texts in Italy cf. E. Lojacono, *Immagini di René Descartes nella cultura napoletana dal 1644 al 1755*, Lecce, Conte Editore, 2003, p. 11–76.

[4] Cf. E. Giancotti, "Philosophie et méthode de la philosophie dans les polémiques sur Descartes en Italie entre le XVIIème et le XVIIIème siècle", in *Problématique et réception du* Discours de la méthode *et des* Essais, edited by H. Méchoulan, Vrin, Paris, 1988, p. 283–299, in particular p. 285.

[5] F. D'Andrea, *Avvertimenti ai nipoti*, edited by I. Ascione, Naples, Jovene, 1990, p. 203.

Tommaso Cornelio, doctor and physiologist,[6] has long been considered by historiographers as a man who made a significant contribution to the renewal of southern culture, encouraging it to open up to an anti-dogmatic, rationalistic and experimentalist science.[7] His admiration for Descartes is obvious as far back as his pamphlet *De Cognatione aëris et aquæ*, written in 1646, and only published for the first time in 1663, and in his *Progymnasmata physica*, in which the author of the *Discours de la méthode* is presented as an "unrivalled philosopher".[8] In this text, urged on by his friend Marco Aurelio Severino,[9] Cornelio tackles the problem of the affinity between water and air, explaining the change in state of the water, not in the light of the effects of an external agent – that is to say the effects of a certain quality, such as heat or cold –, but on the basis of a variable arrangement of the atoms of the matter itself (water and ether), brought about, according to the principles of Cartesian physics, by movement alone.[10]

In the *Epistola qua motuum illorum qui vulgo ob fugam vacui fieri dicuntur, causa vera per Circumpulsionem ad mentem Platonis explicatur* (1648), which revolves around the Platonic theory of circumpulsion, as formulated in *Timaeus* (79 b–c),[11] we clearly see the emergence of Tommaso Cornelio's epistemological orientation, which had already taken shape in the *De Cognatione*. This is founded upon the idea that a knowledge of natural processes is based on the quantitative unification of matter, on the continuity of the physics of the heavens and that of the earth.[12]

---

[6] Cf. on this subject F. Trevisan, "Between Ancients and Moderns: Tommaso Cornelio's Medical Teaching and Unpublished Comment by Him on the Galenic 'Ars Parva'", in *Nouvelles de la République des Lettres*, II, 1983, p. 59–73.

[7] It is possible to trace this tradition back to P. Napoli Signorelli, *Vicende della coltura nelle Due Sicilie, o sia storia ragionata della loro legislazione e polizia*, 5 volumes, Naples, V. Flauto, 1784–1786, V, p. 213, and to P. Giannone, *Istoria civile del regno di Napoli*, Naples, Lombardi, 1805, l. XXXVIII, p. 430. Cf. also A. d'Atri, "Tommaso Cornelio nella storiografia", in *Omaggio a Tommaso Cornelio*, G. Mocchi edition, 2 vols, Soveria Mannelli, Rubbettino, 2004, vol. I, p. 172–186.

[8] "Facilis videbitur explicatio, si usurpemus hypotheses incomparabilis Philosophi Renati des Cartes" (*Thomae Cornelii Cosentini De Cognatione aëris et aquæ epistola ad Marcum Aurelium Severinum*, in Id., *Progymnasmata physica*, Neapoli, Ex Typographia Jacobi Raillard, 1688, p. 387). Cf. also Id., *Epistola qua motuum illorum qui vulgo ob fugam vacui fieri dicuntur, causa vera per circumpulsionem ad mentem Platonis explicatur*, in Id., *Progymnasmata physica*, *op. cit.*, p. 351: "[…] præsertim si principiis perspicacissimi atque incomparabilis Philosophi Renati des Cartes insistentes, opinemur corpora gravia, non proprio liberoque, sed coacto potiùs, atque extrinsecùs adveniente motu versus centrum propelli". In *De ratione Philosophandi. Progymnasma I* Descartes is described as a "vir incomparabilis" (Id., *Progymnasmata physica*, *op. cit.*, p. 80).

[9] Cf. M. Torrini, *Lettere inedite di Tommaso Cornelio a M. A. Severeino*, in "Atti e Memorie dell'Accademia Toscana di scienze e lettere La Colombaria", XXXV, 1970, p. 146–147.

[10] Cf. M. Torrini, *Tommaso Cornelio e la ricostruzione della scienza*, Naples, Guida, 1977, p. 22–23. The experimental example and the analogies to which Cornelio refers in his explanation (the water particles compared to tiny, living eels scattered at random) bear further witness to the Cartesian influence, in this case, the *Météores* in particular. Cf. Tommaso Cornelio, *De Cognatione aëris et aquæ*, *op. cit.*, p. 389–390.

[11] According to this thesis, the expelled air displaces the neighbouring air, and so on as far as the point from which it emerged; this occurs because there is no void. Cf. E. Lojacono, *op. cit.*, p. 24–25.

[12] Cf. T. Cornelio, *De Cognatione aëris et aquæ*, *op. cit.*, p. 384–385; Id., *Epistola qua motuum illorum … circumpulsionem ad mentem Platonis explicatur*, *op. cit.*, p. 321.

The central theme of this treatise, in which Cornelio "reaches the point of maximum consonance with Descartes",[13] is a rejection of escape from the void as an explanation of movement. The chief idea revolves around the principle that nature acts in a uniform manner; it is, therefore, possible to come up with a hypothesis that will explain, not a single isolated phenomenon, but the entirety of natural phenomena; the hypothesis in question is that of the ether, a space full of bodies in which anything which moves pushes away its neighbouring body which, in turn, pushes aside another body and so on, until the last body to be displaced takes the place of the body which moved in the first place.[14]

Tommaso Cornelio's *Progymnasmata physica*, which is a collection of scientific writings on a number of themes, although not a systematically structured theoretical text in its own right,[15] is a profound and illustrious testament to the Italian philosophico-scientific intellectual climate, particularly in the south, and to its reformatory tendencies. The volume, which was printed in Venice in 1663,[16] is the fruit of 15 years of intense research and the intellectual exchange of ideas occurring between Cornelio's stay in Rome and his return to Naples where, in 1653, he took up a post teaching mathematics at the university. The *Progymnasmata physica* endeavour to impose the model of a new *in fieri* knowledge upon the contemporary philosophical debate; this model is perfectible and has a practical dimension, specifically in the sense that it aspires to be shared by an ever wider circle of end-users.[17]

The first seven *Progymnasmata* (*De ratione philosophandi*, *De rerum initiis*, *De universitate*, *De sole*, *De generatione hominis*, *De nutricatione*, *De vita*) as well as the posthumous *De sensibus*, added to the edition of 1688, put together under the supervision of Tommaso Cornelio's friends, aspire to implement an epistemological reform programme. The initial dialogue between Stelliola, Trusiano and Bruno is, for example, a clear taking of the modernists' side against pharmaceutical medicine, scholastic philosophy and official medical organization.[18] The author separates metaphysics from the natural sciences and constructs an itinerary that kicks off with mathematics, the language of astronomical phenomena and mechanical physics, and concludes with physiology, dealing with complex phenomena in which one

[13] M. Torrini, *op. cit.*, p. 79.

[14] Cf. T. Cornelio, *Epistola qua motuum illorum ... circumpulsionem ad mentem Platonis explicatur*, *op. cit.*, p. 307–308. Cf. also *ibid.*, p. 311–312.

[15] In this regard, see P. Cristofolini, "Tommaso Cornelio et l'histoire du matérialisme", in *Gassendi et l'Europe (1592–1792)*, Proceedings of international meeting *Gassendi et sa postérité (1592–1792)*, Paris, Sorbonne, 6th–10th October 1992, edited by S. Murr, Paris, Vrin, 1997, p. 335–346, in particular p. 342.

[16] Various other editions were published after this one, culminating in the posthumous edition, Naples 1688.

[17] Cf. the dialogue between Stelliola, Trusianus and Bruno at the beginning of the work: *Dialogus in Præmii locum suffectus*, *op. cit.*, p. 2–3. On this point cf. M. Torrini, *op. cit.*, p. 104.

[18] Cf. *Thomae Cornelii Cosentini Dialogus in Præmii locum suffectus*, in Id. *Progymnasmata physica*, *op. cit.*, p. 49–50, 53.

comes up against the limitations and the imperfection of the senses. Tommaso Cornelio's position emerges clearly in this work: although he is sensitive to the novelty of Cartesian philosophy, he does not accept it without criticism in its entirety and he does not, for example, fully fall heir to its mechanism as the basis for interpreting the phenomena of life.[19] After all, a not entirely mechanical bearing is a general characteristic of Neapolitan Cartesianism,[20] which borrows from Descartes those concepts and aspects which are particularly close to Renaissance philosophy.[21]

6.1.1 Traces of *L'Homme* Treatise and the Critical Reception of Cartesianism in Tommaso Cornelio's *Progymnasmata physica*

Tommaso Cornelio deals essentially with Cartesian physics and with Descartes writings on optics and dioptrics. It is more difficult to find precise traces of the *Traité de l'Homme* in the *Progymnasmata physica*. And yet it is possible to identify certain consonances between the two works, and we may thus suppose that Cornelio was familiar with an edition of this Cartesian treatise which, published in Latin in Leiden in 1662 by Florent Schuyl, and in French in Paris in 1664, under the supervision of Clerselier, would only become part and parcel of the *corpus* of Cartesian works studied in Naples towards the end of the century.[22] In any case, it is well known that, in the fifth part of the *Discours de la méthode*, one of the Cartesian texts that was certainly doing the rounds in Italian and Neapolitan intellectual circles, Descartes evokes the unpublished treatise,[23] although he does not put forward a

[19] On these points cf. F. Crispini, *Metafisica del senso e scienze della vita. Tommaso Cornelio*, Naples, Guida, 1975, p. 71–103.

[20] This is the opinion of G. Belgioioso, *La variata immagine di Descartes. Gli itinerari della metafisica tra Parigi e Napoli (1690–1733)*, Lecce, Milella, 1999, p. 57. Referring to Di Capua and to Cornelio himself, the scholar also reveals: "[...] i cartesiani di Napoli hanno spesso una conoscenza indiretta dei testi cartesiani – emblematico il caso di Vico – e, quando ne hanno una conoscenza diretta e corretta, ne prendono le distanze. Ad un esame ravvicinato, l'appellativo di 'cartesiani' per questi 'letterati', che sono scienziati e coltivano esperimenti, appare improprio [The Cartesians of Naples often have an indirect knowledge of Cartesian texts – Vico is a case in point – and whenever they have a direct and accurate knowledge thereof, they distance themselves from them. Upon closer inspection, the name 'Cartesian' when applied to these academics, who are scientists and who perform experiments, would appear to be erroneous]" (*Ibid.*, p. 57–58).

[21] Cf. N. Badaloni, *Introduzione a Vico*, Milan, Feltrinelli, 1961, p. 78.

[22] Cf. the memoirs of Francesco Maria Spinelli, prince of Scalea and pupil of Gregorio Caloprese (1654–1715), who, in his teachings, made use of a great number of Cartesian texts, including the treatise *De homine*. Cf. *Vita e studi di F. M. Spinelli Principe della Scalea [...]*, in A. Calogerà, *Raccolta di opuscoli scientifici e filosofici*, Venice, Simone Occhi, 1753, vol. XLIX, p. 463–521, in particular p. 470–478.

[23] Cf. *Œuvres de René Descartes*, Adam-Tannery edition, Paris, Vrin, 1996, volume VI, p. 45 (henceforth referred to as: AT).

summary faithful to the contents of the *Monde* and *L'Homme*; nevertheless, in the *Discours*, he offers a detailed explanation of the workings of the heart, which actually picks up on some of the points found in *L'Homme.*[24]

A famous passage from *L'Homme* seems, in particular, to have provided inspiration for a page of *Progymnasma II*. This is the paragraph in which Descartes draws attention to the perfection of machines built by men, which are able to move on their own:

> Nous voyons des horloges, des fontaines artificielles, des moulins, et autres semblables machines, qui n'étant faites que par des hommes, ne laissent pas d'avoir la force de se mouvoir d'elles-mêmes en plusieurs diverses façons; et il me semble que je ne saurais imaginer tant de sortes de mouvements en celle-ci, que je suppose être faite des mains de Dieu, ni lui attribuer tant d'artifice, que vous n'ayez sujet de penser, qu'il y en peut avoir encore davantage.[25]

In *Progymnasma II*, Cornelio too, in his own way, highlights the wonders of artificial machines and compares them with the works of nature;

> Indeed, just as craftsmen use iron to make knives to cut things, clocks to mark the passage of time, ploughshares to till the earth, chains, arrows and countless other instruments of different uses and varying applications depending on position, size, shape, movement and the number of parts the tool is made up of, so too do I believe that all natural things are manufactured and produced using the same universal matter, and differ one from the other according to the arrangement of the particles of matter, movement, size, number and shape.[26]

In the rest of the text, Cornelio goes on to claim that the Cartesian *ratio*, in itself and unobstructed, leads us into the heart of physiology, along wide, open, gentle roads, thereby exposing the clearest signs of nature.[27] And yet Cornelio does not believe that all natural phenomena may be adequately explained according to the principles of Cartesian philosophy. He highlights the fact that, for Descartes, every function of an inanimate body, as well as its movement, depends on the varying alterations of the matter and occurs entirely according to the laws of mechanics. Living beings themselves should, therefore, move and become animated as if they

[24] Cf. A. Bitbol-Hespériès, *Le Principe de vie chez Descartes*, Paris, Vrin, 1990, p. 86–91; Id., "Introduction", R. Descartes, *Le Monde, l'Homme*, edited by A. Bitbol-Hespériès and J.-P- Verdet, Paris, Seuil, 1996, p. XL–XLII, XLVI.

[25] R. Descartes, *Le Monde, l'Homme*, *op. cit.*, p. 119 (AT, XI, p. 120).

[26] "Nam quemadmodum ex eodem ferro artificis opera elaborantur cultri ad res incidendas, horologia ad intervalla temporis designanda, vomeres ad terram perstringendam, catenæ, sagittaæ, & alia ferè innumerabilia instrumenta, quorum varia est. utilitas, ac diversa operatio pro ratione situs, magnitudinis, figuræ, motus, ac numeri partium ipsius ferri: ita pariter verosimillimum censeo ex eadem Universi materia effingi conflarique res omnes naturales inter se dissimiles, propter diversam particularum materiæ dispositionem, motum, magnitudinem, numerum, atque fíguram" (T. Cornelio, *De Initiis Rerum naturalium, Progymnasma II*, in Id. *Progymnasmata physica*, *op. cit.*, p. 94).

[27] "Facilis mehercule atque expedita est. hæc philosophandi ratio Cartesiana, quae ad interiora Physiologiæ penetralia, nonquidem per salebras, sed per planam patentemque viam, nos dirigit, & clariora in dies naturæ indicia depromit" (*Ibid.*, p. 94–95).

were robots. The majority of their actions, especially those pertaining to sensations and desires, are not, however, tied to the laws of mechanics and would seem to be separate, in some way, from the actions of the body. Although Descartes has ingeniously demonstrated how the origins of the varying natures of things may be traced back to matter, and how it is moulded in various ways, the concept of "vis efficiens" does not fully convince Cornelio, especially when it comes to explaining biological phenomena.[28]

In summary, for Cornelio, Cartesian mechanical physics is not able to give a satisfactory explanation of how living beings function. He recognizes the complexity of the physical and biological world, which cannot be explained on the basis of a simple geometric model. For Descartes, life operations, even the most complex ones, depend solely on the arrangement of the organs ("la disposition des organes"),[29] that is to say, on a system of apparatus and mechanical connections[30]; Cornelio, on the other hand – in keeping with a biological way of thinking that tends to oppose the radicalism of Cartesianism and to perceive, within animal bodies, a number of functions that cannot simply be the result of the interplay of springs, tubes and filters[31] –, expressing an anti-reductionist condition which is typical of vitalism,[32] identifies a multiplicity of mechanisms in the physical-biological sphere. The process of food assimilation, for example, is a particular mechanism of the living being[33]: in the nutrition process, the movement is the basic explanatory law, but it cannot be attributed exclusively to the automatic devices of the body-machine, neither can it be fully explained on the basis of purely anatomical data.[34] The entire body system is, in fact, subjected to a force of movement which, although mechanical in nature, adapts to the peculiarities of the organic sphere, in which multiple functions contribute to the growth process and to the conservation of the living being.[35]

---

[28] "Verumtamen ne Cartesio plus tribuere videar, quàm res, & veritas ipsa concedat, non dissitebor desiderari quippiam in ejusdem Philosophia, nec per illius principia explicari satis commodè posse omnia naturæ phænomena. Fac enim omnes inanimorum corporum functiones, motusque pendere a diversa materiæ modifícatione, & omnino juxta leges Mechanicæ fieri, quis tamen nobis persuaserit ipsas etiam animantes instar automatan cieri; quum pleræque earum actiones maximè quæ ad sensum appetitumque attinent, ita Mechanicis legibus non sint obstrictæ, ut potiùs ab actione corporis sejunctæ quodammodo videantur. Ad hæc Renatus quanquam expressit accuratè materiam, ostenditque qua ratione ex illa variis conformata modis, possint diversæ rerum naturæ proficisci, vim tamen efficientem nequaquam pro dignitate satis explicasse videtur" (*Ibid.*, p. 95–96).

[29] R. Descartes, *L'Homme*, *op. cit.*, p. 119 (AT, XI, p. 119).

[30] Cf. G. Canguilhem, "Machine et organisme", in Id., *La connaissance de la vie*, Paris, Vrin, 2009 [1965], p. 129–164, in particular p. 146–147.

[31] Cf. F. Dagognet, *Le corps*, Paris, PUF, 2008 [1992], p. 60–61.

[32] Cf. Mayr, *This is Biology: The Science of the Living World*, Cambridge, Harvard University Press, 2001 [1997], p. 9.

[33] Cf. T. Cornelio, *De Nutricatione*, *Progymnasma VI*, in Id. *Progymnasmata physica*, *op. cit.*, p. 213.

[34] *Ibid.*, p. 257–260.

[35] On these points and, in particular, on the concept of the plurality of mechanisms, cf. F. Crispini, *op. cit.*, p. 97–101.

In the Cartesian model, then, the order of ideas is, as is well-known, distinct from that the body and is regulated by a strict determinism. Cornelio, on the other hand, shifts the attention onto that which unites from within, the physical world which obeys the laws of matter and motion. A monist position emerges here and, in its harking back to hylozoism and the Renaissance-style energetistic model, seems to provide a bulwark against the domination of mechanical philosophy, although it is by no means a rejection thereof.[36]

Cornelio distances himself from Descartes on a number of specific points, such as the question of animal souls, as we have seen, or on blood circulation.[37] Nevertheless, even when he does not fall in with Cartesian theories or explanations, he does make ample reference to Descartes, such as in the *Progymnasma VII, De vita,*[38] and demonstrates his familiarity with the French philosopher's explanatory model and the wide, centuries-old debate on which it has had such an explosive impact. Let us, then, consider a particular aspect of Cornelio's critical reception of Cartesianism as regards the subject of biological phenomena.

In *L'Homme*, Descartes claims that, in the pores of the heart, there is a fire without light ("un de ces feux sans lumière") which cooks the tissues and makes them blazing hot: Thus, as the blood enters one of the two cavities, it immediately swells and expands. The purpose of the fire in the heart of the machine is to expand, heat and thin the blood ("et le feu qui est dans le cœur de la machine [...] n'y sert à autre chose qu'à dilater, échauffer, et subtiliser ainsi le sang [...]"), which falls, non-stop, drop by drop, via a thin tube of the vena cava, into the right cavity, from which it is exhaled into the lung. From the vessel of the lung, which anatomists call arteria venosa – it proceeds to the other chamber and, from here, is distributed all around the body.[39] Respiration, which is used to thicken the vapours of the body-machine in the world "imagined" by Descartes, is necessary to keep this fire going, just as respiration on our part is necessary for the preservation of life.[40] Descartes highlights, therefore, the connection between life and cardiac heat, and he came back to this point in his later writings, from *Discours de la méthode* to *Passions de l'âme.*[41]

---

36 Cf. F. Crispini, *op. cit.*, p. 110.

37 Cf. on this subject V. Jullien, "De la *fortuna* du cartésianisme napolitain", in *Les Méditerranées du XVII*[e] *siècle*, Actes du VIe colloque du Centre international de rencontres sur le XVIIIe siècle, Monopoli, Bari, 13–15 avril 2000, edited by G. Dotoli, Tübingen, G. Narr, 2002, p. 337–347, in particular p. 344.

38 Cf. for example, this passage in which Cornelio introduces the Cartesian thesis according to which the beating of the heart is stimulated by the entrance of the blood into its cavities: "Cartesius palpitationem cordis excitari censet a fervore fanguinis: inquit enim sanguinem in ventriculos cordis illapsum, insito vel potiùs innato illius calore effervescere, & rarefieri; atque adeò in majorem molem distendi; donec cordis cavernulæ ab intumescente sanguine præter modum distractæ extensaæque, dilatari ampliùs non queant, sed comprimantur tandem, conclusumque sanguinem actutum propellant: itàque alterno pulsu fieri systolem, atque diastolem. Hæc autem ad mentem Aristotelis dicta esse nonnuli, reclamante Cartesío, arbitrantur" (T. Cornelio, *De Vita Progymnasma VII*, *op. cit.*, p. 270–271).

39 R. Descartes, *L'Homme*, *op. cit.*, p. 120 (AT, XI, p. 122).

40 *Ibid.*, p. 121 (AT, XI, p. 123).

41 On this subject we return to A. Bitbol-Hespériès, *Le Principe de vie chez Descartes*, *op. cit.*

Cornelio also tackles these arguments in his *Progymnasmata physica*. Indeed, in *De Vita*, he writes that heat in the blood is stimulated by the mixing of the breath of life, or of the liveliest spirit that continually enters the ventricles of the heart and that, as long as it is mixed with the blood, shakes up the particles and increases the temperature ("[...] excitari [...] ab admistione vitalis halitus, seu spiritus acrioris in cordis ventriculos jugiter influentis, qui dum sanguini permiscetur ejus particulas agitet, & in fervorerm adigat").[42] Life, then, consists in the continuous and incessant movement of the blood ("[...] facilè intellectus est vitam in continuata & perenni sanguinis motione consistere"). Indeed, we will live as long as the blood, which is heated in our hearts and which is animated by the breath of life, is transported to our limbs via our arteries ("Quippe tamdiu vivimus, quamdiu sanguis in corde calescens vitalique halitu animatus per arterias in membra diffunditur").[43] In the remainder of the text, while arguing that nothing is as necessary to the preservation of life as respiration, Cornelio dwells on the function of air in the respiratory process and on the connection between this process and the movement of the blood.[44] He claims, among other things – and almost echoing the Cartesian text – that the blood pushed into the lungs via the vena arteriosa from the right ventricle of the heart, cannot flow into the left ventricle unless the air borne by the spirit inflates and expands the branches of the trachea.[45]

These passages of the *Progymnasmata physica*, among others,[46] show that Cornelio does not stop at faithfully reproducing Cartesian explanations as formulated primarily in *L'Homme* and in the *Discours*: when it comes to explaining the phenomena of life and of its preservation, he brings up, without repudiating the mechanism itself, a further element, a breath, a driving life force that animates and guides the work of the mechanical system. Cornelio highlights the fact that a common vitality co-exists with the apparatus and mechanical workings of the organism and that living matter is not purely passive and cannot be reduced entirely to geo-

---

[42] T. Cornelio, *De Vita. Progymnasma VII*, *op. cit.*, p. 281.

[43] *Ibid.*, p. 282. For which reason, if a part of the body is no longer moistened by the blood, it grows cold, dies and finally rots ("Quapropter si membrum aliquod, vel corpus universum hujusmodi influxu deficitur, protinus inalgescit, emoritur, ac demum putrescit") (*Ibid.*).

[44] *Ibid.*, p. 282–283 *et seq.*

[45] "Quippe sanguis ille, qui e dextero cordis ventriculo in pulmones per venam, ut vocant, arteriosam propellitur, nequit in sinistrum ventriculum permanare, nisi aër spiritu ductus arteriæ asperæ surculos inflet, atque distendat: Hinc enim fit, ut venæ arteriosæ ramuli comprimantur, atque adeò conclusus in his sanguis protrudatur in surculos arteriæ venosæ" (*Ibid.*, p. 283).

[46] See, for example the final paragraph of *De Vita*, in which it emerges that biological phenomena depend upon a life principle which cannot simply be a matter of the laws of mechanics ("[...] ita pariter haæc ipsa animabilis spirabilisque natura ad confervationem vitaæ neccessaria est.: hinc enim vitalis, & salutaris animalium spiritus perenni successione viviscit", *ibid.*, p. 291).

metric extension,[47] but includes something else to the extent that it is active, sensitive and animated.[48]

In *De Sensibus, Progymnasma Posthumum* too, Cornelio distances himself more explicitly from the Cartesian concept of space, and from a conception which, by reducing matter to extension, excludes any intrinsic activity. Believing that life phenomena cannot be explained solely by virtue of the mechanical movement of the corpuscular particles, Cornelio hypothesizes a spontaneous and unique activity of matter that accounts for the physiological dynamics. In this regard, he distinguishes, for example, between natural things ("res naturales") that are stirred by constant movement and cannot spontaneously find themselves in a state of stillness ("earum quædam perenni motu, ac perpeuta agitatione cientur, nec umquam sponte quiescunt") and those things that never move of their own accord; although these are, through their own force and by their own nature, always at rest, they can be stimulated to move ("Aliæ vero numquam per se ipsæ moventur; sed quum sua vi, et natura quiescant, possunt aliunde ad motum concitari").[49] Cornelio thus dwells on the self-propelled corpuscles ("ea corpuscola, quæ per sese moveri"), igneous, ethereal elements of which some bodies, such as fire, which move of their own volition and spontaneity, consist.[50] In the remainder of the text, Cornelio tackles the question of the part of the body in which the ability to feel exerts its force ("Sed adhuc tamen obscura est quæstio de loco vel corporis parte, in qua animus sentiendi vires exerceat"). In this regard, he does not believe that there is a single origin of feeling in this or that part of the body (heart, brain etc.) as has been claimed by some philosophers, but rather that the "animus sentiendi" is decentralized and distributed horizontally, both in the centre and at the periphery.[51] In any case, and again from a rather non-Cartesian perspective, he would seem to be drawing attention to an active dimension of feeling, a mobilization of spirits, forces and internal movements, an ability to self-direct in the individual parts of the living organism.[52] In other words, both feeling, in its various forms (sense and intellect), and the movement of physi-

[47] For Cornelio, it would not, therefore, be possible to structure physiology according to the model of mathematics, as Niels Stensen (1638–1686), for example, suggested doing, at least in the early days of his career, in the wake of Descartes. On the subject of the followers of Descartes who apply the mechanism to the medical field, and their general mechanistic interpretation of life phenomena, cf. M. D. Gremek, "A Survey of Mechanical Interpretations of Life from the Greek Atomists to the Followers of Descartes", in *Biology, History, and Natural Philosophy*, edited by A. D. Breck and W. Yourgrau, New York, Plenum Press, 1972, p. 181–195, in particular p. 189–191; S. Carvallo, "Pourquoi ne pas être cartésien en médecine?", in *Qu'est.-ce qu'être cartésien*, edited by D. Kolesnik-Antoine, Lyon, ENS Éditions, 2013, p. 437–463, in particular p. 441–444.

[48] Cf., for example, T. Cornelio, *De Generatione. Progymnasma V*, in Id., *Progymnasmata physica, op. cit.*, p. 179–183.

[49] T. Cornelio, *De Sensibus. Progymnasma Posthumum*, in *Thomæ Cornelii Cosentini Opera quædam posthuma, Numquam antehac edita, Ad Nobilissimum Virum Francescum Ab Andrea*, Naples, Ex Typographia Jacobi Raillard, 1688, p. 8–9.

[50] "[…] nam ignis, seu flamma, quæ magnam partem ex corpusculis sese moventibus constitui videtur, per se ipso ac sua sponte indefínenter movetur […]" (*Ibid.*, p. 10).

[51] Cf. *ibid.*, p. 70–71.

[52] Cf. *ibid.*, p. 74–75.

ological processes depend "on the force and the power of the spirits" and it is thanks to the spirits' power and ability to self-direct – those spirits who, among the physical spirits, preside over the functions of the living – that such cognitive processes and life process are carried out and brought, time after time, to their conclusion.[53]

6.2 Concluding Notes

Tommaso Cornelio is certainly someone who contributed to a wind of change in Italian culture and in Neapolitan culture in particular, in the seventeenth century; indeed, in his works, we perceive a striving for knowledge of the phenomena of nature and of her various elements.[54] As we have explained from the very beginning, he has been credited with introducing Descartes' method and bearing to Naples. From his books we get a very clear idea of the debt he owed to the author of the *Discours de la méthode* as well as his awareness that the Cartesian programme implicated a reorganization of the entire field of knowledge (it is interesting, in this regard, to consider how broad the horizon of Cornelio's own interests was); precise references to Descartes' texts are, though, somewhat thin on the ground.[55] It is by taking this information into account, however, that we may hypothesize that Cornelio was familiar with *L'Homme*, probably also on the basis of the fifth part of the *Discours de la méthode*, where, as is well known, there is a summary and a completion of the treatise that Descartes declined to publish. Sensitive to the new Cartesian anthropology,[56] which is based on the idea that the human body is just another component of the world, regulated by the laws of nature established by God, Cornelio takes a wide perspective, in the sense that he makes a critical examination of various philosophico-scientific traditions, from the Pre-Socratics to the Renaissance thinkers (Telesio, Bruno, Campanella) and his own contemporaries (Bacon, Galileo, Harvey, Hobbes, Malphighi, Redi, and so forth). In this way, the philosopher and physician from Cosenza would seem to be the embodiment of the need to introduce the novelty of Cartesian teachings and positions to a wider context, in the light of which we may, time and again, assess and make use of the tools and the ideas that the French philosopher has given us.

---

[53] "Convenit profectò inter Physicos spiritus esse opifices administrosque functionum, & operum omnium, quæcumque ab animalium vita proficiscuntur. Enimvero & motus, & sensus, & intellectio, & quicquid denique vitali facultate peragitur, id omne a vi, & potestate spirituum pendet, atque perficitur" (*Ibid.*, p. 79). Cornelio then adds that the question of the origin and nature of spirits is exceedingly complex and has not yet been fully resolved (*Ibid.*, p. 79–80).

[54] As we find in one of his elegies: "Atque juvat varios hominum perquirerc mores,/Et rerum causas noscere quippe juvat,/Stellarum Cœlique vias,Terræque recessus,/Lataque cœrulei scire theatra maris" (*Thomæ Cornelii Cosentini Elegiraum Liber Posthumus*, in *Opera quædam posthuma, op. cit., Elegia I*, p. 90).

[55] Cf. E. Lojacono *op. cit.*, p. 21.

[56] On this subject cf. in particular D. Antoine-Mahut, *L'Homme cartésien. La "force qu'a l'âme de mouvoir le corps": Descartes, Malebranche*, Rennes, Presses Universitaires de Rennes, 2009.

Chapter 7
The Reception of *L'Homme* Among the Leuven Physicians: The Condemnation of 1662 and the Origins of Occasionalism

Domenico Collacciani

Abstract There is abundant commentary in the literature on the successive censures of Descartes's philosophy by the Leuven theological faculty in 1662 and the Holy Office in 1663. This essay seeks to demonstrate that the 1662 condemnation struck in particular an occasionalist interpretation of the mind–body problem that had been developed in the Leuven medical faculty. The first part studies the traces of occasionalism in the works of the Leuven physicians Guillaume Philippi and Gerhard van Gutschoven. The second part examines the *Compendium omnium præcipuarum actionum automaticarum*, a thesis defended under the direction of Professor Pierre Dorlix 5 months prior to the theologians' censure. I will show that the thesis cites long passages from the manuscript of *L'Homme* in van Gutschoven's possession and that it also contains the first occurrence, 4 years before La Forge, of the term *causa occasionalis*. I conclude that the 1662 censure was prompted by the spreading of a variant of Descartes's philosophy that drew on the occasionalist solution to account for the functions of the soul (attention, voluntary action) exclusively with the movements of the machine of the body.

7.1 Introduction: The Condemnations of Descartes's Philosophy

Undoubtedly, Descartes's condemnation by the Roman Church is the most momentous event for the reception of Descartes' philosophy at Leuven University. In fact, however, the series of events culminating in Descartes's works being placed on the Index in 1663 was triggered a year earlier at Leuven University. There, the theological faculty, prompted by the internuncio, Girolamo de' Vecchi, condemned a thesis set to be defended in the medical faculty that was largely inspired by the new

D. Collacciani (✉)
République des savoirs (USR 3608), ENS-CNRS-Collège de France,
45 rue d'Ulm, Paris 75005, France
e-mail: domenicocollacciani@yahoo.it

D. Antoine-Mahut, S. Gaukroger (eds.), Descartes' *Treatise on Man* and its Reception, Studies in History and Philosophy of Science 43,
DOI 10.1007/978-3-319-46989-8_7

philosophy. Georges Monchamp's work from more than a century ago and the recent discovery of Holy Office documents combine to present a good picture of what took place.[1] The implications of the event for the history of Cartesian philosophy, however, have so far not been fully established. On the one hand, the reasons, both theological and philosophical, for which anti-Cartesianism manifests itself so forcefully in Leuven are certainly as complex as those that motivated the spread of various Cartesianisms across Europe.[2] On the other hand, judging by their effects, the censure *donec corrigatur* (i.e, forbidden until amended) and the condemnations by the French universities that followed were rather weak: after 1663, the 'Cartesian party' continued to publish Descartes's posthumous texts along with other philosophical works.[3] The quarter century or so that separates Descartes's death from the publication of Spinoza's *Opera postuma* and Malebranche's *Recherche de la verité* was marked by rich debates, and Cartesian philosophy underwent significant evolutions vis-à-vis Descartes's thought. One of the reasons for this evolution no doubt lies in the posthumous publication of works by the philosopher himself (the *Compendium musicæ*, *Le Monde*, *L'Homme*, and the *Letters*), another in the new reflections on method, logic, and physics put forward by Johann Clauberg, Antoine Arnauld and Pierre Nicole, Nicolas Poisson and Jacques Rohault.

In the same year as the censure by the university in Leuven, Florent Schuyl published the first Latin edition of the *Treatise of Man*, which greatly impacted the way Descartes's philosophy spread. In particular, it changed the history of the reception of Descartes oeuvre as a whole, for his other works were now read in light of this treatise.[4] What I set out to show here, however, is that the 1662 condemnation resulted above all from a premature distribution of the physiological treatise. When we look at the causes of the condemnation rather than its effects, the censure appears in a whole new light. It then becomes possible to discern the philosophical richness of Cartesianism in Leuven, which proves to be an important site for the theoretical elaboration of Descartes's philosophy.

The examination of certain texts rarely evoked in the critical literature, of course, needs to be prefaced with a word of caution: just like the condemnation in France, anti-Cartesianism in Leuven is far from a purely philosophical stance and its political and religious stakes must not be overlooked. The condemnation that struck

---

[1] Georges Monchamp, *Histoire du cartésanisme en Belgique* (Bruxelles: Hayez, 1886); Jean-Robert Armogathe and Vincent Carraud, 'The First Condemnation of Descartes's *Œuvres*: Some Unpublished Documents from the Vatican Archives', *Oxford Studies in Early Modern Philosophy* (2005:1), 67–109. See also Roger Ariew, *Descartes among the scolastics* (Leiden-Boston: Brill, 2011), 217–266.

[2] On the complexity of "Cartesianism" as a historical category, see Sophie Roux, 'Pour une conception polémique du cartésianisme: Igace-Gaston Pardies et Antoine Dilly dans la querelle de l'âme des bêtes' in Delphine Kolesnik-Antoine, ed., *Qu'est-ce qu'être cartésien?* (Lyon: ENS Éditions, 2013), 315–337.

[3] On the commercial success of Descartes's works and the publication history of the posthumous oeuvre, see Matthijs van Otegem, *A Bibliography of the Works of Descartes (1637–1704)* (Utrecht: Zeno, 2002).

[4] For the influence of the treatise on Spinoza's Descartes reception, see Emanuela Scribano's recent book, *Macchine con la mente: Fisiologia e metafisica tra Cartesio e Spinoza* (Rome: Carocci, 2015).

Cartesian theses was only one among a number of censures issued by the Leuven theological faculty in the 1660s. An examination of the old university's archives in fact shows that the spreading of Descartes's philosophy was a secondary problem compared to the great theological debates about Jansenism, on which the theological faculty's interest and attempts at containment focused.[5] One might therefore be tempted to claim that in Leuven as in France, Cartesianism and Jansenism had entered into an alliance. Scholarship, however, has not as yet produced sufficiently convincing evidence to bear out this point.[6] In Leuven, unlike in France, the new philosophy seems not to have inscribed itself in a polarized debate between the Society of Jesus and the followers of Cornelius Jansen. As we will see, the views of Jansen's successor, Libert Froimont, long served *anti*-Cartesians as a model.

The condemned theses advocated the application of mechanicist physics in medicine. According to the theologians, their author's enthusiasm for mechanicist explanations had led him to express his contempt for certain Aristotelian dogmas only all too clearly. According to the list of the censured propositions, the theologians condemned in particular: (1) the text's polemical spirit (it accuses physicians of following each other like sheep); (2) the rejection of secondary qualities, which are said to be fully explained by extension and movement; (3) the thesis that the existence of the soul is more certain than the existence of the body; (4) the negation of the idea that animals are endowed with souls; (5) the negation of real accidents in favor of an explanation based solely on "the modification of motion, of rest, of situation, of shape, of the magnitude of the small parts of matter"; (6) the affirmation that animals live in more sound a manner than human beings; and (7) the view that mortality is a consequence of human ignorance.[7]

7.2 Gutschoven et Philippi, Two Cartesian Physicians

Unfortunately, no copy of the theses has survived even though they were "publicly displayed and divulged amidst an extraordinary crowd of listeners." What we know of the text comes from the excerpts cited by the theologians.[8] Nonetheless, we are

[5] See Lucien Ceyssens, 'Le cardinal F. Albizzi (1593–1684): son autobiographie et son testament' (Rome: Academica Belgica, 1977). Cf. the case of the theologian Christian de Wulf (Lupus) and the condemnation of John Baptist Sinnich's 'Saul ex re'.

[6] The documents published by Ceyssens show that Philippi was considered a moderate Jansenist; Lucien Ceyssens, *Sources relatives à l'histoire du jansénisme et de l'antijansénisme des années 1661–1672* (Louvain: Publications universitaires de Louvain, 1968), 532. Compare F. Claeys Bouuaert, *L'ancienne Université de Louvain: Études et Document* s (Louvain: Publications universitaires de Louvain, 1956). 178–179. Francisque Bouiller claims that the Cartesian theory of the will resembles the Jansenists'; see his *Histoire de la philosophie cartésienne*, 3rd ed. (Paris: Delagrave, 1867), vol. 1, 432–35. See also, Tad Schmaltz, 'What Has Cartesianism To Do with Jansenism?', Journal of the History of Ideas, Vol. 60/1, 1999, 37-56.

[7] *Histoire du cartésanisme en Belgique*, 618–21 and 357–67. *Sources relatives à l'histoire du jansénisme*, 83–100.

[8] The internuncio Girolamo de' Vecchi owned a copy of the theses; cf. *Histoire du cartésanisme en Belgique*, 352.

not lacking in sources to characterize the Cartesianism of the Leuven physicians. There are, first of all, the printed works of Guillaume Philippi, the professor of medicine to whom, following Monchamp, the theses are to be attributed. He is the author of a *summa philosophiæ* in three volumes, published between 1661 and 1664. These three "medullas"—in order of publication, *Medulla logicæ*, *Medulla metaphysicæ*, and *Medulla physicæ*—do not constitute a coherent philosophical work; they clearly bear the mark of the author's change of mind after the events of 1662. Nonetheless, Philippi manages to draw some remarkable Cartesian ideas from the traditional *quæstiones*. Chief among these is a superb formulation of occasionalism as an explanatory model of the relationship between the soul and the bodily machine dating from 1661. We shouldn't be surprised to find a thesis of this kind in a logic manual: for Philippi, the union of soul and body is merely a case of a logical relation defined by the concept of the sign. On the very first page, the author establishes the sign as "that which represents something to a cognitive being." Signs can thus be either a concept, that is to say, an image produced by the mind, or a sign that determines *occasionaliter* the cognitive power to produce a concept.

> A sign is that which represents something to a cognitive being … Through its knowledge, the sign occasionally [*occasionaliter*] determines the faculty of knowledge such that it forms the concept of something.[9]

Yet in the course of elaborating his logic/semiotics, the author comes to inquire into the origin of sensations, which amounts to investigating the production of mental signs that begins with the natural signs generated by the body. Philippi's answer is derived from Cartesian neurophysiology, which he reinterprets with the help of a classic metaphor:

> Q. Whether there are given qualities that are beings really distinct, in the strict sense, from substance, and whether the first of these is sensation.
>
> A. To answer, note that our soul is bound to the body in such a way as to be affected by the movements of the body and that it is being determined to have different sensations by the trembling of the nerves[, that is,] when the tubes of the nerves are inflated by the animal spirits and the fibers they contain are duly extended such that when one extremity of the nerve is stirred, the other extremity in the brain (from which all nerves lead off) moves as well, the way the two ends of a rope move when one of them is struck. Moreover, this movement determines the soul … to have sensations like a natural idiom, the way a spider perched in its web is being determined to pursue or to flee. That the soul is determined by this movement alone is proven by the fact that when a blind man is

[9] 'Signum est quod aliquid enti cognitivo repræsentat. … Signum mediante sui cognitione potentiam cognitivam occasionaliter determinet ad formandum conceptum de aliqua re' Guillaume Philippi, *Medulla logicæ* (Leuven: Petrus Sassenius, 1661), 1.

> being determined, by means of a cane, to know the different bodies he touches, movement alone intervenes.[10]

As early as 1661, Philippi's logic thus displays the ambivalence that was increasingly to characterize the development of occasionalism.[11] While the sign model aimed above all at protecting the spirituality of the soul from the causal action of the body, its application ultimately neutralized the soul, leaving nothing to subsist but the purely physiological description of the bodily machine. This is evident in the transposition of the example of the blind man just cited: what in the *Dioptrics* serves to explain the mechanical action of light on the sense organ[12] becomes, in Philippi, a proof for a direct, still mechanical, action of the organ on the soul.

The passage from Philippi clearly shows that the critique of secondary qualities unfolds entirely within a discourse that concerns the application of mechanical principles to physiology. Initially, it consists in an extension of the critique to pulsific quality, which turns into a general critique of all secondary qualities.[13] In the view of the theologians, of course, this generalization is one of its most dangerous aspects because it at least implies a critique of the traditional explanation of the dogmatic view of the Eucharist.[14]

---

[10] 'An dentur qualitates quæ sunt entitates stricte dictæ a substantia realiter distinctæ? et primo an tales sint sensus?

Pro responsione, Nota, quod anima nostra sic sit alligata corpori, ut a motibus corporis afficiatur, et ad varias sensationes determinetur ope tremoris nervorum dum tubuli eorum a spiritibus animalibus sunt inflati et filamenta illis inclusa debite extenduntur, ita ut quando una nervi extremitas movetur, etiam moveatur altera in cerebro (a quo omnes nervi derivantur) eo modo quo movetur utraque extremitas funis extensi quando altera percutitur. Determinat autem iste motus animam in cerebro residentem ad sensationem instar naturalis idiomatis, sicut aranea insidens telæ determinatur a variis filamentorum motibus ad prosecutionem aut fugam. Quod anima per solum istum motum determinetur probatur, quia quando cæcus ope baculi determinatur ad cognitionem diversorum corporum quæ tangit, nihil aliud intervenit nisi motus. ... Atque ita merito concludi potest animam a solis motibus etiam determinari ad sensationes aliorum sensuum, ita ut ad hoc non indigeat specie intentionali, maxime cum vix possit explicari quid sit illa species, quo modo prodeat ab objecto etc. et cum illud quod animam determinat ad cognitionem non debeat esse simile objecto, ut patet in vocibus et scripturis, quæ nullam habent similitudinem cum re quam nobis significant' *Medulla logicæ*, 105.

[11] I am referring here to occasionalism as it relates to the mind–body problem. See Steven Nadler, *Occasionalism, Causation Among the Cartesians* (Oxford: Oxford University Press, 2011), esp. the first chapter, 'Occasionalism and the Mind-Body Problem', 6–29; Véronique Le Ru, 'La réception occasionaliste de Descartes: des Malebranchistes à l'Encyclopédie', *Recherches sur Diderot et sur l'Encyclopédie* 38 (April 2005): 191–202.

[12] *Dioptrique*, *AT* VI, 85, 4–13.

[13] The anti-Cartesian definition of the *qualitas pulsifica* can be found in Plempius, *Fundamenta medicinæ* (Leuven: Nempæi, 1654), II, section 6: *de facultatibus animæ*, cap. 5, 150. See also Annie Bitbol-Hespériès, *Le principe de vie chez Descartes* (Paris: Vrin, 1990), 49–57. On criticism of the *qualitas pulsifica* and for a detailed analysis of the correspondence between Descartes, Plempius, and Fromondus, see Lucian Petrescu, 'Descartes on the Heartbeat: The Leuven Affair', *Perspectives on Science* 21/4 (2013), 397–428.

[14] At almost the same time, Louis de la Ville (Le Valois) articulates a criticism of the Eucharist from a Cartesian point of view in his *Sentimens de M. Descartes touchant l'essence et les propriétés du corps, opposés à la doctrine de l'Église* (Paris: Estienne Michallet, 1680); see also Jean-Robert Armogathe, *Theologia Cartesiana, l'explication physique de l'Euchristie chez Descartes et Dom Desgabets*' (La Haye: Martinus Nijhoff, 1977).

I won't dwell here on the 1662 metaphysical work that contained, according to Monchamp, statements close to the condemned theses[15]; I will only point out that in his metaphysics, Philippi follows the order of demonstration in Descartes's *Principia*, including the placement of metaphysics before physics, and thereby rejects the scholastic-Aristotelian organization of knowledge.[16]

In turn, it may be useful for evaluating the effect the condemnation had on the thinking of the author to examine briefly the *Medulla physicæ* published in 1664. In the first chapter on the essence and existence of the body, the author adds a long digression to the definition of the body as *substantia extensa*. The digression discusses the body of Christ in the Eucharist as well as the souls of animals, and it obviously constitutes an abjuration of the views expressed in 1661 that had so troubled the theologians. In his first book, Philippi had considered the possibility of a definite refutation of secondary qualities based on an entirely mechanical explanatory model, that is, based on movement alone. In 1664, the perspective is entirely the inverse: now it is movement that needs to be explained as an object of sense perception. More carefully this time, Philippi goes back to the example of the blind man and distinguishes more clearly between, on the one hand, the material chain that links the object to the brain and, on the other, the soul that is said to be determined on the occasion of the movement.

The *Physics* adds a remarkable fact concerning the reasons that led some of Descartes's earliest readers to highlight the rare occurrences of the word "occasion" in passages where the philosopher speaks of the union of the two substances.[17] In all these instances, Descartes employs the notion of occasion to solve the problem of how movement can be known via the senses. He states that in each sense perception, there is more than just the sensations produced by extension and movement. Movement in particular has an ambiguous status: it can at the same time be both a principle of explanation and the object of perception. According to the Cartesian gnoseology, it becomes impossible, in fact, to account for the perception of movement itself, which is sensible, to be sure, but not material. The idea of sensible movement, Philippi tells us, does not correspond to the data of sensibility: it is produced and added by the soul on the occasion of the series of perceptions of colors and shapes "as if [*ac si*] it were known by the outer senses."

> Sensible movement is the movement that on occasion of an external sensation or of its absence can be known by the understanding as easily as if it were known by outer sense. We call Peter's walking sensible on the occasion of the external occasion of some things connected with this action of walkin, that is to say, on the occasion of seeing, for example, Peter's colors. As we see those colors change their distance from something we conceive to

[15] *Histoire du cartésanisme en Belgique*, 368–370.

[16] Roger Ariew has shown how significant a role the inversion of the order of the treatises on metaphysics and physics played in the scholastic reception of Descartes; see his *Descartes and the First Cartesians* (Oxford: Oxford University Press, 2015), 113–117

[17] *Dioptrique*, *AT* VI, 114; *Responsiones sextae*, *AT* VII, 437.

be immoblie, we easily judge that Peter too is moving locally and changes his distance from this immobile thing as if we perceived Peter's local movement via the external sense.[18]

Philippi proposed an occasionalist solution to explain the idea of movement, which seemed to him to exceed the contents of sense perception, whereas 5 years earlier, his colleague Gerhard van Gutschoven had been thinking about an apparent defect of the soul when compared to the movements of the body. This early follower of Cartesianism, a direct student of Descartes, had written a series of objections to his anti-Cartesian colleague Vopiscus Fortunatus Plempius's *Ophthalmographia*. These objections, along with Plempius's responses, had been added to that book starting with the third edition of 1659. Gutschoven sets out to answer a question complementary to the one we saw had occupied Philippi: what happens when it is the soul that acts on the body and not the other way around? In several cases it seems that the simplicity of volition cannot be the cause of a series of complex movements that take place inside the machine. Gutschoven reminds his readers that the problem comes up in the third discourse of the *Dioptrics* at the point at which Descartes counts the movements of the pupil among the voluntary movements. He writes: "this movement must be called voluntary, although those who make it are usually unaware of it, because it does not for all that cease, as it depends on and follows their desire to see well."[19] From this limit case, Gutschoven deduces a general principle of the action of the soul on the body. He writes:

For the movement of a part to be called voluntary it is not necessary that our volition first strive for the movement of this part; it suffices that the movement of this part follow [*sequi*] whatever other volition, as different as you like from the part to be moved.[20]

The volition to see well (*voluntas bene videndi*) evoked by Descartes, Gutschoven says, does not differ from the volition to make a fist, which does not address the muscles of the arm. All it takes for all the movements required by an action to be called voluntary is a *contraendi voluntas*, and by the same token, the object of the volition to catch a ball is not our body but the ball itself. The domain of Cartesian physiology is thus preserved intact by rejecting an explanation of the soul's action

[18] 'Motus sensibilis est, qui occasione sensationis externæ, aut defectu ejus, tam facile potest cognosci existere ab intellectu, ac si fuisset cognitus per sensum externum. ... Dicitur autem illa ambulatio Petri sensibilis, occasione sensationis externæ aliquorum coniunctorum isti ambulationi, scilicet occasione visionis, v.g. colorum Petri, quos quia videmus mutare distantiam ab aliquo, quod concipimus tanquam immotum, tam facile iudicamus Petrum etiam moveri localiter et mutare distantiam ab eodem, ac si motus localis Petri sensu externo perciperetur' Guillaume Philippi, *Medulla physicæ* (Leuven: Petris Sasseni, 1664), 71.

[19] The passage quoted is the following: 'Et observandum, hunc motum voluntarium esse dicendum, licet, ut plurimum, a nobis ignorantibus peragatur; neque enim ob hoc minus dependet aut minus sequitur ex voluntate quam habemus bene videndi' *Dioptrice*, *AT* VI 596, AT VI 107, 25–29.

[20] 'Ut motus alicujus partis dicatur voluntarius, non necesse est, ut voluntas nostra prius tendat in motum istius partis: sed satis est, ad quamcumque voluntatem, quantumvis plane a motu partis movendæ diversam, illius partis motus sequi' (Gerhard Gutschoven 'Animadversiones in Ophtalmographiam', in Vopiscus Fortunatus Plempius, *Ophthalmographia, sive Tractatio de oculo*, 3rd ed. (Lovanii: Nempæi, 1659), 250.

by direct causal relations and by limiting the explanation to noting the chronological priority volition has over movement. This interpretation of Descartes's thinking has the merit of aligning the *Dioptrics* with a statement found in the *Secundæ responsiones*, more precisely, in the *Rationes more geometrico dispositæ*. Having defined thought as "everything that is so much in us that we are immediately aware (*consci sumus*) of it," the philosopher writes: "I have added *immediately* to exclude the things that follow and depend on our thoughts; the principle of voluntary movement, for example, surely is volition; nonetheless, it is not itself a thought."[21] Yet, in Gutschoven, "to follow" and "to depend on" are synonymous: in the body, everything that follows volition depends on it and therefore is voluntary. Remarkably, the word "follow" here has lost any connotation of causality and signifies nothing but a chronological order.

Gutschoven's exegetical effort nonetheless neglects certain details Plempius does not fail to point out. He emphasizes that admitting such an ultra-Cartesian definition of voluntary action leads to qualifying every biological phenomenon as voluntary. If the pupil opens thanks to the volition to see well, Plempius asks, why do we not call voluntary the development of the fetus, which follows after the volition to copulate, or digestion, which follows after the volition to feed oneself?[22]

In 1659, Plempius further develops the criticism his teacher Libert Froimond had leveled at Descartes's philosophy. In his *Responsiones*, the physician responds to Gutschoven's Cartesian theses with pithy formulas such as "nature does not imitate; art, rather, imitates nature, and all of nature's works cannot be explained by mechanics,"[23] or makes ironic statements of the kind that Descartes the mathematician may be excused for not understanding what a voluntary movement is, but a physician and anatomist like Gutschoven sharing these views is unacceptable.[24]

Over the years, Plempius had witnessed in Leuven the spreading of the ideas of Descartes (whom he calls *Renatus Democritus*) and been worried about seeing his university "infected."[25] He went as far as invoking a censure and rallied his colleagues to this end. In a venomous passage of his *Responsiones* he writes: "Soon we will hear the condemnation by the primate of the theologians, who will thrust into

[21] *Rationes more geometrico dispositae, AT* VII, 160. CSM, II 113.

[22] 'It is not true that it suffices for someone to have the volition to see well for the movement of the pupil to be called voluntary. In fact, for that same reason the formation of the foetus would have to be said to be voluntary because it follows the volition to unite and to procreate' *Animadversiones*, 251.

[23] Plempius, *Animadversiones*, 247.

[24] Ibid., 251.

[25] Cf. the preface, addressed to the "judgment of learned men" in *Fudamenta Medicinæ*, 4th ed. (Lovanii: Nempæi, 1664). The medical jargon in which opinions were diseases and infections was commonplace. Compare *Sources relatives à l'histoire du jansénisme*, 532; See also *Histoire du cartésanisme en Belgique*, 347.

hell Cartesian philosophy, which is contrary to Scripture, in dissent from the Fathers, far removed from the sentiments of the Church."[26]

7.3 The *Compendium omnium præcipuarum actionum automaticarum*

The violence of Plempius's rejoinders becomes more understandable when we consider his position as the only professor in the medical faculty who had not converted to mechanical medicine. His account confirms that the decision to condemn Cartesian philosophy had been taken at least 3 years before 1662. One had to wait, however, for the scandalous theses of August of that year for the censure to be actually pronounced. Only 5 months before the condemnation, between March 18 and April 1, 1662, three theses were defended before the medical faculty. These documents, unknown until now, provide valuable information and complete previous research on a number of important points. Among these theses, we find in particular a text by Pierre Dorlix, that is, the fourth professor in the medical faculty of whom previously nothing was known but his name. This professor, whom Monchamp mentions only once, turns out to have been an enthusiastic follower of Cartesian medicine. The following pages provide a study of a qualifying thesis, *A Compendium of the Specific Automatic Actions in Human Beings with Their Failures and Their Remedies, Set Out According to the Principles of Descartes's Mechanics and Philosophy*, defended on March 31 with Dorlix presiding.[27]

This academic exercise proves to be of capital importance in more than one respect: first of all, from the perspective of the history of Descartes's condemnation, this text is the last account of Leuven Cartesianism and gives us a rather clear idea of the views of the Cartesian physicians. Second, it undoubtedly constitutes the first reception of Descartes's *Traité de l'homme*. And finally, Pierre Dorlix's text furnishes the first occurrence of the expression *causa occasionalis* in a Cartesian context.

An investigation of the way the *Treatise of Man* circulated among the first Cartesians always runs the danger of remaining inconclusive. What we do know is

[26] 'Brevi audiemus censuram Theologorum Primatis, qui illam et S. Scripturæ repugnantem, S. Patribus dissentientem et ab Ecclesiæ sensu aberrantem ad inferos usque deijcet' *Animadversiones*, 252.

[27] *Compendium omnium præcipuarum actionum automaticarum in homine cum earundem vitijs et horum curationibus secundum mechanicæ et Philosophiæ Carthesianæ principia digestum*, quod pro licentiæ gradu obtinendo sub præsidio clarissimi viri domini Petri Dorlix Med: Doct: et Profess: Prim: publice repetet Vincentius Philippeaux bruxellensis. Die ultima Martij ab hora 9 usque ad 11 1662, Lovanii, apud Cornelium de Blehen.

that the work was widely distributed, first in manuscript, then in the two print editions by Schuyl and Clerselier. And yet it always remains possible that an author employs Cartesian concepts in describing the human body without having read the *Treatise.* We also know that a good part of the treatise was reworked by Descartes and used in his published works. Many of the theories about physiology were accessible, even before the publication of the *Treatise*, in the fifth part of the *Discourse*, the fourth part of the *Principia philosphiæ*, the *Passions of the Soul*, the first two volumes of the *Letters* (published in 1657 and 1659), and in Henricus Regius's *Fundamenta Physices*.[28]

There can, meanwhile, be no doubt in the case of Dorlix's thesis. A comparison of the texts confirms that Descartes's posthumous work is indeed its source. A first as it were macroscopic sign of the resemblance between the *Compendium* and *L'Homme* is the order of their arguments. Dorlix, too, describes the machine of the body by following the path of the blood's circulation, beginning with the heart and moving successively to the lungs, pulmonary circulation, arteries, liver, stomach, followed by the path the spirits take in going up from the heart to the brain and the effects this produces.[29] Taking into account that none of the works published by Descartes follow this order of demonstration, we may already claim that the thesis is conceived and written as a paraphrase of *L'Homme* accompanied by parallel remarks on diagnosing the diseases that may affect the machine. From the physician's point of view, the description of the machine is not of interest per se but only insofar as it is useful for medical practice. Dorlix examines not only the functioning of the human body but possible malfunctions and means for restoring normal conditions as well. This back and forth between theory and practice constitutes the true focus of his philosophical medicine; the description of respiration in the second paragraph of the thesis, for example, is thus but a paraphrase of its counterpart in the *Discourse on the Method*[30]:

[28] On Regius, see Theo Verbeek, ed., *Descartes et Regius: Autour de l'Explication de l'esprit humain* (Amsterdam: Rodopi, 1993) as well as Delphine Bellis, 'Empiricism without Metaphysics: Regius' Cartesian Natural Philosophy', in M. Dobre and T. Nyden, eds., *Cartesian Empiricisms* (Dordrecht: Springer, 2013), 151–183.

[29] On the order of demonstration in traditional medical treatises, see *Le principe de vie chez Descartes*. On the relationship between the order of demonstration in Descartes's physiology and the *Regulæ ad directionem ingenii*, see Vincent Aucante, *La philosophie médicale de Descartes* (Paris: PUF, 2006), 88–89.

[30] The *dissertatio de methodo*, in turn, is a slightly revised reprise of *L'Homme* (*AT* XI 124 8–22; René Descartes, *The World an Other Writings*, ed. Stephen Gaukroger (Cambridge: Cambridge University Press, 2004), 102. In 1637, Descartes reorganizes the argument by placing the passage on the cooling of the air in the lungs at the beginning of the proposition. It is this order that we find in the *Compendium*.

Compendium omnium præcipuarum actionum, Conclusio 2.	*Dissertatio de Methodo*, AT VI 569–570 (AT VI, 53–54)
Ex dextro cordis sinu sanguis rarefactus exhalat in pulmones, quorum substantia est adeo rara et mollis ab aëre tamen inspirato adeo refrigerata, ut simul atque vapores sanguinis egredientes ex cordis cavitate dextra intrant in arteriam pulmonarem illi ibi incrassentur, rursusque in sanguinem conversi guttatum in sinistrum recidant; quam conversionem si non subirent, igniculo qui ibi est nutriendo inepti essent, saltem in ijs hominibus qui partu seclusi sunt; nam fœtui in utero existenti de alijs ductibus ad eum effectum provisum est quod petenti dicemus.[31]	Ex eo cognoscitur verum respirationis usum esse, satis recentis aeris in pulmones inferre, ad efficiendum ut sanguis qui eo ex dextro cordis ventriculo defluit, ubi rarefactus et quasi in vapores mutatus fuit, ibi incrassescat et denuo in sanguinem convertatur, priusquam in sinistrum refluat; sine quo, alendo qui illic est igni aptus esse non posset. Idque ex eo confirmatur, quod videamus animalia pulmonibus destituta, unicum tantum cordis ventriculum habere: quodque in infantibus qui eo uti non possunt quamdiu sunt in matrum uteris inclusi, foramen quoddam deprehendamus per quod sanguis e vena cava in sinistram cordis cavitatem defluit.[32]

Cartesian physiology is then extended to include a pathology concerning the causes of respiratory diseases:

> It is thus easy to understand that the causes of diseases in the lungs have their origin in a vice either of the air inhaled, or in the lungs themselves not inhaling in the way prescribed, or in the blood issuing from of the right-hand ventricle.[33]

This pragmatic approach, which aims to identify the mechanical causes of diseases and to find effective treatments, determines Dorlix's interpretation of Cartesian physiology. In *L'Homme* as in later writings, Descartes proceeds carefully in the study of physiology: the study of the body according to mechanical principles depends entirely on the metaphysical foundations of physics. From the very first

[31] 'The rarified blood issuing from the heart's right-hand cavity disperses in the lungs, whose substance is so rare and soft and cooled by the air inhaled that the vapors of the blood that issue from the right-hand cavity and enter the pulmonary artery thicken there, and as soon as they have been converted into blood, they fall to the left-hand side. If they did not undergo this conversion, they would not be able to fuel the fire, at least in human beings who have been born, for fetuses do not possess any tubes to this end.'

[32] 'The true function of respiration is to bring enough fresh air into the lungs to cause the blood entering there from the right-hand cavity of the heart, where it has been rarefied and almost changed into vapours, to thicken immediately into blood again before returning to the left-hand cavity. For if this did not happen the blood would not be fit to serve as fuel for the fire in the heart. This is confirmed by seeing that animals without lungs have only one cavity in their hearts, and that unborn children, who cannot use their lungs while enclosed within their mother's womb, have an opening through which blood flows from the vena cava into the left-hand cavity of the heart, and a tube through which blood comes from the arterial vein into the great artery without passing through the lungs.' CSM, I, 138

[33] 'Hinc facile est intelligere causas morborum in pulmonibus ex eo oriri quod sit vel vitium in aere inspirato, vel in ipsis pulmonibus non halentibus conditiones quas hic præscripsimus, vel in sanguine ex dextro ventriculo exhalante' *Compendium omnium præcipuarum actionum automaticarum*, Conclusio 2.

lines of his text, Dorlix, on the contrary, seems to dismiss Descartes's metaphysics; he thus appears to renounce the anthropological aim and stakes of the *Discourse*:

> The principles of Descartes's mechanics and philosophy being presupposed, as well as their notions by which it is shown that all bodies are made from movement, rest, shape, place, and the size of their parts, this thesis attempts quickly to understand the automatic actions of already formed human beings as well as their failures and their remedies.[34]

The thesis discusses already formed human beings, that is, human beings such as they are, with a view to understanding their automatism and diseases. This is the exact opposite of the opening of the *Treatise of Man*, where the verb in the future tense introduces the fiction of a human being yet to be made—"These men will be composed, as we are, of a soul and a body," Descartes writes—and at the same time marks the origin of physiology in the physical science of *The World*, as Clerselier and La Forge point out as well.[35]

Apart from this difference at the beginning, the text faithfully repeats the principles of Descartes's physiology:

> The first and particular cause of automatic actions is the natural heat, which is a kind of real fire without light housed in the heart: this fire is heated by the humors called into the heart, maintained and restored by the fermentation and the heat that remains after each pulsation in the heart's ventricles.[36]

From the very beginning, Dorlix's *conclusiones* set out to respond to Libert Froimond's objections to the physiological part of the *Discourse on the Method*. The opening lines of the thesis explicitly operate a synthesis between the mechanistic explanation of the heart's heat and the explanation of psychological phenomena; they thus do precisely what Froimond had feared, who did not want *noble actions* to be explained *by so ignoble and brutish a cause as heat*.[37]

---

[34] 'Suppositis mechannicæ et philosophiæ Carthesianæ principiis eorumque cognitione quibus omnia corporea per motum, quietem, figuram, situm et magnitudinem partium eorumque inter se proportionem fieri probantur; hominis iam formati actiones automaticas earumque vitia et curationes hac thesi cursim comprehendere conabimur' *Compendium omnium præcipuarum actionum automaticarum*, [1].

[35] *L'Homme*, *AT* XI, 119; *The World an Other Writings*, 99; *L'Homme de René Descartes* (Paris: Fayard, 1999), 181–82.

[36] 'Harum actionum automaticarum prima causa et præcipua est calor naturalis, qui est species quædam veri ignis sine lumine in corde hospitantis: hic fovetur ab humoribus in cor appellantibus, continuatur et restauratur a fermento et calore post singulos pulsus in cordis ventriculis residuo' *Compendium omnium præcipuarum actionum automaticarum*, Conclusio 1.

[37] To Plempius for Fromondus, 3 october 1637, *AT* I 403/*CSM* III, 61.

One of the most explicit quotations from Descartes's treatise can be found in *conclusio* 6, where Dorlix demonstrates that saliva originates in the vapors that come up directly from the heart. On this particular point, which is not discussed anywhere else by Descartes, Dorlix takes up the arguments of the *Treatise on Man* point by point and in the same order.

Compendium omnium præcipuarum actionum automaticarum, Conclusio 6	*L'Homme*, AT XI, 127, l. 14–22
Ulterius partes sanguinis aliæ sic propulsæ, a corde in lienem, vesiculam fellis, et tam ab illis quam immediate ab arterijs regurgitant aliquæ in stomachum et intestina, digestioni illic faciendæ, adinstar aquæ fortis, prompta dissolventia; nam cum satis immediate a cordis foco propulsæ veniant, non possunt non summe æstuare. Ex ventriculo autem per œsophagum in os latæ per modum vaporis una cum materijs ex vasis lymphaticis et arterialibus eo desinentibus salivam constituunt.[38]	De plus il y a quelques-unes des parties du sang qui se vont rendre dans la rate, et d'autres dans la vésicule du fiel; et tant de la rate et du fiel, comme immédiatement des artères, il y en a qui retournent dans l'estomac et dans les boyaux, où elles servent comme d'eau forte pour aider à la digestion des viandes; et parce qu'elles y sont apportées du cœur quasi en un moment par les artères, elles ne manquent jamais d'être fort chaudes; ce qui fait que leurs vapeurs peuvent monter facilement par le gosier vers la bouches, et y composer la salive.[39]

The explanation of respiration given in *conclusio* 21, too, appears to be very close to its analogue in the *Treatise on Man*. This is one of Dorlix's Cartesian theses that undoubtedly originates in the manuscript of the *Treatise*, the one according to which the flow of subtle matter in the nerves is regulated by valves that allow for the exchange of spirits between two opposed muscles. This system explains involuntary movements (such as respiration) by the conversion of a continuous flux of spirits emanating from a nerve that is always open into an alternating contraction of opposite muscles.

[38] 'The other parts of the blood are pushed further from the heart into the spleen and the gallbladder and some of these parts coming from these organs as well as immediately from the arteries overflow in the stomach and in the bowels. They act as a solvent, like aqua fortis, available to execute digestion. And because these parts are pushed directly from the fire of the heart they are very hot. The parts are carried from the cavity through the gullet toward the mouth where they make up saliva combined with the matter left there from the lymphatic vessels.'

[39] 'In addition, there are some parts of the blood that proceed into the spleen, and others to the gall bladder, and, via the spleen and the gall bladder as well as directly from the arteries, there are some parts that re-enter the stomach and the bowels, where they act like *aqua fortis*, helping in the digestion of food. And because they are carried here from the heart almost instantaneously through the arteries, they are always very hot, which enables their vapours to rise easily through the gullet toward the mouth, where they make up the saliva' *The World and Other Writings*, 104.

Compendium omnium præcipuarum actionum automaticarum, Conclusio 21	*L'Homme*, AT XI, 139, l. 3–25
Spiritus vero lati in tubulos illos a cavitatibus cerebri oriundos, qui continuo naturaliter aperti sunt, quales sunt respiratorij, valvulam deprimendo musculum inflant certis aliquibus membranis tectum taliter dispositis, quæ antequam omnes spiritus ex opposito musculo versus illum transierint, cursum eorum sistunt et quasi regurgitare faciunt in alium tubulum naturaliter etiam apertum musculo opposito insertum: atque sic deinceps reprimuntur ab alio, donec sublato illo impetu spirituum et pellicularum dictarum alij tandem ex cerebro cursum suum exorsi idem incipiant. Quodipsum de omnibus illis motibus continuis intellige.[40]	[Pensez que] les esprits animaux qui sont dans la concavité de son cerveau marquée m, coulant par le pore ou petit canal marqué n, qui demeure naturellement toujours ouvert, se vont rendre d'abord dans le tuyau BF, où abaissant le petite peau F, ils font que ceux du muscle E viennent enfler le muscle d. Pensez après cela, qu'il y a certaines peaux autour de ce muscle d, qui les pressent de plus en plus à mesure qu'il s'enfle, et qui sont tellement disposées, qu'avant que tous les esprits du muscle E soient passés vers lui, elles arrêtent leur cours, et les font comme regorger par le tuyau BF, en sorte que ceux du canal n s'en détournent; au moyen de quoi, s'allant rendre dans le tuyau cg, qu'ils ouvrent en même temps, ils font enfler le muscle E, et désenfler le muscle d; ce qu'ils continuent de faire aussi longtemps que dure l'impétuosité dont les esprits contenus dans les muscle d, pressés par les peaux qui l'environnent, tendent à en sortir. Puis, quand cette impétuosité n'a plus de force, ils reprennent d'eux-mêmes leur cours par le tuyau BF, et ainsi ne cessent de faire enfler et désenfler alternativement ces deux muscles. Ce que vous devez juger aussi des autres muscles qui servent à même effet.[41]

[40] 'The spirits that are being carried in the tubes, which come from the cavities of the brain that are continuously and naturally open, like those of respiration, inflate the muscle by lowering a valve by means of certain membranes arranged in a roof-like shape, which stop the course of these spirits before spirits pass from the muscle opposite the other, and make them flow into another tube, which, too, is open naturally, inserted in the muscle opposite. In this way, this latter muscle is in turn compressed by the other until, the force of the spirits and membranes having been lifted, other spirits issuing from the brain begin anew with the same thing.'

[41] 'The animal spirits that are in the brain cavity marked m, running through the pore or little channel marked n, which is by its nature constantly open, proceed first to the tube BF where, lowering the little membrane F, they cause those from muscle E to come and inflate muscle d.

Reflect next that there are certain membranes around this muscle d, which press on it increasingly as it is inflated, and which are arranged in such a way that, before all the spirits from muscle E have passed through it, they stop in their course, and it causes them to be regorged, as it were, through the tube BF, so that those from channel n are re-directed; in this way, they proceed to cg, simultaneously forcing it open and causing the inflation of muscle E and the deflation of muscle d. And they continue to do this for as long as they endure the impetuosity of the spirits contained in muscle d, which, squeezed by the surrounding membrane, tend to be discharged from it. Then, when this impetuosity has been exhausted, they resume their course through the tube BF, so that they are unceasingly forced to inflate and deflate alternately. You should also take this to hold for the other muscles that serve the same end' *The World and Other Writings*, 115–16.

The valve hypothesis, as we know, circulated among his followers already in Descartes's lifetime. The part just cited of the manuscript of the *Treatise* is exactly the one whose being "stolen" Descartes decries in a series of letters to Mersenne and Elizabeth. Disobeying his master, Regius in the eleventh chapter of his *Fundamenta physices* had copied the section dealing with the muscle valve hypothesis from a manuscript that was "confusing" and almost unreadable, as Descartes himself admits in one of his letters to Mersenne. Without Descartes's explanatory figure at his disposal, Regius provided an incorrect one.[42] The author of the *Compendium* must have found himself in the same predicament, yet he chose the opposite strategy. There is no explanatory image in the thesis and accordingly, all of the references to images found in Descartes's manuscript have been taken out. The decision was no doubt prompted by the academic context and the literary genre of the qualifying thesis. As the written support of an oral defense, the thesis cannot refer listeners to images the way the author of a book can refer readers to illustrations.

If we exclude Regius, who was more an interlocutor than a student of Descartes', as a special case, the *Compendium* is the first account proper of the reception of Descartes's physiological treatise. As we saw earlier, the structure of the text as a whole evidently replicates the structure of the *Treatise on Man*. This brings us, finally, to an examination of a remarkable passage of the thesis, which will allow us to show that its source cannot lie either in other works published by Descartes or even in Schuyl's Latin edition, *De Homine*, but only in the French manuscript. The passage in question is the *conclusio* 18, which is here compared to the texts found in Clerselier's 1664 French edition and Schuyl's 1662 Latin translation:

[42] Descartes to Mersenne, 23 November 1646, *AT* IV, 567/*CSM* III 301–302. Descartes to Elisabeth, March 1647, *AT* IV 627/*CSM* III 315. Delphine Kolesnik-Antoine examines the stakes of this erroneous adaptation in her *L'homme cartésien: 'la force qu'a l'âme de mouvoir le corps'* (Rennes: PUR, 2009), 57–75. See also Tad Schmaltz's essay in this volume, chapter 5.1.

Compendium omnium præcipuarum actionum automaticarum, Conclusio 18	*L'Homme* AT XI, 128 l.3–129 l.23	*De homine*, p. 13–15
Sed maxime notatu dignum, est, quod vivacissimæ, subilissimæ et fortissimæ sanguinis partes ex corde erumpentes, secundum leges mechanicæ, via a corde quam rectissima in cerebrum propellantur, dum interea partes minus his agitatæ, tum propter minorem arteriarum ad caput ascendentium capacitatem, tum etiam quia a dictis ad latera in motu deturbantur	Mais ce qu'il faut ici principalement remarquer, c'est que toutes les plus vives, les plus fortes, et les plus subtiles parties de ce sang, se vont rendre dans les concavités du cerveau; d'autant que les artères qui les y portent, sont celles qui viennent du cœur le plus en ligne droite de toutes, et que, comme vous savez, tous les corps qui se meuvent tendent chacun, autant qu'il est possible, à continuer leur mouvement en ligne droite	Sed quod hoc loco comprimis observandum, magis vividæ et vehementius actæ sanguinis particulæ in cerebri concavitates sive ventriculos ascendunt. Quia nimirum arteriæ per quas eo deferuntur, in rectiori linea sunt sitæ, quam ceteræ, quæ ex corde promanant; atque, ut manifestum, omnia corpora, quæ moventur, quantum possunt motum suum secundum lineam rectam continuare conantur
	Voyez, par exemple, le cœur A, et pensez que, lorsque le sang en sort avec effort par l'ouverture B, il n'y a aucune de ses parties qui ne tende vers C, où sont les concavités du cerveau; mais que, le passage n'étant pas assez grand pour les y porter toutes	Exempli gratia: Ecce Cor A, et cogita cum sanguis vi prorumpit ex apertura B, nullas esse ejus partes, quæ non connitantur ad C, ubi nimirum sunt cerebri ventriculi: sed quoniam via angustior est, quam ut omnes eo deferri
illæ alias minus agitatas ad latera deturbandovia etiam quam rectissima ad genitalia pro eorum usu tendunt	les plus faibles en sont détournées par les plus fortes, qui par ce moyen s'y vont rendre seules. Vous pouvez aussi remarquer en passant, qu'après celles qui entrent dans le cerveau, il n'y en a point de plus fortes ni de plus vives, que celles qui se vont rendre aux vaisseaux destinés à la génération	possint, solæ validiores se eo conferunt, debilioribus a fortioribus impeditis. Per transennam quoque notare licet, nullas sanguinis particulas, iis exceptis, quæ cerebrum ingrediuntur, esse vel fortiores, vel vividiores iis, quæ in vasa generationi dicata deferuntur
	Car, par exemple, si celles qui ont la force de parvenir jusqu'à D, ne peuvent aller plus avant vers C, à cause qu'il n'y a pas assez de place pour toutes, elles retournent plutôt vers E, que vers F ni vers G, d'autant que le passage y est plus droit. En suite de quoi je pourrais peut-être vous faire voir, comment, de l'humeur qui s'assemble vers E, il se peut former une autre machine, toute semblable à celle-ci; mais je ne veux pas entrer plus avant en cette matière	Nam, exempli gratia, illæ quibus sufficienns vis est perveniendi ad D, sed ulterior transitus ad C propter angustiam loci negatur, revertuntur potius ad E, quam ad F, aut G. Quandoquidem via rectior eo ducit. Et consequenter, possem fortassis ostendere, quomodo ex illo humore, qui circa E congregatur, alia machina huic prorsus simili formari possit. Sed non lubet jam prolixius exspatiari

(continued)

Compendium omnium præcipuarum actionum automaticarum, Conclusio 18	*L'Homme* AT XI, 128 l.3–129 l.23	*De homine*, p. 13–15
Priores iam dictæ cum sanguine suo ad caput latæ non tantum substantiæ cerebri nutriendæ inserviunt, sed præcipue producunt ventum aliquem subtilissimum vel potius flammam vivacissimam et purissimam spirituum animalium nomine insignitam ; quando quidem arteriæ quæ illam a corde advehunt, postquam in indefinitos ramulos divisæ sunt, et composuerunt rete istud quod aulæi instar cavitatum cerebri fundo insternitur, colligantur circa quandam glandem in cerebro in principio cavitatum ejus, ibique infinita foraminula habeant, perque partes subtiliores sanguinis quas continent in illam glandem effluere possunt, adeo tamen stricta quæ nullis crassioribus transitum concedant (quantumvis hisce subtilioribus dictis paulo crassiores et debiliores immediate ab ipso plexu carotidum in cerebri ventriculos prorumpant) indeque partes aliæ sanguinis ulterius pergunt in vas quoddam seu euripum, per quem tota cerebri superficies irroratur[a]	Pour ce qui est des parties du sang qui pénètrent jusqu'au cerveau, elles n'y servent pas seulement à nourrir et entretenir sa substance, mais principalement aussi à y produire un certain vent très subtil, ou plutôt une flamme très vive et très pure, qu'on nomme les esprits animaux. Car il faut savoir, que les artères qui les apportent du cœur, après s'être divisées en une infinité de petites branches, et avoir composé ces petits tissus, qui sont étendus comme des tapisseries au fond des concavités du cerveau, se rassemblent autour d'une certaine petite glande, située environ le milieu de la substance de ce cerveau, tout à l'entrée de ses concavités; et ont en cet endroit un grand nombre de petits trous, par où les plus subtiles parties du sang qu'elles contiennent, se peuvent écouler dans cette glande, mais qui sont si étroits, qu'ils ne donnent aucun passage aux plus grossières. Il faut aussi savoir, que ces artères ne s'arrêtent pas là, mais que, s'y étant assemblées plusieurs en une, elles montent tout droit, et se vont rendre dans ce grand vaisseau qui est comme un Euripe, dont toute la superficie extérieure de ce cerveau est arrosée[b]	Illa vero sanguinis portio, quæ ad cerebrum usque assurgit, non tantum nutritioni et conservationi substantiæ cerebri opitulatur: sed imprimis quoque valde subtilem auram, vel potius flammam admodum agilem puramque, quam spiritus animales vocamus, ibidem generat. Notandum vero arterias, per quas illa sanguinis portio ex corde ad cerebrum deducitur, postquam in infinitos ramusculos fuerint divisæ et compusuerint admiranda illa reticula, quibus velut tapetis ventriculorum cerebri pavimentum instratum est, congregari circum glandulam quandam præter propter in cerebri meditullio, juxta introitum ventriculorum, sive concavitatum ejus; ibique infinitis fere poris constare. Ex quibus, negato propter pororum angustiam crassioribus senguinis partibus transitu, subtiliores duntaxat, quibus distenduntur in glandulam profluunt. Sciendum etiam has arterias ibi non desinere, sed invicem sibi magis junctas recta ascendere ad amplum illud vas, quod instar Euripi totam cerebri superficiem irrigat

[a] 'But what is worth being stated most of all is that in erupting from the heart, the most energetic, finest, and strongest parts are pushed from the heart to the brain in the straightest possible line. At the same time, the parts that are less animated than these are pushed to the side, be it because of the movement of the more energetic parts, be it because of the lower capacity of the arteries going up to the head. As some of the parts pushed aside are more agitated, energetic, and mobile, they in turn push the less animated parts aside and hasten in the straightest possible line toward the genitals for the latter's purposes. The parts mentioned first that are carried with the blood to the head serve not only to nourish the substance of the brain, they also and above all produce a very fine wind or, rather,

(continued)

a very energetic and pure flame, which has been given the name of animal spirits; moreover, having been divided into an indefinite number of small branches and having built up the net that is spread across the bottom of the cavities of the brain, the arteries that carry this flame from the heart come together around a certain gland at the entrance of its cavities, and there they have an infinity of small holes through which the finest parts of the blood they [the arteries] contain can flow into the gland and which are yet so narrow that they do not let any of the thicker parts pass through.'

[b] 'But what must be noted above all at this point is that all the most energetic, strongest, and finest parts of this blood proceed to the cavities of the brain, inasmuch as the arteries bearing them there are in the most direct line from the heart; and as you know, all moving bodies tend as much as they are able to continue their motion in a straight line.

Consider the heart A, for example, and consider that when the blood is forced from it through the aperture B, all its parts tend toward C, that is, toward the cavities of the brain; but because the passage is not sufficiently large to bear all of them there, the weakest are turned back by the strongest, which in this way proceed there alone.

You should also note in passing that the strongest and most energetic parts, other than those which go directly to the brain, go to the vessels destined for reproduction. For if those that have the force to reach D, for example, cannot progress on to C, because there is no room for all of them there, they turn instead toward E, rather than toward F or G, in so far as the passage toward E is straighter. Beyond this, I could perhaps show you how, from the humour that gathers at E, another machine which is similar to this can be formed, but I do not wish to enter further into this matter. As for those parts of the blood that penetrate as far as the brain, they serve not only to nourish and sustain its substance, but above all to produce there a certain very fine wind, or rather a very lively and very pure flame, which is called the 'animal spirits'. For it should be noted that the arteries that carry these from the heart, after having divided into countless small branches and having composed the little tissues that are stretched out like tapestries at the bottom of the cavities of the brain, come together again around a certain little gland which lies near the middle of the substance of the brain, just at the entrance to its cavities; and those in this region have a large number of small holes through which the finest parts of the blood can flow into this gland, and these are so narrow that they do not allow the larger ones to get past.

You should also know that these arteries do not stop there, but being gathered up into a single one, they go straight up and enter that great vessel which, like Euripos, bathes the whole external surface of the brain' *The World and other Writings*, 104–106.

While in other paragraphs, the theses typical of the *Treatise of Man* are paraphrased, there can be no doubt that the *conclusio* 18 is a direct quotation of paragraph 13 in *L'Homme*. A word-by-word comparison shows that it is this same French text that has been translated into Latin. We cannot completely exclude the possibility that Dorlix was working with a copy of *De homine*, which was in fact published the same year, because we have no way of knowing if the book had already been printed in March 1662. In any case, however, this external fact is not essential since a comparison of the texts in column 1 and 3 shows clear differences in semantics and syntax. In turn, there can be no doubt that 1 and 3 are two distinct translations into Latin of the same French original, an example of which is reproduced here in column 2. Yet it is very likely that 1 (Dorlix) and 3 (Schuyl) refer to two different sources. The text Dorlix used does indeed come from a tradition that differs from the tradition that provided the manuscript copy on which Schuyl based his edition. In the family tree of the circulation of Descartes's manuscript, the *Compendium* and the 1662 Latin edition are located on two separate branches.[43] It is implausible to suppose that Dorlix relied on but heavily redacted Schuyl's edition. A comparison with the French text instead confirms the hypothesis that he was directly familiar with the original text. The one he used is certainly the manuscript in the possession of his colleague van Gutschoven who, during precisely this time, took on the task of drawing the illustrations needed for the Clerselier edition published 2 years later.[44]

To conclude, it can be shown not only that Dorlix did not limit himself to unknown sources but also that he reworked Descartes's physiology by elaborating on it in at least two essential respects: the *Compendium* stands out by employing the notion of particles of different shapes in the description of phenomena and by a technical and well-defined deployment of the notion of occasional cause.

As for the first point, Dorlix improves on the explanatory model on which the *Treatise of Man* is based by inserting hypotheses about the shapes of parts taken from the *Principia Philosophiæ*. As we know, Descartes at the time of composing the *Treatise of Man* had not yet elaborated the distinction between "round" parts and

[43] The most complete *stemma codicum* of Descartes's manuscript can be found in Franco Meschini 'Filologia e scienza: Note per un'edizione critica de *L'Homme* de di Descartes', in *Le opera dei filosofi e degli scienziati* (Florence: Olschki, 2011), 165–204, which extends the one established by Otegem.

[44] In his preface to the second volume of Descartes's correspondence, Clerselier had called on other scholars to contribute. Pierre Guisony and René de Sluse were the intermediaries between van Gutschoven and Clerselier. See the *préface* in René Descartes, *L'Homme et la formation du foetus* (Paris: Girard, 1677), [17].

"ramified" parts. Not until the *Meteors* does he enrich his mechanics with an analysis of different figures that, although they are invisible, produce macroscopic phenomena such as the clouds described in the book. He himself, however, never applied this important theoretical tool to the description of the bodily machine. Yet in *conclusio* XII, Dorlix develops an explanation of kidney stones based on different figures of particles. According to him, renal calculi form when the ramified parts of the urine accumulate.[45]

The theoretical stakes as well as the *Treatise*'s value for the spreading of Cartesianism, meanwhile, are certainly higher when it comes to the second point. As we have seen, the purely medical approach to Descartes's texts came with a purging of all references to the metaphysical foundations of physiology. The simple description of the human body in the *Compendium omnium præcipuarum actionum automaticarum* also erases the fictional fable of the world that preceded Descartes's *Treatise on Man* and which his editors La Forge and Clerselier considered to be of central importance.

Yet when the mechanicist model manages to explain even cognitive and motor functions, the problem of the union of soul and body arises. Dorlix's solution inscribes itself entirely in the wake of his colleagues' Cartesianism. It takes the rudiments of the occasionalist theory we saw in Philippi and Gutschoven all the way:

> The pores on the gland, which are caused by the spirits coming immediately from this gland and the spirits issuing from it, are the occasional causes thanks to which the soul united with the gland has ideas of the things that happen to the body. And if these ideas in the gland appear very often, they are the occasional causes by means of which the soul that pays attention to them adheres to them. They appear thanks to either the senses excited by the same object, or the vestiges of the sensorial fibers, or the fact that the gland remains turned to one side because of its rigidity, its hardening, or other similar causes sufficient to block it. And on the contrary, if these ideas we observe on the gland remain for so short a time that we go immediately from one to the next, they are the occasional causes by means of which the soul that pays attention to them very quickly, without any reflection, has the idea of one thing or another. … And from this it possible to deduce all kinds of follies.[46]

[45] Delphine Antoine-Mahut, 'La machine du corps', in Frédéric de Buzon, Élodie Cassan, and Denis Kambouchner, eds., *Lectures de Descartes* (Paris: Ellipses, 2015), 229–252.

[46] 'Pori autem quos spiritus immediate ex glandula veniendo in ejus superficie constituunt simul cum spiritibus egredientibus, sunt causæ occasionales quare anima glandulæ unita habeat ideas eorum quæ in corpore fiunt. Et hinc si ideæ illæ in glandula nimis sæpe, fortiter, vel continuo obversentur: erunt causæ, occasionales quare anima ad illas attendens nimis illis inhæreat: sive illæ obversentur gratia sensuum sæpius ab eodem objecto excitatorum, sive vestigiorum aliunde quam per sensorias fibrillas immediate valide nimis impressorum: sive fiat ex eo quod glandula in unam partem obversa maneat, propter suam rigiditatem indurationem et similes causas quæ ipsam fixare valent. E contra vero si ideæ illæ quæ glandi observantur tam pauco tempore persistant, ut statim ad alias eatur, erunt causæ occasionales quare anima ad illas attendens nimis citam et subitam modo de hoc, modo de illo objecto citra ullam debitam reflexionem, ideam habebit … Et hinc omnium generum stultitiæ deduci possunt.' *Compendium omnium præcipuarum actionum automaticarum*, Conclusio 23.

Dorlix stabilizes, as it were, his fellow physicians' intuitions. His colleagues had explained cognitive functions (Philippi) and motor functions (Gutschoven) by highlighting the value of the notion of "occasion." Dorlix endows the examples used by Descartes himself with the stable status of a cause and, 4 years before La Forge, is the one to coin the term *causa occasionalis*.[47] The invention of this notion is anything but reckless; Dorlix, quite to the contrary, wants it to account for the interactions between soul and body.

The first of the occurrences of the term in the last quotation all by itself exhausts the question of the union of soul and body. The movement of the spirits is the mechanical cause of the pores that open onto the surface of the pineal gland. These pores are the occasional causes of the ideas the soul has of everything that happens in the body. Dorlix then draws a momentous consequence from the parallel between the movements in machines and psychological phenomena established thanks to the occasional cause. The connection is made via attention: the persistence of an idea in the soul is an effect "occasioned" by the persistence of the flow of spirits that keeps the pore open. In the same way, in the third occurrence, the absence of attention is said to be caused by the instability of the flow. Here as in the preceding paragraphs, Dorlix is interested in the phenomena tied in with the union because he seeks to find in them the causes of diseases. The instability of the flow produces a chaotic movement of the gland, which causes what is here called *stultitia*, an incapacity to focus one's attention.

It is worth noting that Dorlix's reflections pick up on articles 77–78 of Descartes's *Treatise*. The *causa occasionalis* is born from a formalization of the "weak" version articulated by Descartes.

> We can assume that what makes tube 8 turn towards point b rather than toward some other point is simply that the spirits that issue from this point tend toward it with a greater force than do any others. And the same thing will cause [*donnerait occasion*] the soul to sense that the arm is turned toward object 8, if it is already in this machine, as I shall later suppose it to be.[48]

A few lines down, Descartes reminds his readers that the idea of the movement of the arm "would be formed at the same time … if one's attention were not diverted, that is to say, if gland H were not prevented from leaning toward 8 by some different, stronger action."[49] Yet in the context of a physiological treatise, this notion of attention might seem to imply a limitation of the freedom of the soul. That, at least, is what La Forge sees in it, and in his commentary, he is at pains to distinguish

---

[47] Louis de La Forge, *Traité de l'esprit de l'homme*, in *Œuvres philosophiques* (Paris:PUF, 1974), 177. Vincent Carraud, *Causa sive ratio: la raison de la cause, de Suarez à Leibniz* (Paris: PUF, 2002), 347.

[48] AT XI 181, 14–21/*The World and other Writings*, 153.

[49] AT XI 181, 7–14/*The World and other Writings*, 154.

between two senses of the word "attention:" a proper sense allegedly not at issue here, in which attention is "the act of the will by which one wills or at least accepts that the same idea continues to be present to our apperceptive faculty," and a purely physical sense that refers exclusively to the movements of the gland.[50]

The parallel between the physical and the mental established by occasional causality, however, leads Dorlix to envelop one notion of attention in another. The *Compendium* in fact concludes with a study of the passions, the limits of the will, and the powerlessness the will may experience given certain disruptions in the functioning of the machine.

Conclusio 24 spells out the occasions that may limit the will:

> Every sensation and every voluntary movement of the body is canceled out if, given a sufficient dilation of the ventricles, the flow of spirits across the gland into the brain is prevented, which can happen when the gland lets parts of blood enter into its pores that suffice to clog them. … Voluntary movement can also be vitiated by another source, namely when the spirits cannot enter into the muscle one seeks to move by a decree of the soul (*imperium animæ*) because a nerve is clogged, compressed, or cut, or when they can enter but the valves are deformed.[51]

The spreading of the *Treatise of Man* in Leuven has certainly had a remarkable impact on the reception of Descartes's philosophy. This study of its context has shown that the condemnation of August 1662 was not an isolated incident. The ultra-Cartesian views censured by the theological faculty now appear as the result of a collective labor on the texts of the master, an effort of which the *Compendium omnium præcipuarum actionum automaticarum* represents the penultimate step. It is on the basis of their absolute confidence in the explanatory power of the princi-

---

[50] *L'Homme de René Descartes*, 338–339.

[51] 'Omnimoda autem sensatio et motus corporis voluntarius tolletur si posita sufficienti ventriculorum cerebri dilatatione impediatur trans glandulam spirituum in cerebri cavitates influxus. Qui casus fieri potest dum in porulos suos glandula partes sanguinis obstructioni eorum sufficientes admittit. … Motus tamen voluntarius ex alio adhuc capite vitiari potest, si scilicet ad imperium animæ spiritus in musculos quos movere vult per nervum utpote obstructum compressum, resectum intrare non possint, vel si saltem intrent sit defectus in valvulis quod hæ scilicet per exesionem sint plane sublatæ, distortæ, intro fora spectent etc.' *Compendium omnium præcipuarum actionum automaticarum*, Conclusio 24.

ples of Descartes's physics that the physicians developed a philosophical medicine built around the the occasionalist solution of the mind–body problem.

In keeping with the tradition at the University of Leuven, the *Compendium* ends with *impertinentia*. These short concluding statements in which the author speaks freely on a variety of subjects once more bear witness to the radicalness of their author's brand of Cartesianism:

> Descartes's dogmas are not new because they are true.
>
> One should have pity rather than get upset with those who refute Descartes's philosophy because either they have not read it or they have not understood it.[52]

[52] 'Impertinens: Non sunt nova dogmata Cartesiana, sunt quippe vera. Aliud: Qui philosophiam Cartesij universim convellunt sunt potius digni commiseratione quam indignatione, quippe qui illam vel non legerint vel lectam non intellexerint' *Compendium omnium præcipuarum actionum automaticarum*, Impertinens following Conclusio 32.

Chapter 8
Machine and Communication of Corporeal Dispositions in Descartes and La Forge: The Mysterious 'Article 83' of *L'Homme* and La Forge's Comments

Philippe Drieux

Abstract In the *Remarques* he added to *L'Homme*, La Forge lays the emphasis on a few lines that Clerselier had isolated in his Table as an "article 83". For la Forge, Descartes didn't tell enough to be completely understood, though the subject was in itself full of beauty. What is at stake are the relations between machines – that kind of fictitious machines that enabled Descartes to describe and explain animal as well as human corporal behaviour, solely and apart from mental influence. What Descartes really meant remains rather unsure, and even confused, but La Forge seems to consider that he may go further. He intends to make up for the missing explanation. What kind of relations are at work between corporeal machines, especially when similar behaviours are observed ? What consequences are to be expected for society? Is La Forge faithful to Descartes in this operation, or is he sliding to a different stage, which has more to do with occasionalism than cartesianism ?

In preparing for *L'Homme* a Table of Contents like that he did for the *Dioptrique*, Clerselier isolates as 'article 83' an elliptic passage of the treatise, in which Descartes declares that:

> the memory effect that seems to me the most worthy to be considered, consists in this: that without any soul present in this machine, it can naturally be disposed to mimic [imiter] every movement that real men, or other similar machines, will do in its presence.[1]

[1] AT XI 185.

P. Drieux (✉)
IHRIM, UMR 5317 CNRS, Université de Rouen-Normandie,
Place Emile Blondel, 76130 Mont Saint Aignan, France
e-mail: philippe.drieux@wanadoo.fr

D. Antoine-Mahut, S. Gaukroger (eds.), Descartes' *Treatise on Man* and its Reception, Studies in History and Philosophy of Science 43,
DOI 10.1007/978-3-319-46989-8_8

[Mais l'effet de la Mémoire qui me semble ici le plus digne d'être considéré, consiste en ce que, sans qu'il y ait aucune âme en cette machine, elle peut naturellement être disposée à imiter tous les mouvements que de vrais hommes, ou bien d'autres semblables machines, feront en sa présence.]

Following the *Remarques sur L'Homme de René Descartes*, La Forge sees in this passage an allusion to corporal dispositions communication from a machine to another. He even considers that this transfer consists in an 'involuntary imitation' which

can be clearly seen in those who are yawning, only because they can see others yawning ; or coughing and splitting when seeing others doing so.[2]

[Or cette sorte d'imitation involontaire se voit manifestement en ceux qui baillent, pour en voir bailler d'autres ; ou qui toussent et crachent pour voir faire les mêmes actions.]

The term 'imitation' can be confusing as long as it isn't distinguished from 'psychological' or even 'mixed' meanings (including both mind and body), that are irrelevant in *L'Homme*. Since the treatise proposes to treat of a 'clay machine', namely a material device devoid of any mental perception, intentionality is instantly precluded. Communication of corporeal dispositions may appear in this presentation only if they can be explained without any interference from the soul, as previously quoted. Thus a strictly physical layout might link a machine to another of the same kind, both animal and human.

Such an opening in *L'Homme* would be surprising, since Descartes' later works don't properly thematize the topic.[3] Should we consider that this reading comes from a faithful interpretation or, on the contrary, that it belongs to the long story of clever mistakes?

At the beginning of the *Remarque* devoted to the passage, Louis de La Forge estimates that 'there's something missing in this article' and that 'it is too short to be so beautiful'.[4] Though the article is indeed somewhat elliptical, La Forge's assessment is not less. For sure, the hypothesis of the automaton may arouse an aesthetic judgment, as it combines simplicity of principles with ingenuity of construction, deletion of superfluous entities with a promise of continuity between art and nature. The prospect of expanding those same principles to the sphere of relations between machines, namely distinct bodies, would no doubt increase its interest. But the ultimate reason for La Forge's statement remains unclear. What kind of expectation can be found on his part, which his reading of 'article 83' seems to have fulfilled, and may account for his aesthetical emotion?

---

[2] *L'Homme de René Descartes (...) avec les Remarques de Louis de La Forge* (...), Paris, chez J. Le Gras, 1664, rééd. T. Gontier, Fayard, Paris, 1999, p. 342.

[3] D. Kambouchner showed that *Epistola ad Voetium* contains passages devoted to passions transfer, but they seem to have little in common with this paragraph of *L'Homme*. See D. Kamboucher, 'Descartes et la communication des passions', Rue Descartes, n°12–13, Paris, 1995.

[4] *Op.cit.*, p. 340.

It is feared that this expectation may have influenced him, and that his judgment goes too far and too quickly, which may also happen to any reader of *L'Homme* who is familiar to its posterity, which La Forge yet contributed to shape.

8.1 La Forge's reception

As is appropriate to the elliptical beauty of the text, La Forge's comment in the *Remarques* is bound to set out what was involved in it, and first to identify the proper nature of relevant phenomenon, before proposing a progressive and complete explanation. This is not however without several shifts.

La Forge first substitutes for memory, as the cause of imitative disposition, the mere action of present objects on the senses. He alleges a grammatical mistake in doing so:

> It seems to me that there's something missing in this article (…). I think that what can dispose our machine to mimic the actions that real men, or other similar machines, may do in its presence is rather the action of objects on the senses than a memory effect: otherwise it would have been necessary to say 'would have done in its presence', because memory is only of things past; (…)
>
> [Il me semble qu'il y a quelque chose qui manque dans cet article ; (…) je pense que ce qui peut disposer notre machine à imiter les actions que de vrais hommes, ou autres semblables machines, feront en sa présence, n'est pas tant un effet de la Mémoire, que de l'action des objets sur les sens ; autrement il aurait fallu dire, *auront fait en sa présence*, parce que la Mémoire n'est que des choses passées ; (…)]

More than a clarification, this is in fact a first essential shift. Confined to a strictly reiterative role, memory is now devoid of any causal role in imitative disposition in itself.[5] Descartes would have mentioned it only because it enables to understand how an action, once imitated, may subsequently be reproduced. But no reason remains why it should still be given as the cause of imitation.

The situation of the paragraph becomes also rather difficult to explain. As a matter of fact, it belongs to a series of three paragraphs devoted to memory action, specifically related to the 'first cause' of gland movement,[6] which immediately precedes the description of 'action of objects on the senses'. This is renowned as the 'second cause that may determine[7]' gland movements.

---

[5] 'Therefore it seems to me that we must primarily attribute to the presence and action of object the disposition of our machine that enables and inclines it to mimic actions made in its presence (…).' [C'est. pourquoi il me semble que l'on doit primitivement attribuer à la présence et à l'action de l'objet la disposition de notre machine, qui la rend capable et qui l'incite à imiter les actions qui se font en sa présence (…).] *Remarques*, *op. cit.*, p. 341.

[6] AT XI 184.

[7] AT XI 185. The first of those two causes consists in unequal parts of spirits, as an internal determination of their course. For a further analysis of the consequences of this distinction on Descartes' theory of perception, see our study in *Perception et sociabilité, La communication des passions chez Descartes et Spinoza*, Garnier Classiques, Paris, 2015.

Following La Forge, a more 'correct' and complete presentation of imitative disposition would imply another amendment. Where Descartes claimed that machine 'can be disposed' to imitate 'without any soul in this machine',[8] La Forge suggests to start from a conscious process, and to see it as a finalized effect of union rules: 'No one can however deny that the action of the soul contributes to this very much', in so far as 'we feel generally inclined to imitate actions of our fellow-men'.[9] And as union wouldn't be as perfect as it is if it were lacking, a corporeal device goes with this feeling.

> For, as experience may teach everyone, it is one of the accepted articles between soul and body, when they first allied, that soul perceptions coming to it through the senses are such that they incline the will to want the same things to which the idea, excited by this action, disposes body at the same time; which is so constituted that the less discordant it is, the tighter this union might be between those two substances.[10]
>
> [Car comme l'expérience peut enseigner à un chacun, c'est un des articles accordés entre l'âme et le corps, lorsqu'ils se sont alliés, que les perceptions de l'âme qui viennent en elle par l'action des sens, sont telles, qu'elles incitent la volonté à vouloir les mêmes choses auxquelles l'idée que cette action excite sur la glande dispose en même temps notre corps ; ce qui s'est ainsi fait, afin que l'union de ces deux substances fût d'autant plus étroite qu'elle serait moins discordante.]

This presentation of the phenomenon as an effect of a union goes with another sliding, that brings imitative disposition to the state of an *inclination*, namely to an inner feeling, or even to a *passion*, if passive determination of will is its main feature, following Descartes' definition.[11]

But that's not all: the most obvious explanation of that disposition would ultimately be premised on a spontaneous comparison between the present situation of our mind-body with that of our peers,

> as if we could dimly see that there is some sort of failure in our body, when it can't do what other bodies of the same kind can do, and as if our soul were inclined by this to experiment on what its own body is able to.[12]
>
> [comme si nous apercevions confusément qu'il y a quelque espèce de défaut en notre corps, quand il ne peut faire ce que d'autres corps qui lui ressemblent exécutent ; et comme si notre âme était incitée par là à expérimenter de quoi son corps est capable.]

The formula twice mentions the analogical mode of '*as if*', which is particularly striking. On the one hand because it explicitly involves an intersubjective relationship, whereas the original text only mentioned similar behaviour. On the other hand, because involuntary imitation is somehow treated as a 'remaining' movement, or an *analogon*, of a conscious attempt to experiment *freely* on the powers of body. It

---

[8] Or « any knowing principle » [aucun principe de connaissance] as La Forge turns it in *Remarques*, *op. cit.*, p. 341.

[9] *Remarques*, p. 341

[10] *Ibidem*.

[11] *Passions de l'âme*, art. 40, AT XI 359.

[12] *Remarques*, *op. cit.*, pp. 341–342.

appears as a corporeal counterpart, or material substrate for an intentional and free dispensation of union abilities.

The last item in the description lays emphasis on the fact that will may by itself start the mechanical process of imitation, or inhibit launching.[13] Conversely, it may happen that this process runs without any mental control at all. This involuntary imitation is ultimately said to be the proper object of Cartesian concern.[14] But taken in all its dimensions, the imitation process is described as a typical effect of contingent lawlikeness which rules soul-body union. Moreover, it runs through a binding frame that primarily relates us to our peers.

One can easily predict what kind of metaphysical use La Forge is to make of such rewriting, and the aesthetic motives of his opinion become also clearer. But before coming back to it explicitly, let us examine the whole content of the Remarque and its several attempts to explain ‘article 83’. For the moment, we can only state that imitation, related by Descartes to memory and mere physiological structure, resorts now to actual perception and to a free mental comparison experiment, despite the original framework of *L’Homme*.

8.2 Vain attempts of physical explanation

But this twist is far from leading La Forge to ignore physical explanation. On the contrary, it requires three stages. The first one is dedicated to liken imitation to sound resonance.[15] As the latter is ‘not so hard’ to understand, it may help to lead reflexion. It occurs as a mere physical phenomenon of wave propagation, such as (*mutatis mutandis*)

> a string of lute that is not touched, imitates the sound of another string that is affected nearby, when tuned to the same tone.[16]
>
> [elles les imitent de la même manière que la corde d’un luth qui n’est point touchée, imite le son d’une autre corde qui est touchée auprès d’elle, lorsqu’elle est montée sur le même ton.]

But it seems really uneasy to extend that sort of explanation to a complete behaviour. A sound wave is first communicated to surrounding brain pore fibers, through which spirits travel to the muscles of tongue and throat. The movement of spirits coming in ‘perfectly mimics the opening which was in the network’. It carries

---

[13] « (...) such as it is true that our will sometimes contributes to this, and that it is the cause why we tend to imitate certain action that our machine wouldn’t imitate without it, our will often hinders it, retaining our inclinations back.’*Ibidem*, p. 342. [(...) comme il est vrai que notre volonté y contribue quelquefois, et qu’elle est la cause que nous nous portons à imiter certaines actions que notre machine n’imiterait point sans cela, elle empêche aussi assez souvent que cela ne se fasse en tenant la bride à nos inclinations.]

[14] *Ibidem*.

[15] *Remarques*, p. 344.

[16] *Remarques*, 344.

something that may communicate to the air, coming out of the throat, 'the same movement, with the same intervals', received from the sound which 'opened the network'.[17] Repeating the same sound would thus be a mere physical effect of sound wave propagation through the structure of a machine-body. But how can this be properly transferred to a whole behaviour, which moreover has to be *seen* in another machine?

The second step relates to memory, but eventually reduces imitation to a mere reiterative associational effect. Memory can give efficiency to the 'not considerable' action of the 'presence of objects'. For mimetic behaviour to take place, movements have to be like our own former ones, and memory ought to be involved.[18] We ought to have done these actions before. External action works only as a mere catalyst to our own reiteration.

> I don't think that we can explain otherwise that principal effect of memory, which allows the machine we describe able to mimic actions of other men, or similar machines, when it has previously performed the same action in a similar occasion. I don't fear to add those last words to the text, because without it, it seems to me not to be so understandable.[19]
>
> [Je ne crois pas que l'on puisse autrement expliquer ce principal effet de la mémoire, qui fait que la machine que nous décrivons peut imiter les actions que de vrais hommes, ou autres semblables machines, font en sa présence, quand elle en a autrefois fait de semblables dans une pareille occasion. Je ne crains point d'ajouter ces dernières paroles au texte, parce que sans cela il ne me semble pas si intelligible.]

These failing attempts to explain imitation features, denying any specificity to the phenomenon, are a bit like surrendering. But the third step tries to produce a proper explanation.

It lies on one of the highlights of *L'Homme*, namely the causal reciprocity between ideas an limbs movements. In Descartes' description, the movements of animal spirits coming out from a given point of the gland may turn the network pores 'towards the places they come from, if they don't find them already turned this way'. This effect generates a command of the limb 'towards the places related to' the relevant points of the gland.[20] The idea on the gland is also an idea of the movement of eyes or arm towards a certain point in external space. Conversely, a given disposition of limbs determines a given position on the gland, via the corresponding orientation of pores. Thus 'the movements and ideas may be reciprocally caused by each other.[21]'

The specific action of external bodies is then convened by La Forge, as the 'second cause that may determine gland H movement'.[22] In a fairly striking shortcut, he affirms that

[17] *Ibidem*, p. 345.

[18] *Ibidem*, p. 345.

[19] *Remarques*, 346

[20] AT XI 181

[21] AT XI 182

[22] AT XI 185

> when this object B comes to move, the diverse impressions that it makes on the eyes, according to the various places where it is, are the cause of the gland imitating its movement; and consequently the meshes of the network change their orientation successively, and dispose members that depend on them also to imitate the movement of object B.[23]

> [(...) quand cet objet B vient à se mouvoir, les différentes impressions qu'il fait sur les yeux, selon les divers lieux où il est, sont cause que la glande imite son mouvement ; et conséquemment que les mailles du résil changeant leurs regards successivement, disposent les membres qui en dépendent à imiter aussi le mouvement de l'objet B.]

It is true that in Descartes' description, action of objects can change the position of the gland, and contribute to a muscular accommodation process, or adjustment, in order to optimize perception. In this sense, it's correct to say that this action 'can change the gland situation, if ever it finds it not already disposed as it claims', and that 'when its position changes, it causes the holes in the network to change the way they are orientated to it'[24]

But a huge difference exists between this carefully settled accommodation process, which takes six long paragraphs by Descartes, and what La Forge tries here to explain, namely how to trigger a behaviour similar to one seen previously.

If this explanation were correct, any perception would involve imitation of moving objects, which he's himself aware, since the following sentence goes on with:

> this disposition being most weak, and common to all ideas, can hardly ever have any considerable effect, if memory doesn't contribute.[25]

> [cette disposition étant fort faible, et commune à toutes les idées, ne peut Presque jamais avoir d'effet considerable, si la mémoire n'y concourt.]

In order to explain the non-existence of such facts, an *ad hoc* argument is proposed to minimize potential effects of this hypothetical mechanism. It works only if memory, and its reproductive associational abilities, supports it. Proceeding this way, La Forge expects to explain memorial aspects of imitation. In fact, he substitutes an explanation to another, or even deals with something else. If imitative disposition is nothing but mere associational repeating, there's no use to go further. It has no specific interest left and can be likened to mere animal breeding.[26]

Most of all, according to Descartes, any external object movement can't match to a similar movement of the gland. On the contrary, any transfer of an object on a line has to touch different places on the gland to be perceived.[27]

La Forge's construction here is more an extrapolation of sound resonance than a real explanation of Descartes statements. In both cases, imitative device is reduced to a physical transfer of movements.

---

[23] *Remarques*, p. 347.

[24] AT XI 186.

[25] *Remarques*, p. 347

[26] 'And in this consists the whole discipline of all disciplinable animals', *Remarques*, p. 348

[27] See how corporeal machine can convey an idea of distance in AT XI 183.

When he says that 'article 83' aims to explain how an individual machine can copy another, La Forge is led to question Cartesian coherence, or to override it. How the fact that imitative disposition takes place *before* external action of objects on the nerves be explained, or that it appears as a memory effect but in the presence of perception? This problem, emerging from Cartesian writing, is indeed far from easy to settle. La Forge deserves to be seen as the first to identify this trouble, and to suggest an hypothesis to solve it which could have been expected from Descartes himself. However he may fail to be faithful to Descartes, he has the proper merit not to conceal it.

But did Descartes really intend to talk about imitative disposition, as far as La Forge feels allowed to read it in this passage? Is it really this kind of behaviour, whose most common type is yawning, that Descartes intended to deal with?

8.3 Back to Descartes

In this regard, the sown seeds among post-Cartesian philosophers, such as Malebranche or Spinoza, shouldn't preempt examination. The reasons to hold a safe suspension of judgment are at least as strong. First of all, the literal text of unfaithfully named 'Article 83' doesn't put imitative behavior as an effective *relation* between two machines, but only as a natural and very general disposition.[28] The 'most remarkable effect' of memory appears much more as an undetermined ability, than a real communication of any *particular* behaviour. Such as memory can explain chimeras, or any other fictitious effect of *phantasy*, like spontaneous remembrance,[29] it may be the cause of unexpected occurrence of a particular behaviour in a body-machine, similar to one previously observed. Those effects are difficult to explain without the will. That's an *a fortiori* reason why it deserves a merely corporeal explanation, in order to confirm the general aims of the treatise[30] Nor are intentional thoughts necessary to explain that kind of amazing phenomenon. As with any of the same kind, it belongs to corporeal causality.

This purpose is still mentioned in the Fifth Part of the *Discours de la Méthode*, which is reputed to be a kind of abstract of *L'Homme*. When it comes to knowing what he means by 'memory, which retains' ideas, Descartes immediately mentions 'phantasy, which is able to change them and compose new ones', just like what he said in the former paragraph of *L'Homme*. Hence he adds that phantasy

> by the same means, moving animal spirits into muscles, can move the limbs of this body in as many diverse ways, and as much about objects that come to its senses, or internal passions that are in it, as ours are able to, without any leading of the will.[31]

---

[28] « It can naturally be disposed to mimic all movements…' AT XI 185

[29] AT XI 184

[30] They run « just like clocks or any other automat (…).» AT XI 202

[31] AT VI 55

[par même moyen, distribuant les esprits animaux dans les muscles, faire mouvoir les membres de ce corps, en autant de diverses façons, et autant à propos des objets qui se présentent à ses sens, et des passions intérieures qui sont en lui, que les nôtres se puissent mouvoir, sans que la volonté les conduise.]

Therefore, the nature of the relation between machines described at the end of the articles devoted to memory doesn't seem essentially different from this external imitation of human corporal functions – or if you like from this perfect counterfeiting –, by the statue of clay. So in this effort to discriminate which functions can be imputed to the body, Descartes is bound to explain how a machine may

utter words, and even utter some about corporeal actions that will cause some change in its organs: as if, touched in some place, it asks what we mean to it; shouts when it is hurt, and so on.[32]

[(proférer) des paroles, et même qu'elle en profère quelques-unes à propos des actions corporelles qui causeront quelque changement en ses organes : comme, si on la touche en quelque endroit, qu'elle demande ce qu'on lui veut dire ; si en un autre, qu'elle crie qu'on lui fait mal et choses semblables (...)]

The most perfect counterfeit ought to go as far as possible, and explain how, without any intentionality, a machine can reproduce the behaviour of other similar machines. To give an example, the parrot is able to repeat *at the right* time the words he has heard. But strictly speaking, a perfect counterfeit behavior is not necessarily an active imitation. It may occur accidentally –so to speak, as an automatic effect of the organization of the machine – just like two dogs seem to hunt together.

Secondly, counterfeiting seems to fall into a determined category of perception which remains rather stable in the corpus, from *L'Homme* to the *Passions de l'âme*,[33] and even to the *Correspondence*.[34] This category includes fortuitous imaginations 'whose cause is the sole body'. Animal counterfeiting can therefore be likened to the illusions of dreams, daydreaming, or a situation of 'nonchalance' of fantasy that relates it to the field of fiction. In so far, counterfeiting animals, as the parrot, may simulate actions that are not actually his, in a way comparable to those of a sleepwalker.

More than a lack of attention, 'nonchalance' refers to a status/disposition of the machine when some of its 'ideas', similar to those formed under action of external objects (through senses), are spontaneously produced by the brain itself. Sensitive perceptions are then counterfeited, and may deceive the soul:

It's useful to note here that all the same things that the soul can perceive through the nerves may also be represented by the fortuitous course of the spirits, without any other difference other than that the impressions coming into the brain through the nerves are more vivid and clear than those that are there excited by the spirits; which made me say in Article 21 that they are like a shadow or painting of the others.[35]

[32] AT V 56

[33] AT XI 344

[34] *Letter to Elisabeth*, October 6th of 1645, AT IV 311

[35] *Passions de l'âme*, art. 26, AT XI 348.

[Il reste ici à remarquer, que toutes les mêmes choses que l'âme aperçoit par l'entremise des nerfs, lui peuvent aussi être représentées par le cours fortuit des esprits, sans qu'il y ait autre différence, sinon que les impressions qui viennent dans le cerveau par les nerfs, ont coutume d'être plus vives et plus expresses, que celles que les esprits y excitent. Ce qui m'a fait dire, en l'article 21, que celles-ci sont comme l'ombre ou la peinture des autres.]

Memory is not a simple stock of determined images. Its functions are not simply dedicated to remembrance.[36] It can get the machine to feel something like outer and inner feelings, even while we are awake, though no vital necessity or any command of will are concerned.[37]

What is true of *sensitive* function of the cerebral pores has also to be true of their role in *motion*. The figure that is plotted on the internal surface of the brain is both a perceptive datum and a motion control. Counterfeit perception shall develop into a complete sleepwalker-like behaviour, even while machine is awake.

Memory appears as an ability to suggest both perceptions and behaviours. Machines can be set in motion and engage in many operations independently, without any impulsion from outside. There is no movement a machine can do that a similar machine wouldn't be able to perform by itself without being subject to the same external stimuli. Its internal organization and its own inner activity allows it to counterfeit actions of similar machines. By definition this is not on purpose, but may accidentally be effective.

It is therefore by itself 'capable' of the same actions a similar machine may perform in its presence. 'Article 83' ought to remain as a general observation about abilities of the machine. In this sense, one can easily understand why Descartes didn't have to explain how *current* perception of similar individuals may launch imitative behaviour. It's not for *L'Homme* to do more than finally describe

> the outer movements of all limbs, so properly following from objects that touch the senses, as from passions, or impressions that occur in Memory, that they mimic as perfectly as possible those of a real man.[38]

> [(...) et enfin les mouvements extérieurs de tous les membres, qui suivent si à propos, tant des actions des objets qui se présentent aux sens, que des passions, et des impressions qui se rencontrent dans la mémoire, qu'ils imitent le plus parfaitement qu'il est possible ceux d'un vrai homme.]

It's time to notice that, as far as *passions* are concerned, counterfeit phenomenon seems to be excluded by Descartes. Passion *in itself* seems impossible to counterfeit:

> It should also be noticed that it sometimes happens that this painting is so similar to the thing it represents, that we may be deceived concerning perceptions that relate to objects that are outside of us, or those relating to some part of our body, but we can't be deceived

---

[36] See J. Sutton, *Philosophy and memory traces, Descartes to connectionism*, Cambridge University Press, 1998, p. 57

[37] 'Thus often when we are asleep, and sometimes even when we are awake, we imagine things so strongly that we think we are seeing them in front of us, or we are feeling them in our body, though they are no'. AT XI 348–349 [« Aussi souvent lorsque l'on dort, et même quelquefois étant éveillé, on imagine si fortement certaines choses, qu'on pense les avoir devant soi, ou les sentir en son corps, bien qu'elles n'y soient aucunement ; »]

[38] AT XI 202.

in the same way concerning passions, especially since they are so close and internal to our soul that it is impossible for it to feel them without their being just what it feels they are.[39]

[Il faut aussi remarquer qu'il arrive quelquefois, que cette peinture est si semblable à la chose qu'elle représente, qu'on peut y être trompé touchant les perceptions qui se rapportent aux objets qui sont hors de nous, ou bien celles qui se rapportent à quelques parties de notre corps ; mais qu'on ne peut pas l'être en même façon touchant les passions, d'autant qu'elles sont si proches et si intérieures à notre âme, qu'il est impossible qu'elle les sente sans qu'elles soient véritablement telles qu'elle les sent.]

Passions are different from any other feeling. Outer feelings may deceive the soul, since they fundamentally provide an *external* information. But passions don't run this way. It's not that corporal basic movement of passions is excluded from imitation at all. As any other corporal behaviour, it can be launched without proper necessity, or only *as if* it were necessary to have it – though it's not.

But once a passion is launched, it belongs to *our* soul, and it's impossible to consider it as a fake one. Accidental passion is actually one of *our* passions, even if it has no proper cause specific to us in particular. Dreaming anger is actually being angry, as Descartes says in the following lines.[40]

A passion can't really take place in us without us, so to speak. Passion implies an inclination of will, whatever it is, that plainly belongs to us, and we are responsible for it. One can be abused where disabuse is possible. You can decline responsibility when you are not fairly engaged. But this is not possible with passion. What is impossible is not passion counterfeiting, but that passion wouldn't be mine at all.

Ego as a substance is concerned here and not only my possession of a body. The mere definition of self is here at stake.

This might be the most obvious reason why Descartes renounced any development about spontaneous communication of corporeal dispositions as a social process between machines in the treatise of the *Passions of the Soul*, despite its closeness to *L'Homme*. The example of yawning, so obvious to La Forge in order to explain Descartes' intentions, is completely absent from Descartes descriptions.

8.4 The story of a well-founded mistake

Though Descartes may appear, due to this article, as a precursor of social communication of passion theories supported by Malebranche or Spinoza, he would probably oppose such theories. Descartes seems to refuse imitation as a regular determination of ordinary perception, in the name of a strong and strict difference between external and internal perceptions of the soul. If passion occurs as a 'fortuitous imagination', it can't be deceitful, and passion as such just can't remain fortuitous.

[39] AT XI 348.

[40] *PA*, art. 26: 'But though we are asleep and dreaming, one can't ever feel sad, or moved by any other passion, that it shouldn't be utterly true to say, that soul has really got this passion for herself.'(['mais, encore qu'on soit endormi et qu'on rêve, on ne saurait se sentir triste, ou ému de quelque autre passion, qu'il ne soit très vrai que l'âme a en soi cette passion.']) AT XI 349.

La Forge's astonishment and aesthetic emotion rather belong to his reading viewpoint than to Descartes' own intentions. The Cartesian text nevertheless offers an unequalled opportunity to construe it as an explicit reference to social interaction. This mechanical, accidental and spontaneous effect of counterfeit is actually read by La Forge as a regular effect of natural institution and of mind-body union articles. He reads their perfection in such a well-combined effect of nature, which goes beyond isolated individuals and makes them parts of an integrated whole. The *singularity* of this effect comes as a proof and confirmation of an underlying *order*. It's now easier to understand La Forge's judgement about this article, both aesthetic and metaphysical.

But this metaphysical ambition is not *L'Homme's one*. The treatise tries to reduce animal behaviour to physicality, and to confirm by this way the substantial distinction between body and mind, which is also a metaphysical issue, but avery different one.

Reading this passage as a proof of a transcendent order that would account for the origins of society by providing its physical basis is a double metaphysical movement, which Descartes probably didn't foresee.

As far as Descartes himself is concerned, human social life is related to the fact that any human being is bound to consider himself as part of a whole.[41] It's not a result of dedicated corporeal devices. How animal communication is mentioned in subsequent texts is meaningful. While human beings essentially share meanings, animals don't.[42] Considering oneself as a part of a whole relates much more to a meaning than to a physical fact. Society must be based on these shared meanings, and upon thoughts rather than physical determinations.

Moreover, the possibility of a passion being transferred from one to another seems to be excluded, as soul can exercise a complete authority on its judgments. This is a second 'metaphysical lock' that may prevent La Forge's views. Consequently, the subsequent reserve that Descartes showed on the subject of the communication of passions, which may appear deceptive for his readers, is not really surprising after all.

But the opportunity was given, and La Forge is the first to seize it. His fascination is quite logical. A new way is opened up to explain the origins of society. Malebranche follows him on this path, as he takes corporeal behaviour imitation as a secondary cause of those ties which bind us to other human beings, wisely instituted by the Author of Nature, the sole effective power in this world.

The other aspect of La Forge's explanation, which makes it an irresistible effect of the corporeal machine, has more to do with Spinoza's possible reading of this article, that may have led him to his own explanation of '*affectuum imitatio*'.[43] In a way, those two options are contained in La Forge's *Remarques*, and this is the reason why it provides a precious testimony about what his contemporaries were expecting from *L'Homme*.

---

[41] *To Elisabeth*, September 15th., 1645 AT IV 293; *To Chanut*, February 1st, 1647, AT IV 612

[42] For example, *Letter to Newcastle*; November 23rd 1646, AT IV 573.

[43] Spinoza, *Ethics*, III, 27.

Chapter 9
La Forge on Memory: From the *Treatise on Man* to the *Treatise on the Human Mind*

Emanuela Scribano

Abstract In his remarks on *L'Homme*, La Forge aims at a rigid separation of the functions of the body from the activity of the soul. This project looks authentically Cartesian, but some critical issues reveal how difficult it is taking away any activity of the soul in sensitive experience. In the *Traité de l'esprit de l'homme*, La Forge explicitly limits the cognitive capability of the memory without the active presence of the mind.

La Forge's notes on the posthumous *Treatise on Man* consciously emphasize the Cartesian project in order to account for body's abilities independently of the mind. It is the only part of Cartesian project concerning the study of the man we have. Indeed, the plan was to issue two more parts, one devoted to studying the mind independently of the body and the other to studying the union of mind and body.[1] In his commentary, La Forge tells that he has devoted another treatise, the *Treatise on the Human Mind*, to the missing parts of the Cartesian project—those relative to the mind independently of the body and to the union of mind and body.

First, in relating to Cartesian physiology and, later, in completing the Cartesian program, La Forge takes two things into account: (i) Descartes' writings that, though written later than the *Treatise on Man*, were published before his physiology text; and (ii) texts by other authors published when the *Treatise on Man* had not yet appeared. Both elements are pertinent in evaluating the relevancy, limits, and role of the physiological analysis developed in the *Treatise on Man*. Here, we shall verify the analysis concerning memory and reminiscence.

---

[1] La Forge (1999), 59.

E. Scribano (✉)
Department of Philosophy and Cultural Heritage, University Ca' Foscari Venezia,
Palazzo Malcanton Marcorà, Dorsoduro 3484/D, I-30123 Venezia, Italy
e-mail: emanuela.scribano@unive.it

D. Antoine-Mahut, S. Gaukroger (eds.), Descartes' *Treatise on Man* and its Reception, Studies in History and Philosophy of Science 43,
DOI 10.1007/978-3-319-46989-8_9

9.1 Descartes: Material Memory

Memory, as discussed in the *Treatise on Man*, is the disposition of the brain's internal parts to reopen spaces that had been opened by previous stimulations, affecting the pineal gland (and hence the mind) in the same way as the first stimulations. Descartes compares the brain to a textile through which small metal wedges have passed. The gaps the wedges produce in the textile close after their passage, but the textile is more easily reopened where the wedges crossed than at other points.

Memory plays a major role in cognitive phenomena, especially concerning the association of ideas. Thanks to the association of traces in the brain, memory fills what is missing in sense perception:

> It must be noted that if one were to re-open just some of (the holes) …, this in itself would cause others … to reopen at the same time, especially if they had all been opened together several times and had not usually been opened separately. This shows how the recollection of one thing can be excited by that of another which had been imprinted in the memory at the same time. For example, if I see two eyes with a nose, I immediately imagine a forehead and a mouth, and all the other parts of a face, because I am unaccustomed to seeing the former without the latter. And seeing fire, I am reminded of heat, because I have felt this in the past when seeing fire.[2]

In this respect, material memory overlaps imagination: "if I see two eyes with a nose, I immediately *imagine* … and seeing fire, I am *reminded* …".[3]

The living machine reacts differently to environmental stimuli depending on past experience, and since past experience is never the same, living machines react differently to the same stimulus. Memory connects brain traces and inserts any new stimulus into a brain network already marked by previous experiences. In this way, a dog who has been beaten while a violin was playing will be scared any time a violin plays.[4] Thanks to material memory, the body-machine can "recognize" places and things belonging to its past experience and react to the reoccurrence of circumstances in ways as different as its reactions to previous occurrences of those circumstances. Traces can work as a true material memory, inducing behavior by their very presence.

These features of material memory were examined by Descartes again in a letter to Meyssonnier on January 29, 1640.[5] Descartes also resumes the issue, the same year, discussing it with Mersenne.[6] The disposition of the brain's tissue to more easily reopen traces that have already been opened, i.e. material memory, is a

[2] Descartes, AT XI, 179; (2004), 151–52.

[3] Material memory had already been assimilated to imagination in the *Regulae*. AT X, 416, CMS I, 42.

[4] Descartes to Mersenne, 18 March 1630, AT I, 134. On the importance of material memory in Descartes the seminal essay is Sutton (2007). See also Morris (1969).

[5] Descartes to Meyssonnier, January 29, 1640, AT III, 18–21; p. 20, CSMK, 144: "I think also that some of the impressions which serve the memory can be in various other parts of the body: for instance, the skill of a lute player is not only in his head, but also partly in the muscles of his hands, and so on."

[6] Descartes to Mersenne, April 1, 1640, AT III, 47-8, CSMK, 145–6.

phenomenon concerning the internal part of the brain and does not involve the pineal gland—except in the case of human beings with a slow and dull mind.[7] Indeed, a more mobile gland, and for this reason a less retentive one, corresponds to the smartest minds.[8] Moreover, memory affects other parts of the body too—the whole body has marks, such as the marks that we see impressed on the fetus at birth.[9]

The quote above, according to which a brain trace produces the memory corresponding to an associated brain trace, leads us to think that in the *Treatise on Man* the conscious memory occurs in the mind upon the reopening of a brain trace. If this were the case, memory would behave like sensation, which produces a conscious perception in the mind via the brain modification connected with it.

9.2 Intellectual Memory

During spring 1640, in his letters to Mersenne, Descartes mentions a notion not yet introduced: intellectual memory. On April 1, 1640, Descartes writes to Mersenne that, besides the memory depending on body traces, "I believe there is also another one, entirely intellectual, which depends on the soul alone."[10] The existence of an intellectual memory is reiterated, again to Mersenne, on June 11, 1640: "the intellectual memory has its own separate impressions (*especes*), which do not depend in any way on these folds".[11] Descartes appropriates a notion he finds in the *Conimbricenses* commentary on Aristotle's *De memoria et reminiscentia*, but, by doing so, he introduces a new requirement, which brings him to search within a tradition he is well acquainted with for a different form of memory to that with which he is concerned in the *Treatise on Man*.[12] On August 6, 1640, again in a letter

[7] Descartes to Meyssonnier, January 29, 1640, AT III, 20, CSMK, 144.

[8] Aristotle had already argued that intellectual smartness and memory were inversely proportional. Cf. *Of Memory and Reminiscence*, 1, 449b: "indeed, as a rule, slow people have a good memory, whereas those who are quick-witted and clever are better at recollecting."

[9] AT XI, p. 177, (2004), 150. Repeated in the *Dioptrique*, AT VI, 129. As it is well known, the topic will be resumed by Malebranche, *Recherche de la Vérité*,II,I,VII, OC, I, 232 ss.

[10] Descartes to Mersenne, April 1, 1640, AT III, 48, CSMK, 146

[11] Descartes to Mersenne, June 11, 1640, AT III, 84–5, CSMK, 148.

[12] On the view of intellectual memory in the *Conimbricenses* commentary to *De memoria et reminiscentia*, see Gilson (1979), s.v. *Mémoire*. The passage quoted by Gilson clearly shows that most of Descartes' remarks about memory before 1640 derive from this tradition. The *Conimbricenses* ascribe to intellectual memory the remembrance of universal and immaterial things and deny any difference between intellectual memory and intellect. Descartes repeats a traditional claim even in placing material memory in the back of the brain. Ivi, 78. Conversing with Burman, Descartes says that the remembrance of universals pertains to intellectual memory, AT V, 150: "Verum haec memoria intellectualis magis est universalium quam singularium…"

It is important to stress, besides this, that the *Conimbricenses* add the notion of intellectual memory to the Aristotelian text. The intellectual memory they speak of, as a matter of fact, does not correspond to the Aristotelian theory of reminiscence, which was not, in any way, devoted to preserving the memory of immaterial and universal concepts.

to Mersenne, intellectual memory becomes a further aspect distinguishing human beings from animals. Indeed, it is to intellectual memory that Descartes ascribes the most meaningful part of the human activity of remembrance: "Moreover, in addition to the corporeal memory, whose impressions can be explained by these folds in the brain, I believe that there is also in our intellect another sort of memory, which is altogether spiritual, and is not found in animals. *It is this that we mainly use*."[13] This remark is entirely Descartes' own, with no comparison in the *Conimbricenses* commentary to the *De memoria et reminiscentia.* If there were cues to intellectual memory before 1640, they was never the object of any systematic reflection.[14]

A theory of "intellectual" memory not only surfaced in the year 1640,[15] but, and above all, it was given a central role in human memory—even if that role was not specified.[16] In any case, in 1644 memory still seems to be "intellectual" because it concerns thoughts not produced via brain traces, thoughts representing immaterial things. Indeed, Descartes speaks of peculiar "traces" of "intellectual things", enduring traces that account for the memory of those things, but impossible to exemplify because of their immateriality. This is the reason why, until 1640, intellectual memory stands alongside but does not substitute the material memory presented in the *Treatise on Man*: the former appropriated for thoughts originating from the intellect, the latter for thoughts originating from experience. The few hints at a double memory found in letters between 1640 and 1644 do not stray much from what one can read in the *Conimbricenses* commentary to the Aristotelian *De memoria et reminiscentia*, except for the mysterious statement that we help ourselves *mainly* with intellectual memory. What is new, rather, is the evocation of that second kind of memory. Some years later, due to Arnauld's insistence, Descartes considerably modified the theory of intellectual memory occasionally touched upon between 1640 and 1644, and fully justified the mysterious hint of the August 6, 1640 letter to Mersenne. In fact, the claim becomes more extreme: the human mind mostly uses intellectual memory and only intellectual memory produces human recollection. For this reason, a brand new theory of intellectual memory is required.

---

[13] Descartes to Mersenne, August 6, 1640, AT III, 143, CSMK, 151. Emphasis mine.

[14] On the *Regulae* see Angelini (2000). The existence of an intellectual memory does not tell yet us whether this notion points to a specifically Cartesian theory, as Joyce (1997) seems to think, or whether it overlaps that proper to scholastic commentaries to the *De memoria et reminiscentia*. On the presence and the sense of this notion in Aquinas see Bazán (1990). The hint in the *Studium bonae mentis* that Baillet cites seems dependent on the theory of an intellectual memory extractable from the *Conimbricenses* commentary to *De memoria et reminiscentia*, which identifies intellectual memory and intellect. Cf. Baillet (1987), II, 66: "It seemed to doubt that memory were distinct from intellect and imagination. It did not think that memory could extend or grow, but rather be more or less filled." AT X, 200–01 and Descartes (2013), 134–5.

[15] Once and only once, in a letter to Mersenne, on August 6, 1640, Descartes will call it "spiritual". AT III, 143.

[16] Morris (1969) and Machamer, Mc Guire (2009), 188–193 take this as a reference to innate ideas. The conjecture, however, conflicts with what Descartes tells Mersenne and repeats in a letter to Mesland on May 2, 1644, i.e. that intellectual memory should have its species. AT IV, 114–15, CSMK, 233.

9.3 Recollection and Traces

The discussion with Arnauld that concerns us links back to Gassendi's objections to Descartes and to his replies about thought as the essence of the mind. If this were the case, Gassendi objected, thought would always be actual and the mind would always be thinking. Descartes, fully accepting this consequence, had to reply to the problems originating from it. Gassendi neatly points out one such problem: if it is true that the mind is always thinking, why do we not have any recollection of our prenatal and early childhood thoughts?[17] In the context of the replies to Gassendi, the analysis of memory is then instrumental in clearing the obstacle to thought being continuously actual.

To account for the actuality of thought being compatible with the absence of memory, replying to Gassendi, Descartes ascribes the absence of recollections of prenatal experience to the inability of the fetus' brain (and of lethargic people's brains too) to retain traces:

> So long as the mind is joined to the body, then in order for it to remember thoughts which it had in the past, it is necessary for some traces of them to be imprinted on the brain; it is by turning to these, or applying itself to them, that the mind remembers. So is it really surprising if the brain of an infant, or a man in a deep sleep, is unsuited to receive these traces?[18]

In July 1641, the reply to Gassendi matched an insistent series of remarks by an anonymous person who introduced himself as Hyperaspistes.[19] In his response to these remarks, Descartes held that the unborn baby's mind, closely tied to the body, cannot have purely intellectual thoughts, but only unclear perceptions, such as pleasure and pain, which derive precisely from its close union with the body.[20] Moreover, even if the fetus had purely intellectual thoughts, these would not produce memory because "where purely intellectual things are concerned, memory in the strict sense is not involved; they are thought of just as readily irrespective of whether it is the first or second time that they come to mind".[21] The absence of memory of intellectual things was a new claim, which strengthens what Descartes had said to Gassendi, i.e. that the fetus does not have, and cannot have, any memory of any kind whatsoever.

Some years later, Arnauld's objections again bring up themes from the discussion among Descartes, Gassendi, and the Hyperaspistes. The problem was the same: how is it possible to maintain that thought is the essence of the mind, since, if it were so, one would then have to argue that the mind is always thinking? It is not possible that the inability to remember prenatal thoughts is due to brain traces being

[17] *Objectiones Quintae*, AT VII, 264.

[18] *Quintae Responsiones*, AT VII, 356–7, CSMK, II, 247.

[19] Landucci convincingly argue for identifying the Hypersapistes with Mersenne. Cf. Landucci (2001).

[20] Descartes to X*** [Endegeest, August 1641], AT III, 423–24, CSMK, 189–90.

[21] AT III, 425, CSMK, 190.

surface-based or to their vagueness, as Descartes asserted in reply to Gassendi. If it were so, one would have to maintain that the only memory with which the human mind is endowed is a material memory. Whereas, according to Arnauld, "it seems necessary to accept that our mind has two memory faculties (*vis*), one purely spiritual and one requiring a body organ."[22] The necessity of accepting two memory faculties, Arnauld insists, originates from Descartes' own theses. Descartes theorized a double power of thinking—*duplex cogitandi vis*—, one that is exercised without resorting to any body faculty, and one that applies to images impressed on the brain. If the mind can understand without the brain's help, why can it not remember without that help?" Besides, that there has to be a true memory of pure intellectual understanding is proved by the fact that in an argument the next step depends on memory of the previous ones.[23] Hence, Arnauld concludes, it has to be possible to have memories of pure intellectual understandings and these memories have to be possible independently of brain traces. The absence of memories from the prenatal stage are then presented again by Arnauld as a best objection to the Cartesian thesis according to which the mind is always thinking.

In answering Arnauld, Descartes definitely concedes the existence of two memories: "I agree with you that there are two different powers of memory".[24] But in the fetus there are neither pure intellections, because of the close link between mind and body, nor a purely intellectual memory, if by this we mean the recollection of thoughts independent of brain traces. This remark does not introduce the possibility that adults remember concepts of purely intellectual origin, possibility denied to Hyperaspistes and allowed to Mersenne, but it opens up a different understanding of the intellectual memory that is not active in the fetus. For the first time, Descartes individuates mental activity as the feature that makes memory intellectual—mental activity applies to material traces and is capable of *recognizing* new traces impressed on the brain as similar to those impressed before, or of detecting them as fully new ones. Only thanks to mental activity do brain traces become recollections and it is thanks to this ability that the relevant mental activity comes to be called "memory". In contrast to bodily memory, this mental activity does not retain anything—it is just that: what allows us to turn the brain trace into a recollection. This mental activity, because of the close link between mind and body, does not operate in the fetus:

> I agree with you that there are two different powers of memory; but I am convinced that in the mind of an infant there have never been any pure acts of understanding, but only confused sensations. Although these confused sensations leave some traces in the brain, which remain there for life, that does not suffice to enable us to remember them. For that we would have to observe that the sensations which come to us as adults are like those which we had in our mother's womb; *and that in turn would require a certain reflective act of the intellect, or intellectual memory, which was not in use in the womb.*[25]

[22] Arnauld to Descartes, June 3, 1648, AT V, 186.

[23] Ivi, 187.

[24] Descartes to (Arnauld), June 4, 1648, AT V, 192, CSMK, 354.

[25] AT V, 192–93, CSMK, 354–55, emphasis mine. This aspect of the Cartesian theory is well explicated by Elisa Angelini (2000), 197 and 206. See also Minerbi Belgrado (2006), 850.

This shift—to saying that only a non-bodily activity can produce remembering—radically overturns the problem. If, with Gassendi and Mersenne, missing prenatal memory experiences are justified because brain traces were wanting, Descartes now directly claims that purely intellectual activities are missing. Brain traces—which in *L'Homme* constituted material memory—do not properly allow the mind to remember. Not even a new flow of animal spirits reopening old brain traces suffices to produce a recollection in the mind. Only the intellect's "reading" the brain traces produces recollection and this reading is impossible in the prenatal stage.

But if Descartes' reply signals progress and defines the theory of intellectual memory, it makes, if this is possible, Arnauld's objection even stronger and, at the same time, gives Gassendi's objection new vigor. Having attributed the capacity to turn brain traces into memories only to intellectual memory, Descartes thinks he can acknowledge that brain traces formed in the fetus "remain there for life" without producing a memory. If, as Descartes recognizes, brain traces to which intellectual memory can be applied are there, why is intellectual memory then "not in use in the womb?"

Arnauld immediately grasps Descartes' reply's feebleness and stresses that it differs from the reply to Gassendi's objection. If only intellectual memory is capable of recognizing brain traces and of turning them into memory, why can't we use it during our prenatal life, since Descartes now concedes that brain traces get impressed on the brain, even the brain of the fetus? Let purely intellectual thoughts go, but conceding an intellectual memory distinct from material memory, together with the thesis that thought is always actual and the assumption that traces impressed on the fetus' brain last, by logical consequence would mean that the recollection of pre-natal experience is possible:

> [e]ven if the not yet born baby has no pure intellections, but only vague sensations, why can he not remember them later, since their traces anyway remain impressed in the brain (a thing you seemed, however, to deny in the *Metaphysics* at page 507). You will rely that this it due to the fact that recollection depends on intellect's reflection, which had not at all been exercised in the womb. Yet, concerning reflection, the intellect, that is intellectual memory, looks reflexive by its own nature. There is still to explain what is the reflection out of which you make up intellectual memory, and how it differs from the simple reflection intrinsic in any thought and why one cannot use it at all in the womb.[26]

Descartes seems to have realized that his first reply to Arnauld was counterproductive, and with elegance adjusts it in the next letter, basically giving up the rash step made in the first reply to Arnauld and again focusing on the brain traces. Intellectual memory, in its role as a reader of brain traces, is there in the prenatal phase—hence, it is true that thought is always actual. There are also brain traces, but

---

Understanding intellectual memory as acknowledgement quietly echoes Fernel's account of the recollection of universals known in the past: "When (concepts of kinds) come to the mind, if we are considering the past, the mind anyway recognizes to have already entertained and known them and this certainly is intelligence memory (*intelligentiae memoria*)." Fernel (2003) VI, 14, 500, on which see Céard (2002), 129.

[26] Arnauld to Descartes, July 1648, AT V, 213. Emphasis mine. Arnauld quotes the *Quintae responsiones* from Soly 1641 edition.

intellectual memory cannot act on them because the traces are too messy to be deciphered. What is missing in prenatal development is neither intellectual memory, as Descartes states in the first reply to Arnauld, nor material memory, as he asserts in answer to Hyperaspistes and Gassendi. What is missing is a *decipherable* material memory. When it comes to early childhood brain traces, one can repeat the saying that in well-trodden sand we claim not to detect the trace of any human, because no footprint can be told apart.[27]

The outcome of Descartes' mature reflections on memory is that brain traces, which he searched for by dissecting animal heads, deserve only metaphorically to be called memory, because they do not allow recollection even in the presence of a mind if the mind is restricted to perceiving brain traces without any further interpretative work.[28]

If the reply to Arnauld is a clarification of the mysterious hint in Descartes' letter to Mersenne, in August 1640, that intellectual memory is what human beings *mainly* use, then this hint has to be understood in the following sense: human beings, like any other animal, have mechanical behaviors that transduce past brain traces into actions, as when we pull back our hand from a cactus because seeing it reactivates the traces a previous sting impressed on the brain. In such a case, memory is a mechanical reflex on the part of the body. What is truly and characteristically human is conscious recollection, which happens *only* when a mind is capable of handling and deciphering brain traces. When a mind is connected with the machine, it will not *record* brain traces as conscious recollections, but will *interpret* the traces. Only on this interpretation will it transform them into conscious recollections.

Descartes only once speaks of the intellectual memory of past experience working beyond death, i.e. without applying the mind to brain traces, in a letter aimed at consoling Huygens on the occasion of his brother's death: "Those who die pass to a sweeter and more tranquil life than ours; I cannot imagine otherwise. We shall go to find them some day, and we shall still remember the past; for we have, in my view,

[27] Descartes to Arnauld, July 29, 1648, AT V, 220, CSMK, 356–7: "it is not sufficient for memory that there should be traces left in the brain by preceding thoughts. The traces have to be of such a kind that the mind recognizes that they have not always been present in us, but were at some time newly impressed. Now for the mind to recognize this, I think that when these traces were first made it must have made use of pure intellect to notice that the thing which was then presented to it was new and had not been presented before; for there cannot be any corporeal trace of this novelty. Consequently, if ever I wrote that the thoughts of children leave no traces in their brain, *I meant traces sufficient for memory, that is, traces which at the time of their impression are observed by pure intellect to be new*. In a similar way we say that there are no human tracks in the sand if we cannot find any impressions shaped like a human foot, though perhaps there may be many unevennesses made by human feet, which can therefore in another sense be called human tracks." (Emphasis mine)

Clarke (2003), 203, misses this true reverse of the argument. Clarke himself, coherent with his own empiricist reading of Descartes, deems intellectual memory a theological relic with no role in Descartes' cognitive system. Ivi, 99–105, but see the relevant remarks by Des Chene (2006).

[28] To Mersenne November, or December 1632, AT I, 263, CSMK, 40: "I am now dissecting the heads of various animals, so that I can explain what imagination, memory, etc. consist in."

an intellectual memory which is certainly independent of the body."[29] Descartes writes to Huygens about a memory that preserves memories of the past without brain traces as well as memories of particular events. It should then be different from any other kind of intellectual memory that Descartes wrote about before or later. It looks like Descartes was driven to a conjecture that his own reflection deemed impossible, with the only aim being to console his friend.

9.4 La Forge: Memory After the Treatise on Man

La Forge deals with memory at length in his notes to the *Treatise on Man*, substituting Descartes' preferred metaphors, of textile and paper, for forest or a green where vegetation becomes bowed when someone crosses it.[30] Faithfull to Descartes, La Forge thinks that memory traces are not stored in the small gland and, developing Descartes' hints, he forcefully argues that it is restrictive to place memory traces only in the brain. The whole body is a network of traces in the same way and for the same reason that the brain is.[31]

The fact that the whole organism bears material traces, i.e. memory, adds further value, according to La Forge, to material memory, as the mechanism responsible both for surprising animal behaviors and unconsciously performed human actions.[32] The iteration of bodily movements when the same brain traces are reopened, a reopening that directs bodily learning, is ascribed to memory in the *Treatise on Human Mind*, where animals best instantiate it.[33] The association of ideas is also ascribed to memory, and it is meaningful that, in the *Treatise on Human Mind*, this is exemplified in animals, as it was in the notes to the *Treatise on Man*.[34] The insistence on animals as a privileged exemplification of the mechanism of memory traces that Descartes studied in the *Treatise on Man* is an effect of La Forge's being familiar with Descartes' correspondence, an aspect that now has to be taken into account. A second effect of this knowledge on La Forge is the drastic de-escalation of the relevance of material memory to the human being in the *Treatise on Human Mind*.

---

[29] Descartes to Huygens, October 10, 1642, AT III, 578, CSMK, 216.

[30] La Forge (1999), 332.

[31] Cf. also La Forge (1999), 332.

[32] La Forge (1997), 178.

[33] Ivi, 181: "...after the memory traces have thus retraced the original species on the gland, when the spirits pass again in the same way through the same pores they flow into the same muscles and thus dispose our body to begin the same actions which it performed on the occasion of the object which stimulated it the first time. That never fails to happen in animals, and even in human beings when the power of their soul does not inhibit it." Following Descartes, *L'Homme*, AT XI, 185. Cf. also La Forge (1999), 364.

[34] La Forge (1999), 283–4. Cf. also La Forge (1999), 385-6. Cf. Bordoli (1994), 71–79.

Opening the chapter on memory in the *Treatise on Human Mind*, La Forge seems to entrust recollection to material memory:

> By the term 'corporeal memory' here I understand only a certain lasting ease to re-open those pores of the brain's ventricles which have already been opened by the spirits and in the fibers through which the spirits passed, whatever the cause which had made the opening; for by means of this ease, the pores re-open sometimes of their own accord in the same way as they had been opened the first time and do not resist the flow of spirits towards them as much as other pores, and this can cause the same species to be retraced on the gland *and the same idea to return to the mind*.[35]

Immediately La Forge specifies that, in the human mind, material memory "do[es] not seem capable of much." Material memory does not produce a human recollection, and in the animal and human unconscious actions it only plays a behavioral role:

> Although on this understanding memory is not something which is very active, and although this facility and these traces which the spirits leave in the fibres through which they have passed do not seem capable of much, I want you to realize however that it is the principal cause of all the surprising things which are observed in animals and which cause most people to attribute some thought to them, and even of most of the actions which we perform unintentionally.[36]

Recollection requires active participation on the part of the mind. In the *Treatise on Human Mind*, La Forge appropriates the Cartesian theory of intellectual memory.

Because of the complex narrative of the Cartesian theory of intellectual memory, La Forge is obliged to subdivide the Cartesian theory into four kinds of memory, three of which make up the Cartesian theory of memory: bodily memory, reminiscence—into which merges the Cartesian theory of intellectual memory from the exchange with Arnauld—, spiritual memory—the memory of thoughts we had when living when the mind is no longer joined to the body, i.e. the memory Descartes had written to Huygens about—, and, finally, the only memory which La Forge calls intellectual—the memory Descartes spoke about in the exchange with Arnauld, and to which is entrusted the recollection of purely intellectual thoughts, a memory which has to exist if, as Arnauld stressed, reasoning is based on linking thoughts that follow one another in time. So Chapter XIX, entitled (in a very Aristotelian way) "Memory and Recollection", absorbs the development of the Cartesian theory of intellectual memory, from when it appears in 1640 to the 1648 version in the exchange with Arnauld.

One point, however, is clear to La Forge. In order to account for human memory, he cannot simply refer to the presence of a mind that transduces bodily events in psychical terms, as happens with sensations. The mind is required to actively interpret brain traces. For this reason, La Forge privileges, as a common denominator all kinds of memory, that from the exchange with Arnauld:

---

[35] La Forge (1997), 178. Emphasis mine.

[36] Ibid.

> When some species re-appears on the gland it is always an effect of memory, unless the re-appearance depends completely on the object. But it is not always an effect of remembering. For in order to remember it is not enough simply to perceive a species which comes back again, *if one does not also know that this is a re-appearance* and that it is not the first time one has had this thought. Thus remembering or the power we have of recalling something consists in our faculty of recalling the original species on the gland and being aware that this is not the first occasion on which it gave us the thought which is present to the mind at the time.[37]

Now, this is possible only using "reflection and a completely pure conception without any contribution from the imagination,"[38] i.e. thanks to the intellectual memory of which Descartes writes. Because of this, La Forge thinks he can easily solve, in a Cartesian spirit, the problem Descartes left open, namely of how it is possible to speak of a memory of purely intellectual thoughts: "That is why when we use this faculty merely to recall the thoughts of purely intellectual things, it seems not to differ from understanding in the way we perceive them apart from the fact that, besides the perception of the idea, it also provides a perception of its re-appearance."[39]

Yet, it is not just the knowledge of Descartes' reflections on intellectual memory that backs the marginalization of material memory. There was, besides this, a noteworthy event: the publication of a text largely devoted to memory, the *Traité de l'esprit de l'Homme et de ses fonctions* by Pierre Chanet, in 1649—when the text of the *Treatise on Man* and Descartes' correspondence had not yet been published. It is this work that accounts for the use of the word "reminiscence" instead of "intellectual memory", which is Descartes' term.

The whole second volume of Chanet's text is devoted to memory and the third to reminiscence. The reference Chanet privileges in these volumes is Fracastoro's *Turrius*. Without mentioning him explicitly, Chanet hints at Gassendi, as a "cultivated man" who in a "fourth letter" compares "the organ of memory to a sheet of paper differently folded according to the differences of what it is meant to represent."[40] A reference to Descartes is detectable in the gesture of consent to the theory of representation without likeness, as is the case with words and things, which are linked by convention.—a theory Chanet ascribes to "another modern" author.[41]

---

[37] Ivi, 182. Emphasis mine.

[38] Ibid.

[39] Ibid.

[40] Cf. Gassendi (1642), *Epistola IV*, 194. The consonance between the comparisons of memory and a sheet of paper first made by Gassendi in the fourth Letter on *De apparente magnitudine solis* and later by Descartes in the letter to Meyssonnier, January 29, 1640, is striking. AT III, 20, CSMK, 144. Descartes re-proposes the comparison in the letter to Mesland, May 2, 1644. AT IV, 114–15, CSMK, 233: "It is rather as the folds in a piece of paper or cloth make it easier to fold again in that way than it would be if it had never been so folded before."

I am grateful for Theo Verbeek and Jan-Erick Boss' kindness and competence in pointing to Gassendi as the author hinted at by Chanet in this text.

[41] Chanet (1649), 150. To Chanet the *en passant* hint made by Descartes in the fourth *Discours* of the *Dioptrique* was enough, AT VI, 112. In relation to this, La Forge maliciously remarks that

Chanet says he is disappointed by the Aristotelian subdivision of the memory process into memory and reminiscence. The first would be a spontaneous emergence of a recollection, the second a search guided by the intellect for what one wants to recall. According to Chanet, instead, memory is only for the retention of images of things, a "room in which our phantasy let our ideas sleep, and where they stay resting until that same phantasy retrieves them."[42] This memory is not yet a conscious recollection. To yield a recollection a mental activity is required, an activity which, in the lack of a more adequate terminology, Chanet calls "reminiscence", thus indicating "any action for which memory images are worked up and represented anew to imagination."[43]

Fracastoro had maintained the Aristotelian dichotomy between memory and reminiscence: "Data at hand and immediately available, as if they had already been compared with what is under our eyes, stimulate, we think, memory. Instead data in need of being searched at some length by means of a specific logical process, stimulate, we think, reminiscence".[44] According to Fracastoro, as for Aristotle, reminiscence is an intellectual phenomenon, like reasoning.[45] Chanet, on the contrary, criticizes the way in which Fracastoro uses "reminiscence", as well as his imagining it as a kind of reasoning.[46] According to Chanet, to surface, conscious memory always requires "an action of our mind".

This activity extends "to any action for which memory images are worked up and presented anew to imagination."[47] The search is guided by cues until memories surface, i.e. the reminiscence that Aristotle and Fracastoro discuss is only one possible modality of conscious recollection, the only one accidentally requiring reasoning.[48] In "his" reminiscence, then, Chanet includes all the cases already analyzed by Fracastoro under the category of conscious memory, and he places them in four categories. In the first, the view of something lets us remember that we have had already view at some other time; in the second, the view of something associated with a particular object lets one remember the object itself; in the third, a recollection casually surfaces when one is not searching for it anymore; in the fourth, recollection is the outcome of an intellectual search evaluating similarities and dissimilarities with some images of what we are looking for. Only the last involves

---

Chanet had carefully read Descartes because he repeats his thesis without quoting him, as if those theses were his own. Cf. La Forge (1974), 169.

[42] Ivi, 191.

[43] Ivi, 195.

[44] Fracastoro (2006), 76: "Quae ergo promptissima sunt, et statim sese offerunt, ut aliàs collata cum illo quod occasionem praebet, memoriam facere dicuntur; quae vero indigent perscrutatione et discursu quodam, reminiscentiam."

[45] Ibid.: "We then call reminiscence the act which let us know again, via an inquiry, what we once already knew, but which has failed memory." ("Reminisci enim dicimur id, quod de novo per inquisitionem addiscimus, aliàs quidem notum, sed iam è memoria delapsum.")

[46] Chanet (1649), 239, 246.

[47] Ivi, 195.

[48] Ivi, 247 ff.

reasoning.[49] Chanet introduces, then, a new meaning of the word "reminiscence," identifying it with conscious recollection.

Chanet follows Descartes on a relevant point: memory traces, by themselves, are not recollection. Conscious recollection always is a product of the mind, even if the mental activity producing the recollection is not reasoning. Chanet's text seems appropriate to supply a taxonomical framework to La Forge's contribution and, above all, to show that Descartes, with the theory of intellectual memory, is not vulnerable to Chanet's analysis of the phenomenon of memory, since he is able to account for all the cases of memory Chanet individuates. This has had to be the reason why La Forge uses the Aristotelian—and misleading—term "reminiscence" to individuate the intellectual dynamics of memory sketched by Descartes in his letters, dynamics for which Descartes himself never used the term.

Drawing on Descartes' intellectual memory, "it is not difficult to explain the operation of the four kinds of remembering which Mr. Chanet speaks about in his treatise on the operations of the mind."[50] The first kind of reminiscence—remembering we have seen someone when we see him again—occurs because the same brain traces are opened more easily a second time than on the first, and "it is this ease that provides an occasion for the mind to reflect on its thoughts and to realize that it had ahead seen this person previously."[51] The second kind of reminiscence can be explained via the simple connection between brain traces. The third kind can be accounted for in various ways—either the traces are superficial or they are alike and it is then easy to take the one for the other; the fourth kind, finally, requires a thorough intellectual screening of traces and also envisages a case in which the intellect, which possesses a general notion of what it aims to recollect, pushes the imagination to survey the traces to find the particular instance of the general idea it is looking for: "This type of remembering is completely mental and depends only on the power which the mind has over the body."[52] Notice that this way La Forge succeeds in inserting the recollection of universals discussed by the *Conimbricenses* and later by Descartes (in discussion with Burman) into intellectual memory.[53] In agreement with Chanet and Descartes, La Forge claims that reminiscence is always and only a mental activity. Indeed, even the kinds of memory that most value the presence, relation, and quality of brain traces fall under the general premise according to which there is no recollection without the mind interpreting brain traces.

Already in the notes to *L'Homme* there is a distinction between memory and reminiscence, with a brief hint about the role of the mind, to which is ascribed reminiscence but not memory. In this text, La Forge, aware of Descartes' translation into physiological terms, in the *Passions of the Soul*, of the Aristotelian reminiscence, suggests that evoked reminiscence are precisely the search for memories described

---

[49] Ivi, 194.

[50] La Forge (1997), 183.

[51] Ibid.

[52] Ivi, 184.

[53] Infra, fn 12.

by Aristotle.[54] In the *Treatise on Human Mind*, instead, reminiscence is unequivocally Descartes' intellectual memory. Now, La Forge even denies the name 'reminiscence' to the voluntary search for memory traces that decidedly relates to imagination.[55]

La Forge urgently needed to tackle a problem opened and put aside when, in chapter six, he proved that the mind's essence consists in thought and, as a consequence, that the mind is always thinking. It is, indeed, on the theme of memory and on wanting memories of past experiences—especially in the prenatal phase—that, against Gassendi, Hyperaspistes, and Arnauld, Descartes had to defend the thesis that mind's essence is thinking and, consequentially, that the mind is always thinking. La Forge deals with the implicit objection by trying to bring Descartes' replies together in a coherent way. The fetus' brain is of course capable of receiving traces, but these are so mixed up that it is impossible for the mind to read them. At the same time, though mixed up, the traces are very vivid, blocking the pure exercise of the mind. La Forge collects and organizes Descartes' scattered replies as follows: the fetus' mind does not exercise intellectual memory because it is too strictly linked with the body and hindered by the vividness of brain traces; the fetus' body takes and retains traces, even deep ones, but intellectual memory (if it can exercise at all) cannot decipher them because they are too mixed up.[56] This allows La Forge, moreover, to cleverly keep two Cartesian theses together that cannot be trivially overlapped: the thesis that brain traces in the fetus are not decipherable because they are too confused and the thesis that in early childhood the mind is linked to the body so as not to be able to become autonomous from neurocerebral events. Unsurprisingly, lethargic, apoplectic, frenetic, and sleeping people are associated with infants.[57]

Finally, La Forge decides to neatly separate reminiscence from the role of intellectual memory that Descartes hints at only once, in consoling Huygens about the loss of his brother.[58] This kind of memory is called "spiritual" by La Forge. This proves that this memory, even to careful and sympathetic readers such as La Forge, looks rather dissimilar to the intellectual activity of remembering, which La Forge calls reminiscence. Reminiscence still acts on brain traces, whereas the only memory to which La Forge allots the adjective "spiritual" is the memory of pure minds, which remember ideas from the past, ideas no longer matched by a brain trace. For this kind of memory La Forge goes back to the Cartesian theory from before the exchange with Arnauld, a theory that still had intellectual memory as a recollection of thoughts not matched by brain traces: "we have an intellectual memory because our mind can recall some of its thoughts without any traces of them remaining in the brain."[59] La Forge is silent about the fact that the kind of memory mentioned by

[54] Descartes, *Passions of the Soul*, § XLII; La Forge, (1999), 322.

[55] La Forge (1997), 183–4.

[56] Ivi, 185.

[57] Ivi, 186.

[58] Infra, fn 29.

[59] La Forge (1997), p. 187. On the tensions internal to La Forge's theory of memory see Favaretti Camposampiero's analysis (2009), 390–94.

Descartes to Huygens, contrary to the recollection of purely intellectual thoughts mentioned to Arnauld, would imply remembering thoughts that, during life, were matched by brain traces.

In light of the Cartesian writings published before the *Treatise on Man* and the contributions of authors such as Chanet, who knew neither Cartesian correspondence nor the *Treatise on Man*, but offered a sophisticated theory of memory, it is understandable why La Forge so neatly claims that Descartes' writings on memory in the *Treatise on Man* are not particularly interesting when it comes to understanding human memory. In human beings, bodily memory "does not seem capable of much." At least concerning memory, the *Treatise on Man* is confined to the limits Descartes himself seemed inclined to attribute it when, writing to Elisabeth in 1645, he described it as a treatise "on the nature of animals".[60] Indeed, these are the beings that best exemplify what a mindless body can or cannot do.

Bibliography

Angelini E. Traccia e ricordo. Descartes e le due memorie. Annali dell'Università di Siena. 2000;xxi: 181–7.

Angelini E. Le idee e le cose. La teoria della percezione di Descartes. Pisa: ETS; 2007.

Baillet A. La vie de Monsieur Descartes (1691), vol 2. New York/London: Garland Publishing; 1987.

Bazán BC. Science, Expérience et Mémoire. In: Knuuttila S, Työrinoja R, Ebbesen S, editors. Knowledge and the sciences in medieval philosophy, vol II. Helsinki: Siepm; 1990. p. 127–37.

Bordoli, Roberto. 1994. Memoria e abitudine: Descartes, La Forge, Spinoza. Milano: Guerini.

Céard J. La physiologie de la mémoire, selon le médecin Jean Fernel. In: Kany-Turpin J, editor. "Jean Fernel", 2002. Corpus;41:119–33.

Chanet P. Traité de l'esprit de l'homme et de ses fonctions. Paris: chez A. Courbé; 1649.

Clarke D. Descartes's theory of mind. Oxford: Clarendon Press; 2003.

Fernel, Jean. 2003. Physiologia (1567), translated and annotated by J.M. Forrester, Philadelphia: American Philosophical Society.

La Forge, Louis de. Treatise on the human mind (trans: Clarke D). Dordrecht: Springer; 1997.

La Forge, Louis de. In: Gontier T, editor. L'Homme de René Descartes. Paris: Fayard; 1999.

Des Chene D. Review essay: Descartes's theory of mind, by Desmond Clarke. In: Garber D, Nadler S, editors. Oxford studies in early modern philosophy, III. Oxford/New York: Clarendon Press; 2006. p. 315–40.

Descartes R. Treatise on man, in Descartes, The world and other writings, translated and edited by Stephen Gaukroger. Cambridge/New York: Cambridge University Press; 2004.

Descartes R. In: Carraud V, Olivo G, editors. Étude du bon sens. La recherche de la vérité. Et autres écrits de jeunesse (1616–1631). Paris: PUF; 2013.

Favaretti Camposampiero M. Conoscenza simbolica. Pensiero e linguaggio in Christian Wolff e nella prima età moderna. Hildesheim-Zürig-New York: Olms; 2009.

Favaretti Camposampiero M. Pure intellect, brain traces, and language: Leibniz and the Foucher-Malebranche Debate. In: Garber D, Nadler S, editors. Oxford studies in early modern philosophy, V. Oxford/New York: Clarendon Press; 2010. p. 115–56.

[60] Descartes à Elisabeth, Egmond, October 6, 1645, AT IV, 310.

Fracastoro G. Turrius oder über das Erkenne/Turrius sive de Intellectione Dialogus, herausgegeben von Michaella Boenke. München: Wilhelm Fink Verlag; 2006.
Gassendi P. De apparente magnitudine solis humilis et sublimis Epistolae quatuor, Parisiis: apud Ludovicum de Heuqueville; 1642.
Gilson É. Index scolastico-cartésien. Paris: Vrin; 1979.
Joyce R. Cartesian memory. J Hist Philos. 1997;35:375–93.
Landucci S. Contributi di filologia cartesiana. Rivista di storia della filosofia. 2001;LVI:5–23.
Machamer, Peter; McGuire JE. 2009. Descartes's changing mind. Princeton-Oxford: Princeton University Press.
Minerbi Belgrado A. Uno spazio per l'inconscio? Memoria e passioni in Descartes. Rivista di storia della filosofia. 2006;LXI:837–61.
Morris J. Pattern recognition in Descartes' automata. Isis. 1969;60:451–60.
Sutton J. Philosophy and memory traces: Descartes to connectionism. Cambridge/New York: Cambridge University Press; 2007.

Chapter 10
Light and Man: An Anomaly in the *Treatise on Light*?

Gabriel Alban-Zapata

Abstract It has been quite common to talk about Descartes' text on man as if it were the second part of *The World*, following a first part which would deserve by itself to be entitled *Traité de la lumière*. Yet, from the indications given in the *Discourse on Method*, as well as in the first editions of these texts, it can be established that the text known as "L'Homme" did not correspond, in Descartes' mind, to a separated book, nor to a second and formally distinct part of his work. It is nothing but a chapter – the last one – which forms a part of a work that Descartes conceived of as an uninterrupted discourse. It can also be established that Descartes intended to entitle the whole book *Traité de la lumière*. Thus the theses on man do not follow on from the study of light: they are a part of it, and are essential for its realization. The choice of a title may first appear quite anecdotal. The present paper aims to show that it is not. Considering – or not – the text on man as the last chapter of a treatise on light implies specific ways of reading it. It seems hence natural to pay attention to a particular historical context: the history of Optics, and the relationships between Optics and Physics, especially in Kepler's works. The internal cohesion of the treatise must also be reconsidered – in particular, the key position occupied, within the chapter on man, by the theses on imagination, on memory and on the formation of cerebral "ideas".

10.1 The Title and Its Effects on Reading

The techniques of literary analysis have developed as such that we have become accustomed to paying attention to titles and not to overlook the effects they have on reading.[1] Firstly, the title has a meaning. Through conditioning the expectations of the reader, it prepares a "*mise en relief*" of the work's contents. Then it plays a role

[1] See for instance M. Roy, "Du titre littéraire et de ses effets de lecture", *Protée*, 36 (2008) 47–56.

G. Alban-Zapata (✉)
IHRIM, CNRS, UMR 5317. ENS de Lyon,
15 Parvis René Descartes, BP 7000, 69342 Lyon Cédex 07, France
e-mail: gabriel.alban-zapata@ens-lyon.fr

D. Antoine-Mahut, S. Gaukroger (eds.), Descartes' *Treatise on Man* and its Reception, Studies in History and Philosophy of Science 43,
DOI 10.1007/978-3-319-46989-8_10

in setting the work within a certain field: it enables the work to present itself to a targeted public and through displaying familiar features, makes it recognizable and helps anchor it within a certain tradition. Thus, the title is at once a semiological feature – a statement that is part of the network of statements that constitute a work – and an important factor for writing envisaged as a social or cultural practice.

In this study, we would like to begin from a simple, but often overlooked fact: Descartes' text on man is the last chapter of a treatise that "is named On light".[2] Regarding the chapter on man, the title firstly contributes to a feeling of strangeness faced with what appears – at least at first sight – as an anomaly. The anomaly is primarily internal, or structural. The list of functions that conclude his description of "the body on its own" presents a convincing argument for this:

> … the digestion of food, the beating of the heart and the arteries, the nourishment and growth of the bodily parts, respiration, waking and sleeping ; the reception of light, sounds, odours, smells, heat, and other such qualities by the external sense organs; the impression

[2] To (Vatier), 22 February 1638: "the treatise which contains the whole body of my < physics > is named On Light" (CMSK 87, AT I 562). (I use the following abbreviations for Descartes's works: AT for *Oeuvres de Descartes*, ed. C. Adam and P. Tannery, Paris: Vrin, 1964–1974; "CSM" or "CMSK" for *The Philosophical Writings Of Descartes*, trans. by J. Cottingham, R. Stoothoff, and D. Murdoch – volume 3 including A. Kenny –, Cambridge: Cambridge University Press, 1988; "G" for *The World and other writings*, ed. and trans. S. Gaukroger, Cambridge, New-York, Melbourne: Cambridge University Press, 1998. I follow the cited translations when possible; emended passages are marked in diamond brackets.) The testimony given by the letter to Vatier is confirmed by the foreword by D. R., the editor of *Le Monde... ou le Traité de la lumière* (1664a): "je n'ai vu dans l'Original que ces mots, Traité de la Lumière" (AT XI viii-ix), and by Leibniz, who writes in 1676 : « J'ai été aujourd'hui avec Mons. De Tschirnaus, pour lui donner la connaissance de Mons. Clerselier, et pour lui faire voir les restes de Mons. Des Cartes. (…) Il y a encore un traité de la lumière. Voilà son titre. Mais le traité même est ce que Mons. Des Cartes appelle son Monde, ou Méditations physiques, faites, comme les Métaphysiques, d'un style familier, quoiqu'elles ne disent en substance qui ce qui est dans ses Principes philosophiques » (AT XI, 661–662). They are, what's more, some of the first words of the text: "In putting forward an account of light..." (G 3, AT XI 3). We can therefore not consider that Descartes hesitated between two titles, as has been suggested by J.-P. Cavaillé (*La fable du monde*, Paris : Vrin, 1991, 45). On this point, see A. Nardi, "La luce e la favola del Mondo, Descartes 1629–1633", *Annali dell'Instituto di Filosofia*, 3, Florence: Olschki, 1981, 104, n.1. We should otherwise note that Descartes frequently references "my World" (*mon Monde*) in *The Correspondence*, this is not strictly speaking about the *Treatise on Light*. The *Treatise on Light* contains the World of Descartes, or at least "something like an abridged version" of it ("*quasi un abrégé*") (to Mersenne, 25 November 1630, AT I 179). But this does not constitute its only possible form. Also, the project of a 'World' appearing as of October 1629 (to Mersenne, 13 November 1629), does not take the form of a *Treatise on Light* until one year later (to Mersenne, 25 November 1630). Similarly, in 1642, Descartes will leave aside the specific format of a *Treatise on Light* to favor a *Summa philosophiae* better adapted to the polemic context which motivated the publication of his "World" (to Huygens, 31 January 1642, AT III, 782). On the development of Cartesian Physics between October 1629 and November 1633, see G. Cantelli, *L'Uomo*, Torino: Boringhieri, 1960, 20–23; A. Nardi, *art. cit.*, 103–145; D. Garber, *Descartes' Metaphysical Physics*, University of Chicago Press, 1992, 16–20; A. Bitbol-Hespériès' Introduction to *Le Monde; L'Homme*, Paris: Seuil, 1999, v–viii; or S. Gaukroger, *Descartes' system of natural philosophy*, Cambridge, New-York, Melbourne, Madrid, Cap Town: Cambridge University Press, 2002, 10–24. Note that, from our point of view, most of the reproaches expressed by Nardi against Cantelli's interpretation of the letter on November 1630 are not relevant.

> of the ideas of them in the organ of common sense and the imagination, the retention or imprint of these ideas in the memory, the internal movement of the appetites and the passions; and finally the external movement of all the bodily parts…[3]

There is nothing surprising in the fact that it is a question, in a *Treatise on Light* of "the reception of light" in the external organ of vision but it so happens that it is also a question of the "impression" of the idea of light in the organ of common sense and of imagination and its retention in memory. Then, Descartes deals with the reception of other qualities (sound, smell, etc.) and with the ideas of them. Lastly, and most importantly, he treats "the digestion of food, the beating of the heart and the arteries, the nourishment and growth of the bodily parts, respiration, waking and sleeping, … the internal movement of the appetites and the passions, and the external movement of all the bodily parts". What does all of this have to do with the light?

Furthermore, the title invites us to situate the treatise in the history of optics. Descartes is familiar with the work of Witelo, and likely with that of Alhazen.[4] Above all, he had read Kepler, whom he called his "first master" in optics.[5] In May 1630, Descartes undertook the project of a French *Dioptrics*, or study of optics. Like Kepler, he felt the need to begin from a speculative examination of the nature of light. In Kepler, it was the role of *Paralipomena to Witelo* (1604) that preceded his Latin *Dioptrics* (1611).[6] In Descartes, this position is held by the "discourse on light", that was included in the first draft of the *Dioptrics*,[7] before being separated and replaced by more summary explanations.[8] It is taking into consideration this anchoring in optics that reveals the historical dimension of the anomaly of the *Treatise on Light*: optical studies, to this point, had not been in the habit of examining the digestion of meat.

---

[3] G 167, AT XI 201–202.

[4] Cf. F. Risner, *Opticae thesaurus. Alhazeni Arabis libri septem nuncprimum editi. Eiusdem liber De crepusculis & nubium acensionibus. Item Vitellionis Thuringopoloni libri X...*, Basileae: per Episcopios, 1572. On Descartes' optical sources, see A. I. Sabra, *Theories of light from Descartes to Newton,* Cambridge, New-York, Melbourne: Cambridge University Press, 1981, 72; A. M. Smith, "Descartes' Theory of Light and Refraction: A Discourse on method", *Transactions of the American Philosophical Society* 77 (3), Philadelphia: The American Philosophical Society, 1987, 8–12.

[5] To Mersenne, 31 March 1638, AT II 86.

[6] "I thought it a good idea to look a little more deeply into (*paulo penitius inspicere*) the whole nature of light (*totam lucis naturam*), and to relate to their principles those things that appear (*ea quae apparent*), insofar as possible at present..." (D 16, MA 4). For this work, I use the following abbreviations: "D" for *Optics: Paralipomena to Witelo & Optical part of Astronomy*, edited and translated by W. H. Donahue (Santa Fe: Green Lion Press, 2000); "MA" for *Ad Vitellionem Paralipomena...* (Francofurti: C. Marnium & H. I. Aubrii, 1604). On the respective places of *Paralipomena* and *Dioptrice* in Kepler's thought, see Ph. Hamou, *La mutation du visible. Essai sur la portée épistémologique des instruments d'optique au XVIIe siècle. Volume I: du* Sidereus Nuncius *de Galilée à la* Dioptrique *cartésienne* (Villeneuve d'Ascq: Presses universitaires du Septentrion, 1999) 203–206.

[7] Cf. to Mersenne, 25 November 1630, CMSK 28, AT I 178–179.

[8] These are the first six parts of the *Dioptrics* published in 1637.

10.2 Man as "Spectator"; Man in General

Faced with the problem that the chapter on man is disproportionate, editors and commentators have generally opted for the solution of isolating it from the main body of the optical treatise to which it does not seem to be able to be assimilated. Let us consider two examples.

1. Adam and Tannery's edition of the text (in AT XI) oscillates between two ways of presenting its contents. On the title page, the treatise is presented as a whole (the chapter on man included) under the title *Le Monde de René Descartes ou Traité de la lumière.* This dual title is taken from the 1677 edition. The text on man appears here as the continuation and conclusion of a treatise that includes the name *De la lumière* (On light) in its title. Editors exploit this perspective when annotating the section of the sense of sight, in which Descartes evokes his "subject": "*c'est-à-dire l'explication de la lumière. Nouvelle preuve que ce traité n'est que la suite du precedent*".[9] However, in the foreword of this edition of the text, things are shown from a different angle. It is here a matter of two treatises (the *Treatise on Light* and the *Treatise on Man*) that form the ensemble of *The World.* This second approach in separating the two is evoked by the titles offered in the chapter headings throughout the AT edition: "*Le Monde – Traité de la lumière*" (for chapters 1 to 15), "*Le Monde – Traité de l'homme*" (for the last chapter).
2. A note offered in the recent French edition of *Discours de la méthode* in the *Oeuvres complètes*, regarding the Cartesian decision to study man while treating light, "because he is the spectator of it":

   L'évocation de l'homme comme "spectateur" du monde physique vise à renforcer l'unité du traité, en laissant au second plan l'ambition propre à la physiologie cartésienne : l'explication de toutes sortes de fonctions internes à partir d'une description de la "machine" de notre corps.[10]

As noted here, there seems to be some kind of tension between the coherency of the *Treatise on Light* and Descartes' intention to develop a general physiology. To strengthen the unity of the treatise, it would be necessary to leave this intention aside; to argue for it instead means that the unity of the treatise is of lesser importance. By engaging in the explanation of all kinds of internal functions Descartes would have transgressed the limits of a decision to present his anthropology while studying light. Conversely, if insisting on the unity of his work, the main argument of his physiology should have been parenthetical.

---

[9] AT XI 151, note b.

[10] BK III 648, note 334.

10.3 The First Foundations of the Anthropology of the *Treatise on Light*

In November (or December) 1632, Descartes wrote to Mersenne: “My discussion of man in < my World > will be a little fuller than I had intended, for I have undertaken to explain all the main functions in man”.[11] This important testimony allows us to situate the anomaly in the history of his composing the chapter on man.

Initially, and for a long time, Descartes did not intend to provide a general description of human nature. Here we need to look back on the letter of 25 November 1630 in which Descartes announces his intention to replace the “complete physics” or “World” – a promise made to Mersenne some months earlier, with a “Discourse on light” conceived of as a substitute for it.[12] In these circumstances, “fearing that [he] could not put everything [he] had in mind into [his] discourse”, he favors the expression of his concepts related to light. This priority implies a principle of limitation, which, *a priori*, applies as much to the knowledge of man as it does to that of bodies that make up the exterior world:

> …I undertook merely to expound quite fully what I understood about light, < and >, as the occasion arose, < to > add something about the sun and fixed stars, because almost all light comes from them; about the heavens, because they transmit light; about planets, comets and the earth, beacause they reflect light; about terrestrial bodies in particular, because they are either coloured or transparent or luminous; and finally about man, because < he is the spectator of > it.[13]

At this stage, it is not a matter of explaining “all the main functions in man”, but only those that are in some way related to light. What Descartes intends in the chapter on man is an explanation of the sensation of light. Moreover, it is using these terms that he announces, at the end of Chap. 13, after having presented his conceptions about the nature of light:

---

[11] CMSK 40, AT I 263.

[12] CMSK 28, AT I 179. See also: to Mersenne, 15 April 1630, CMSK 21, AT I, 137.

[13] *Discourse on the Method* V, CMS I, 132, AT VI, 42. Note here that Descartes does not present himself as a painter *(peintre)* of “the nature of material things” but as a painter of *ideas* previously acquired about his subject. The analogy with painting relates to writing as understood to be a material expression of thought (“put into... discourse”). The philosopher that undertakes to present a discourse on the totality of what he has “in mind” (*à la pensée*) on the subject of physics is (or at least, he fears to be) similar to the painter who tasks themselves with representing on a flat surface all of the sides of a solid body. Elsewhere, the fear of not being able to say everything takes its meaning in regard to his personal preoccupations – Descartes feared lacking time, he was unsure whether God would leave him enough to “complete” (*venir à bout*) his task (CMSK 14, 41, AT I 85, 271) – but also in regard to the consideration shown towards the public that Descartes was striving to please. See his use of the paradigm of chiaroscuro at the end of Chap. 8: “I do not promise to set out exact demonstrations of everything I say. It will be enough for me to open up the way for you to find them yourselves, when you take the trouble to look for them. Most minds lose interest when one makes things too easy for them. And so as to present a picture which pleases you here, I must use shading as well as bright colours” (G 32, AT XI 48).

> And so, if it were the eye of a man that was at point E, it would really bu pushed, both by the Sun and by all the celestial matter beween the lines AF and DG.
>
> Now one must know that the men of this new world will be of such a nature that, when their eyes are pushed in this fashion, they have a sensation (*sentiment*) very similar to that which we have of light, as I shall explain more fully below.[14]

The role of the chapter on man is to develop this argument. Questioning the "nature" of the men of this new world is subordinate to the examination of a specific problem: to show how these men must be made so that the mechanical action that is taken for light (in the new world) causes them something similar to the feelings that "we" have of light (in the old world). Descartes had this in view, perhaps, when in June 1632, he informed Mersenne that "I have finished all I had planned to include in it concerning inanimate bodies" and that "It only remains for me to add something concerning the nature of man".[15]

Around June 1632, Descartes began to write the chapter on man. What then would be his point of departure? The state of affairs as expressed in the letter of November 1632 gives rise to only two possibilities: "I have already written of the vital functions, such as the digestion of food, the heart beat, the distribution of the nourishment, etc., and the five senses".[16] Why would Descartes suddenly turn his attention to vital functions? Is it not more plausible for him to begin by resuming the explanation of the light that he left open at the end of Chap. 13, that is to say, through human eyes? Thus, the proposed explanation about the sense of sight echoes directly the pronouncement of Chap. 13.

> There still remains the sense of vision, which I must explain a little more precisely than the others because it is more central to my subject. This sense < also > depends in this machine on two nerves, which must certainly be made up from many tiny fibres, as fine and as easily movable as they can be, for their role is to report to the brain the different actions of the parts of the second element, which, following what we said earlier, will enable the soul, when united with this machine, to conceive the different ideas of colours and light.[17]

From the beginning, Descartes does not go straight to the point (regarding an explanation of the sensation of light). He begins further on, through reinscribing the aforementioned explanation within the heart of a broader explanation of the various sensory qualities. It is thus a question of color, which, like light, belongs uniquely to the sense of sight. But mostly, it a question, *prior to that of light and color*, of the ensemble of qualities that do not belong to the sense of vision (tactile qualities, tastes, smells, sounds)[18]; and *after light and color*, of the ensemble of qualities "which are not related to one sense like those we have spoken of up to now".[19]

The developments on the senses of touch, taste, smell, and hearing make it possible to lay the foundations for a general theory of the senses, that may then be used

---

[14] G 62, AT XI 97.

[15] CMSK 39, AT I 254.

[16] CMSK 40, AT I 263.

[17] G 124, AT XI 151.

[18] G 118–124, AT XI 143–151.

[19] G 131–139, AT XI 159–163.

to complete the explanation left open at the end of Chap. 13. This theory rests upon three pillars.

1. A general explanation of the sensory nerve functions. Descartes describes a pressure that pulls "tiny fibres that make up the marrow of the nerves", and through this traction, the information collected at one end (external sense) is instantly transmitted to the other (internal surface of the brain).
2. A thesis, or rather a series of theses on the union of soul and body:

> Now I hold that when God unites a rational soul to this machine, as I intend to explain later on, He will place its principal seat in the brain and will make its nature such that the soul will have different sensations depending on the different ways in which the nerves open the entrances to the pores in the internal surface of the brain.[20]

3. The demonstration of a principle that allows you to specify the different senses: the ease with which the nerves can be moved.[21] The sense of sight is thus only distinguished from the other senses by a difference of degree: the optic nerves can be moved more easily than others. This difference enables them to be moved by "the different actions of the parts of the second element", in other words – according to the definition advanced in Chap. 13 – by light. The circle seems thus to be closed.

Chapter 13 saw Descartes advance onto a mechanical explanation of light. The recourse to the theory of elements overcame the Keplerian hypothesis of an immaterial movement of light,[22] and brought the laws of optics to those of physics. The light is only one peculiar manner through which the body can interact under the universal laws of movement.[23] The chapter on man exploits the result of this important shift. The opposition introduced by Kepler between light and matter, *leges opticis* and *rationes physicis*, had led, ultimately, to an aporia. Between vision understood as an optical phenomenon (in the narrow and literal sense), and vision understood as a concrete or psycho-physical phenomenon (with the persistence of vision and polarisation), according to Kepler there is a gap that the human understanding is not able to fill. The optician must continue until the retina (without stopping at the lens) but he should not venture beyond (into the nerves and the brain). He "[leaves it] to

[20] G 119, AT XI 143.

[21] "The tiny fibres (…) which serve as the organ of taste in this machine, can be moved by slighter actions than those which serve for touch in general" (G 120, AT XI 145); the ones which serve as the organ of smell "can be moved by even smaller terrestrious parts than those of the tongue" (G 121, AT XI 148); the ones which serve as the organ of the hearing "can be easily moved all together and in the same way, by the little blows with which the external air pushes against a certain very fine membrane" (G 122, AT XI 149). Finally, the optic nerves "must certainly be made up from many tiny fibres, as fine and as easily movable as they can be" (G 124, AT XI 151).

[22] Concerning the debates on materiality or immateriality of light in the early Seventeenth Century, see C. Chevalley, "Sur le statut d'une question apparemment dénuée de sens : la nature immatérielle de la lumière", *XVIIe siècle* 136 (1982), 257–266.

[23] On this new theory of light, cf. A. M. Smith, *op. Cit.*, 13–19, 32–46 ; for a more precise analysis on the formation of this theory, see J. Schuster, *Descartes-Agonistes. Physico-mathematics, Method & Corpuscular-Mechanism 1618–1633*, Dordrecht, New-York: Springer, 2013, 184–214.

the natural philosophers (*Physici*) to argue about" the rest, that is to say, about "the impression (which) itself is not optical, but physical and mysterious (*physica et admirabilis*)".[24]

From Alhazen to Porta, opticians and anatomists projected physiological concepts into the field of optics (fixation of species in the crystalline humour) and optical concepts into the field of physiology (rectilinear propagation of light in the nerves). Kepler puts an end to this confusion of genres. However, by dematerializing the movement of light, he deprives himself the means to achieve a full explanation of vision. Inversely, once light has been subsumed under the laws of movement, Descartes can articulate his theory by a description of physiological mechanisms and metaphysical conditions that contribute to making us see.[25] Without reintroducing light into the nerves, he commits to an understanding of the nervous system as a necessary base for a study of optics.[26]

It is still a matter for the section on the sense of sight to provide a description of "the structure of the eye" on which the formation of images in the base of the eye depends.[27] This description enables him to record the achievements of the Keplerian theory (identification of the function of the refraction lens) while supplementing it with new observations. Descartes keeps it concise: "I shall do so briefly, omitting many superfluous details which the curiosity of anatomists has uncovered here".[28] As elsewhere in "L'Homme",[29] Descartes differentiates his project from that of the

---

[24] D 180–181, MA 168–170. See also the conclusion to Chap. 1 on visual persistence and the end of Chap. 5 on other optical illusions (D 234–236, MA 219–221).

[25] Frédéric de Buzon notes: "Kepler refusait de donner une explication précise de la manière dont les impressions sensibles sont transmises de la rétine au cerveau, où est localisé le sens commun, jugeant cette propriété occulte ou obscure. C'est précisément là que Descartes déplace à nouveau le problème, en l'associant à une psychologie d'un côté et de l'autre côté à une explication purement mécanique du transport des images dans le corps" (introduction to *La Dioptrique*, in *Oeuvres complètes*, ed. J.-M. Beyssade and D. Kambouchner, t. VI, Gallimard, 2009, 144–145). See also G. Simon, *Structures de pensée et objets du savoir chez Kepler*, Service de reproduction des thèses, Université Lille III, 1979, 562–572; Ph. Hamou, *op. cit.*, 274–278.

[26] Descartes insists on the fact that the "detours" through which the actions that move the nerves pass do not prevent these actions from being "easily communicated from one end to the other": in such a manner that the reader himself must not be prevented by them from "seeing clearly how the ideas of objects that strike the sense are formed" (G 146, AT XI 174). He is probably thinking here, in the criticism addressed by Kepler to Witelo on the subject of the crossing of optical nerves: "what can be pronounced by optical laws about this hidden confluence (*hoc occulto commeatu*), which, since it goes through opaque, and therefore dark, parts (*per opacas, ideoque tenebrosas partes*), and is administrated by spirits, which differ entirely in kind from humours and other transparent objects, has already completely removed itself from optical laws?" (D 180, MA 169).

[27] G 124–131, AT XI 151–158.

[28] G 124, AT XI 152. Compared to *Dioptrics* III: "je laisse à dessein plusieurs autres particularités qui se remarquent en cette matière, et dont les anatomistes grossissent leurs livres ; car je crois que celles que j'ai mises ici suffiront pour expliquer tout ce qui sert à mon sujet, et que les autres que j'y pourrais ajouter, n'aidant en rien votre intelligence, ne feraient que divertir votre attention" (AT VI 108).

[29] See the introduction to the body (G 99–100, AT XI 120–121), and the introduction to the description of intracerebral movement within animal spirits (G 140, AT XI 165–166).

anatomists. Already, Kepler had highlighted the lack of mathematical culture by the physicians of his time[30] and their detrimental influence on optics.[31] Descartes makes a similar reproach, which makes sense in the light of one of the important precepts of his method: the excess of anatomical knowledge – a knowledge as majestic as it was useless – came from the fact that the anatomists did not take care "not to assume more than the data and not to take the data in too narrow a sense".[32] Descartes refuses to yield to the "curiosity" that fed the major medical anthropology of his time, from Vesalius to Du Laurens, by way of Fernel and Riolan. His interest in the description of anatomical structures is guided by the prior identification of a specific problem: to show how "these men" must be made so that the action exerted by the luminous body on their eyes caused a similar feeling to that that we have of light.

By the end of the description of "the structure of the eye", the explanation of the sensation of light promised in Chap. 13 could be considered an achievement. However, Descartes did not stop there, and delivered a series of dense explanations on the qualities "common to touch and vision and even in some respects to the other senses". At once detailed and restructured within the *Dioptrics*, these explanations are organized in two stages. Descartes shows how the soul perceives these qualities before establishing that "it can often happen that the soul is mistaken in this". Opticians and psychologists commonly distinguished what is visible per se (light, color) from other qualities known through intellectual (or quasi-intellectual) operations of the rational (or "estimative") soul.[33] Yet, the treatment that Descartes offers is original for multiple reasons.

First, he refuses – as was the case already for qualities in themselves – to treat vision independently of the other senses, and ultimately, of touch. We find the same argumentative formulation taken up three times: "if the hand touches […] Similarly, if the eye … is turned…". Every time Descartes leaves from common qualities of objects we touch to explain, by analogy, the common qualities of vision.

Then, the section on common qualities extends and strengthens the explanation of the sensation of light. Descartes had established that the mechanical variations suffice to cause sensations of light and of color in the soul. It might seem absurd that mechanical variations would not be perceived as such. It was also commonly con-

[30] Regarding Platter: "now compare the true mode of vision proposed by me with that of Platter: you will see that the illustrious gentleman was not farther from the truth than befits one of the medical profession, who is not deliberately treating mathematics" (D 223, MA 208).

[31] He thus reproaches Witelo for letting himself be influenced by the *Physici* (this is to say, in the classical vocabulary, the *Medici*) on the subject of the nutritive function of the vitreous humor, refusing for his part to engage in the argument (D 219, MA 204). See also: D 235, MA 220.

[32] *Rules for the Direction of the Mind*, XIII. See the example of: "the bowl... which had a column on the top of which was a figure (*effigies*) of Tantalus": despite appearances, this *effigies* "is merely a coincidental feature and by no means a factor which defines the problem" (CMS I 54–55, AT X 435–436).

[33] See D. C. Lindberg, *Theories of Vision from Al-Kindi to Kepler,* Chicago, London: University of Chicago Press, 1976; A. I. Sabra, "Sensation and inference in Alhazen's theory of visual perception", in P. M. Machamer and R. G. Turnbull (eds.), *Studies in Perception: Interrelations in the History of Philosophy and Science*, Columbus: Ohio State University Press, 1978, 160–185.

sidered in the Schools that "*(si) les seules figures eussent été suffisantes pour faire connaître toutes les autres qualités, la Nature ne nous eût point donné le sentiment de ces autres qualités*".[34] This objection consists of insisting on the fact that if the mechanistic hypothesis was proven, then we should perceive only the shape of things; we should see neither light nor color. To combat the realist bias, Descartes ventures into the terrain of common qualities to make of it a textbook case. In showing that, even in the case of common qualities (shapes and other properties of space), physiological determinations do not need to resemble the sensations they cause, he accredits the explanation delivered *a posteriori* to the subject of the perception of qualities in themselves.

Lastly, from the section on common qualities the problem emerges, in two stages, of a *quasi* rational knowledge inherent to sense perception. The six first explanations concern qualities which are known in a primitive way: (1) position, (2) shape and (3) size of what is touched, and (4) position, (5) shape and (6) distance (of the central point) of what is seen. Just as in the case of the perception of qualities which depend on one sense, the soul is naturally made in such a way as to sense these qualities from the physiological changes that they correspond to. However, when it comes to common qualities, the psychophysiological connection is based on a code to which it relies on human reason to impose logic. The soul that perceives distance to the central point occupying the field of vision behaves as if it had an innate knowledge of the laws of refraction and of the mechanism of the eye explained above[35]: sensing that the lens becomes more "flatter", it judges, or at least seems to judge that the perceived point is farther, and when the lens becomes "more arched", it concludes or seems to conclude that the perceived point is closer. Then come five explanations that do not concern primitive but derived knowledges: (7) distance from (the peripheral points of) what is seen; (8) size and (9) distance of what is seen; (10) distance from what is touched (with the help of two sticks) and (11) from what is seen (with the two eyes). Here, the perception depends on a judgement (or a *quasi* judgment) that implies a plurality of data, and therefore a form of reasoning. Thus, to perceive the distance of peripheral points, you must: (1) know the distance from the central point, (2) take into consideration the differences in sharpness, (3) take into consideration the differences in intensity so as to interpret the differences in sharpness in terms of proximity or distance, bumps or hollows. Likewise on the topic of the perception of a quality through the means of knowledge of other qualities, but especially on the topic of the almost natural geometry ("*comme par une géométrie naturelle*") implied by the binocular perception of distance.

The first part of the explanation does not pose any problem in particular. The soul behaves, no doubt, as if it knew and used the laws of refraction, when it apprehends the distance from the point which occupies the center of the visual image. But it is

[34] P. Chanet, *Traité de l'esprit, de ses connaissances et de ses fonctions,* Paris: Jean Camusat & Pierre le Petit, 1649, 147: "(if) the shapes themselves were sufficient to know all the other qualities, Nature would not have given us the sensation of these others". Here the argument against Gassendi is formulated.

[35] G 128, AT XI 155–156.

unnecessary that it provide this knowledge. It suffices that it is made (by God) in such a way as to sense distances due to variations on the crystalline. Here, the soul has no need to calculate, because God takes care of this in joining it with the body. On the contrary, the second part of the explanation does pose a problem: here, the sensation does not depend directly on the physiological effect caused by the action of an exterior body. It depends on inferential operations thanks to which the soul is able to derive some new qualities from those ones which have already been known. Then, shall we consider that the soul is expressly reasoning while perceiving such qualities? In the *Treatise on light*, Descartes doesn't provide a clear answer to this question. But while treating of the binocular perception of distance, he takes note of the problem, that is to say, of the insufficiency of a purely intellectualist explanation of derived perceptions. Here, it becomes evident that there can not be a reasoning, but only something like a natural reasoning of perception ("as if by a natural geometry").[36]

Consequently, it is a matter of knowing what, in the mechanical nature of man, will be able to take the place of a reasoning that the soul does not need to expressly provide. How does the inferential aspect inherent to derived perceptions translate on the uniquely physiological plane? To this question, the sixth part of the *Dioptrique* will, in retrospect, provide a clear answer. Significantly expanding the explanation of the binocular apparatus, Descartes invokes "a mental act which, through only a very simple < imagining >, involves < a reasoning > quite similar to that used by surveyors…".[37] Coming back to the problem in the *Replies*, while dealing with the "third degree of sense" (*tertium sentiendi gradum*), Descartes clearly moves away from a purely intellectualist explanation by making an important distinction between "judging and reasoning" (*ratiocinamur et judicamus*) about the things that are present to our senses, and "remembering the judgments already made by us" (*judiciorum jam olim a nobis… factorum recordamur*) about things that are "similar" (*rebus similibus*) to the present ones. In this context, he rejects the invocation of "custom" as a principle (*propter consuetudinem*) that would make our reasonings "so quick" (*tam celeriter*) that we could no more notice them. Habituation does not lead from noticed reasonings to unnoticed reasonings, but from unnoticed reasonings

---

[36] See G. Hatfield, "Natural Geometry in Descartes and Kepler", *Res philosophica,* 92, 1, January 2015, 117–148. The author goes against the consensus of interpretation according to which binocular perception of distances would be assured for Descartes by the judicative faculty of the rational soul (see, for example, Nancy Maull, « Cartesian Optics and the Geometrization of Nature », in *Descartes: Philosophy, Mathematics and Physics*, edited by S. Gaukroger, Totowa NJ: Barnes and Nobles Books, 1980, 23–40). Taking into consideration the cerebral physiology presented in "L'Homme", G. Hatfield proposes "an alternative reading of Descartes account of the triangle of convergence, according to which brain mechanisms carry out any "computations" or responses to the physiology of optical convergence and then cause a perception of direction and distance in the mind, without any underlying mental computations or other cognitive processes" (p.119).

[37] CMS I 170, AT VI 138.

(the rational operations effectuated during childhood weren't reflected) to recollections of reasonings.[38]

Thus, the section on common qualities provided in the *Treatise on light*, which, on one side, is built on the explanation of light, calls on the other side of explanations relating to "imagination, memory, etc." (letter of November 1632), that is to say, on the topic of mechanical associations that are likely to take the place of reasonings that the soul does not expressly provide when perceiving external objects. In doing so, it drives the passage towards the theory of ideas and the second foundation of the anthropology of *Treatise on Light.*

10.4 "My Discussion of Man...Will Be a Little Fuller Than I Had Intended"

The first foundation enabled Descartes to assert a three-term schema: (1) properties of exterior bodies, (2) properties of nerves, (3) properties of the soul. He could have stopped there. This is what he will do in later revisions of his theory on sense perception.[39] But in the *Treatise on Light,* he undertook to go further and complete the

[38] AT VII 438: "*(ea) quae vero a prima aetate, eodem plane modo atque nunc, de iis quae sensus nostros afficiebant judicavimus, aut etiam ratiocinando conslucimus, referamus ad sensum, quia nempe de iis tam celeriter propter consuetudinem ratiocinamur et judicamus, aut potius judiciorum jam olim a nobis de rebus similibus factorum recordamur, ut has operationes a simplici sensus perceptione non distinguamus*". According to G. Hatfield (*art. cit.*, p.41), the status of the act of the imagination evoked by Descartes in the *Dioptrics* remains ambiguous: "mechanically effected act of imagination", or even "unnoticed cognitive act of calculation" – as would have the Latin translation of 1644, which proposed "*licet simplex judicum*" for "*une imagination toute simple*"? But the qualification of this operation ("*une imagination* toute simple") and the illumination provided by the *Treatise on Light* and by the *Replies* seem to be sufficient to lean in favor of the first hypothesis. We should need to consider then that the physiological theory of "traces" plays a key role: it enables the introduction, within a pure mechanistic approach, of compositions and associations of ideas. See AT XI, 179, 184, 197–198, and J. Sutton, *Philosophy and Memory traces: Descartes to connectionnism*, Cambridge: Cambridge University Press, 1998, 50–66. Note that on this subject in "L'Homme", the passage on binocular perception does not entail any reference to the imagination, no remark on the possibility of perceiving distances in movement with only a single eye, and no comparison with surveyors. In this, the proposed rewriting in the *Dioptrics* is paradoxical: Descartes does not present there his mechanical explanation of the imagination; but this does not prevent him from therein drawing consequences, in modifying his explanation of the triangle of convergence in such a way as to insist on the role of time, of habit and certain knowledge (mechanical) acquired in the past. Gary Hatfield cites the passages where Kepler evokes the role of habit in connection with his intellectualist explanation of the triangle of convergence (*art. cit.*, p.120–123), and he seems to consider that it is the same in Descartes. Habituation would be considered by judgements and inferences that should need to be necessarily attributed to the rational soul (p.137, p.140). But is what is true for Kepler equally true for Descartes? Does not the physiological explanation of memory (*vestigia*), which is the foundation for a psychological theory of the imagination and of ideas enable us to think of an acquisition (motor but also mental) which does not imply any intervention of the rational soul as active principle?

[39] Cf. *Dioptrics*, IV-VI, and *Principles of Philosophy* IV, art.188–199.

schema, by introducing, between nervous modifications and mental modifications, a fourth term: intracerebral modifications. In the section on "ideas", taking up the explanation of sensation where he left off in dealing with the senses (inner surface of the brain), Descartes delivers the final step: that which leads to the pineal gland.[40] Before resolving to explain "all the main functions in man", Descartes had likely planned to venture beyond the inner surface of the brain, in the field of intracerebral movements; and this, in order to explain if not the integrality of the "internal senses" of the medical tradition, at least the most primary of them, and the most directly concerned by his study: the "common sense". However, he had in all likelihood not expected all that would imply.

To make sense of the decision announced in the letter of November 1632, we need to consider how the chapter on man is organized. As highlighted at the beginning of this study, we can discern two main explanatory approaches: the explanation of sensations, and the description of the machine in general. The first approach takes its roots in certain previous developments in the *Treatise on Light* (Chaps. 1 and 13). In "L'Homme", Descartes tasks himself with describing the senses. It is the eyes that here constitute the point of departure – from them, the nerves, the internal surface area of the brain, the pineal gland and the traces of memory: this is the approach taken in describing the stages of representation. The second approach, to the contrary, does not directly respond to the leading problematic within the *Treatise on Light*. Here, it is the digestion of meat in the stomach that he takes as point of departure; then it becomes a matter of following the movement of blood and of spirits, explaining at each moment the functions (vital and motor) that depend on them.

To understand why had Descartes felt the need to add a second approach to the first, let us turn our attention towards the end of the chapter, namely, the section on the intracerebral course of animal spirits that, alone, play a role in both scenarios. In this section, the two explanatory approaches – starting from the eye and then from the stomach – finally find a coherency. The same properties (the principles of cerebral physiology)[41] that serve to complete the explanation of sensations (transportation of shapes onto the pineal gland and the formation of ideas)[42] can also be interpreted on a strictly physiological level (distribution of animal spirits in the appropriate nerves, and production of motor responses adapted to the circumstance).[43] We can thus come back to our question, but in a way as to pose it, more specifically, within the scope of the section relating to the intracerebral course of spirits: why is he not content to explain ideas following on from his explanation of the senses? Why connect the psychophysiology of ideas to a pure physiology of motor functions tied to the intracerebral course of spirits?

The first reason is that the explanation of motor functions tied to the intracerebral process had been widely neglected. From Avicenne to Chanet by way of Eustache Saint-Paul, the properties of the brain had at first and foremost been interpreted,

---

[40] G 146–160, AT XI 174–189.

[41] G 140–146, AT XI 165–174.

[42] G, 146–160, AT XI 174–189, and G 165–166, AT XI 197–198.

[43] G, 160–165, AT XI 189–197.

in the *Ecoles*, through a psychological perspective. As to what role the brain plays in the production of muscle movement, the invocation of "*appetitus sensitivus*" was common – enlightened by the knowledge of internal senses, and using its potential to deliberate on the motor response that needs to be adopted -. By also invoking a judicative principle of animal movements, the Schlolastics renounced the knowledge of effects by their efficient causes: because it feels pain or pleasure, or that it estimates what it feels good or bad, the appetite determines whether the spirits course towards certain muscles over others. On the contrary, in Descartes, the same physiological properties can be interpreted on the cognitive plane just as well as on a strictly motor one. The realisation of movements adapted to circumstance is thus not conditioned by the prior acquisition of knowledge over a sensed object. We must moreso consider that the sensation emerges – in beings endowed with a soul – at the site where particular strategies of the motor response develop. The sensation is not what logically precedes movement, but that which accompanies a movement in its nascent state.

But also, and more importantly, the Cartesian explanation of ideas lacks the experiments that could serve as its foundations. This is an often-made reproach in regards to the cerebral physiology presented in "L'Homme", and soon after exploited by Regius, by La Forge or by Malebranche; a reproach that we find just as much in certain anatomists of the brain during the second half of the seventeenth century as we do in Hume's refusal to invoke "traces" and "animal spirits", and in brief, to enter into "an imaginary dissection of the brain".[44] We believe that Descartes was aware of this issue, illustrated by the fact that the only work in which he presents the principles of cerebral physiology at length is the one he designates as "my World", that is, the work in which physics is presented by means of a fable. Steno insisted on this point.[45] And yet, the recourse to the fable, if enabling hypotheses that are insufficiently founded in experience, does not suffice to resolve the problem. It is one thing that speculation is tolerated: but it is still necessary for it to offer something attractive to a reader of the old (or real) World. It is thus evident what can motivate the decision to treat "all the main functions" in man in the midst of a *Treatise on Light.* Hume reproaches Cartesian theory for its *ad hoc* character: they claim to explain the psyche with reference to the physical, when in reality, this physical, they only know via a fiction conceived from the observation of psyche: thus committed to a vicious circle. Yet, Descartes seems to have sought to specifically avoid such a circle. To separate the explanation of ideas from those of motor functions tied to the course of spirits would be to expose that theory to a worthy suspicion: why these hypotheses and these principles over any others that would have the potential to be imagined? On the contrary, through describing the motor

[44] D. Hume, *A Treatise on Human Nature* I: *Of the understanding* (1739), II, 5, ed. T.H Green and T. H. Grose, London: Longmans, Green and co, 1878, 364. For the Humean reception of "L'Homme", see the contribution made by C. Gautier in the present work.

[45] On the critique addressing the cerebral physiology of Descartes made by the anatomists of the 17th century, and the rehabilitation of "L'Homme" proposed by Steno, see the contribution from R. Andrault.

functions tied to the intracerebral course of spirits – and this, after having described the "course of blood" and the motor functions tied to the extra-cerebral course of spirits[46]–, Descartes gives credence to the principles of his cerebral physiology. For lack of being sufficiently founded in experience, they have a seductive power that means they can act upon the resolution two issues (completing the explanations relative to sense perception, and also of muscle movements). Apparently by chance in the frame of a *Treatise on Light*, the explanation of muscle movements, and the general descriptions of the machine that it implies, here plays a crucial role: they guarantee the credibility of certain principles without which the explanation of sensations would not find resolution.

10.5 The Consequences of a Generalization

The resolution to explain "all the main functions in man" has resulted in significant consequences.

The first concerns the developments on plants and animals. The course of the *Treatise on Light* is suddenly interrupted in Chap. 15; elsewhere, Clerselier notes that in the heading of the manuscript on the chapter on man, we see mention of a "Chap. 18".[47] If we add to this that Descartes claims, in the fifth part of the *Discourse*, to have provided explanations on the topic of plants and animals,[48] it is tempting to consider that Chaps. 16 and 17 were intended for this purpose. We believe that the absence of chapters on plants and animals should not be seen as a mark of incompleteness, but more as the consequence of an evolution of the Cartesian project.[49]

Responding to Mersenne, in August 1638, on the topic of "*l'herbe sensitive*", a plant that responds to touch, Descartes insists on the fact that his description of the heart's circulation has the advantage that it could "apply as easily to plants as to animals, if the organs found there are the same". Thus, the problem posed by the

---

[46] Which is the opposite of the rule that recommends the explanation of these movements of blood and spirits "in the proper order" (G 100, AT XI 121). Descartes justifies this inversion by invoking a concern for clarity: the theory of the intracerebral process would be abstruse, if it was not preceded by explanations relatives to motor functions tied to "the composition of the nerves and the muscles" (G 108, AT XI 132).

[47] AT XI, xii.

[48] AT VI 45. On the undetermined status of these chapters that were probably never written, cf. *Oeuvres, op. cit.*, VI 651, note 352; A. Bitbol-Hespériès' introduction to *Le Monde; L'Homme*, Paris: Seuil, 1999, xl.

[49] It is clear that Descartes invokes, in the *Principles* (IV, 188), a lack of experience to justify his silence on the subject of plants, animals and man (AT VIII 315). And we know that he was intensely dedicated to studying them in the later years of his life. However the genre of the fable, unique to *Treatise on Light*, permits the presentation of hypotheses little founded in experience, as proven by the theories present on the subject of intracerebral movement within animal spirits (left out of the *Principles*).

particular case of *l'herbe sensitive* seems to be resolved with little difficulty.[50] Once he starts to turn his attention more seriously to botany, Descartes then considers the explanation (already given) of the movement of the animal heart as a matrix from which we can easily make sense of plant phenomena. Once animal life has been described, it is not difficult to describe that of plants. The same principle applies to the subject of men and of animals: animal knowledge is, in some way, enveloped by that of man, and, even, in the description of "body on its own". Descartes thus retrospectively evokes the description of the body as a document on the nature of the animal.[51] It is unclear, from this point of view, what would have been the need for a chapter on animals within his treatise. On this topic, let us finally note that, among the various critiques through which Descartes makes known that he disavows Regius' *Fundamental Physics* (1646), there is one that often reappears: he mocks the heaviness of twice repeating, "word for word", the explanation of muscle movement.[52] In effect, we find two explanations for muscle movement almost identical to the *Fundamental Physics*: the first in Chap. 10 on animals, the second in Chap. 12 on man.[53] And what is relevant on the topic of muscle movement is relevant for the theory of knowledge: everything that that concerns men and animals is said twice in Regius; in the first instance concerning "simple" operations and a second concerning "reflected" operations.[54] Within the frame of a presentation whose principle rule is to not facilitate the ease of the reader as a means to avoid his distaste,[55] it is difficult to imagine Descartes succumbing to a heaviness that he will soon condemn in Regius.

From the perspective of the division initially established between knowledge of bodies (bodies) and that of man (spectator) the nutrient and animal functions seem to come from the first branch. From the pulse to muscle movements, we are concerned with visible phenomenon that are involved (same as the movements of the

---

[50] To Mersenne, 23 August 1638, AT II 329. Translated and commentated in S. Gaukroger, *Descartes' System of Natural Philosophy*, Cambridge: Cambridge University Press, 2002, 186–187.

[51] Cf. to Elisabeth, 6 October 1645: "the treatise which I once drafted on the nature of animals" (CMSK 270; AT IV 310) ; to Mersenne, 23 November 1646: "it is now twelve or thirteen years since I described all the functions < of the human body, that is, of the animal >" (CMSK 301, AT 566) ; to Elisabeth, March 1647 : "what I had written about < the nature > of animals" (CMSK 314, AT IV 626).

[52] Cf. to Colvius, October 1646: "*ce qu'il répète deux fois touchant le mouvement des muscles..., qu'il a tiré, comme je m'imagine, d'un écrit que je n'ai point encore publié...*" (AT IV 517–518); to Mersenne, 23 November 1646: "as for his way of explaining the movement of the muscles, although this comes from me, and has pleased him so much that has twice repeated it word for word..." (CMSK 301, AT IV 566); to Elisabeth, March 1647: "so fond was he of this passage that twice in his book he repeats, word for word, two or three pages from this section" (CMSK 315, AT IV 626).

[53] Cf. Regius, *Fundamenta Physices,* Amstelodami : L. Elzevirium, 1646, 233 sqq., 293 sqq.

[54] We find already these flagrant repetitions in the third *disputatio* of the *Physiologia* (1641); we will see them again between Books IV and V of the *Philosophia naturalis* (1654). They are characteristic of the original way in which Regius exposes Cartesian physics.

[55] Cf. *supra*, note 13.

Sun, comets, or tides) in the great spectacle of nature. At first sight, therefore, the living being and the animal appear as simple objects for the human soul, parts of the spectacle in the same way as the others. But then Descartes realizes that, to complete the explanation of sense perception, he needs to take into consideration the living being and the animal that are at work in man. Knowledge of the living being and the animal thus crosses over from one field to another. They no longer intervene to complete the description of the external world, but in being – directly or indirectly – involved in the study of the mental representation of the world.

Secondly, the resolution to explain "all the main functions in man" leads to an evolution of a status susceptible to be accorded the text on man. Albeit not one in the sense that it would cease to be the last chapter of a consistent treatise *On Light* – we have tried to show here the opposite. But public works (those which Descartes wanted to bequeath to posterity in the event of posthumous publication) have to be distinguished from the informal practices of disclosure, that is to say, the means by which Descartes had consented, in his lifetime, to circulate the contents of his work among some of his friends. If Descartes had conceived of the text on man as the chapter of a *Treatise on Light,* he never consented, during his lifetime, to have it perceived as such. The unexpected growth of anthropology made possible, in his eyes, a separation: to the *Opticiens* the first fifteen chapters; to the *Médecins-Anatomistes* – Reneri, Pollot and (accidentally?) to Regius – the chapter on man. In other words, at the same time that he consents to make seen what he refuses to publish, Descartes obscures something essential: the junction of a mechanical explanation of the light (deduced from universal laws of nature) with an account on memory traces and intracerebral "ideas" (deduced from a general physiology). Descartes would have evidently been the only one to have in his grasp the two principal parts of his great puzzle. And it would be still the case in the *Principles* since, living out the fable of his World and coming back to the real one, he did not divulge in this work the principles of his cerebral physiology.

By thus separating the content of the *Treatise on Light,* so that the chapter on man is directed only to the physicians, as if it was an autonomous text, Descartes was the first to orient the text towards a medical reception whose influence would gain in posterity. By medical reception, we mean here a certain way of reading the chapter on man, which consists of bypassing the optical context and the structuring role of the examination of the sensation of light, to inscribe it immediately into the field of anatomical and medical study. From Regius' *Physiologia* (1641) to La Forge's *Remarques* (1664b) by way of Schuyl's preface (1662), we can detect a similar trend – in three doctors by profession – to read "L'Homme" as a relatively autonomous writing, and one which calls, as a priority, on anatomical and medical references and observations.[56] Clerselier and La Forge have well insisted on the inclusion of the chapter consisting of the treatise on physics, and the demonstrative

[56] On this medical reception, see for instane D. Antoine-Mahut, Delphine, "Les voies du corps. Schuyl, Clerselier et La Forge lecteur du traité de *L'Homme* de Descartes", *Consecutio temporum: Rivista critica della postmodernità* (consecutio.org) 2 (2012).

opening ("these men")[57]: in fact, they continue to read it as if it was a separate text. They carefully avoid noting that "these men" are those that were mentioned, in Chap. 13 and in accordance with the indication in the fifth part of the *Discourse*, as "spectators" of light. The *Anatomistes* are omnipresent in the *Remarques* by La Forge (Bartolin above all) but we do not see any traces of Alhazen or Kepler.

Translated by Collette Alexander

Bibliography

Antoine-Mahut D. Les voies du corps. Schuyl, Clerselier et La Forge lecteur du traité de L'Homme de Descartes, Consecutio temporum: Rivista critica della postmodernità, 2. 2012. consecutio.org

Cavaillé J-P. Descartes. La fable du monde. Paris: Vrin; 1991.

Chanet P. Traité de l'esprit, de ses connaissances et de ses fonctions. Paris: Jean Camusat & Pierre le Petit; 1649.

Chevalley C. Sur le statut d'une question apparemment dénuée de sens : la nature immatérielle de la lumière. XVIIe siècle. 1982;136:257–66.

Descartes R. De homine, figuris et latinitate donatus a Florentio Schuyl. Lugdunum Batavorum: Petrum Leffen & Franciscum Moyardum; 1662.

Descartes R. Le Monde … ou le Traité de la lumière et des autres principaux objets des Sens…, edited by D. R. Paris: Michel Bobin & Nicolas le Gras; 1664a.

Descartes R. L'homme … et un traitté de la formation du foetus du mesme auteur. Avec les Remarques de Louis de la Forge…. Paris: Charles Angot; 1664b.

Descartes R. L'homme … et la formation du foetus, Avec les Remarques de Louis de La Forge. A quoi l'on a ajouté Le Monde ou Traité de la lumière…. Paris: Theodore Girard; 1677.

Descartes R. L'Uomo. (trans. and ad. G. Cantelli). Torino: Boringhieri; 1960.

Descartes R. Oeuvres, Adam C, Tannery P, editors. Paris: Vrin-CNRS; 1964–1974.

Descartes R. The philosophical writings of Descartes. (trans. and editors Cottingham J, Stoothoff R, Murdoch D, Kenny A). Cambridge: Cambridge University Press; 1988.

Descartes R. The world and other writings. (trans. and editors Gaukroger S). Cambridge/New-York/Melbourne: Cambridge University Press; 1998.

Descartes R. In: Bitbol-Hespériès A, Verdet J-P, editors. Le Monde; L'Homme. Paris: Seuil; 1999.

Descartes R. In: Beyssade J-M, Kambouchner D, editors. Oeuvres complètes. Paris: Gallimard; 2009–2013.

Garber D. Descartes' metaphysical physics. Chicago/London: University of Chicago Press; 1992.

Gaukroger, S. Descartes' system of natural philosophy, Cambridge: Cambridge University Press; 2002.

Hamou P. La mutation du visible. Essai sur la portée épistémologique des instruments d'optique au XVIIe siècle. Volume I: du Sidereus Nuncius de Galilée à la Dioptrique cartésienne. Villeneuve d'Ascq: Presses universitaires du Septentrion; 1999.

Hatfield G. Natural Geometry in Descartes and Kepler. Res philosophica. 2015;92:117–48.

Hume D. In: Green TH, Grose TH, editors. A treatise on human nature I: of the understanding (1739). London: Longmans, Green and co; 1878.

[57] See Clerselier's Preface (AT XI xi-xii) and La Forge's first comment in *L'Homme de Descartes et un Traitté de la formation du foetus du mesme auteur. Avec les Remarques de Louis de la Forge...*, edited by C. Clerselier, Paris: Charles Angot, 1664b, 171–172.

Kepler. Ad Vitellionem Paralipomena, Quibud Astronomiae pars Optica traditur.... Francofurti: Claudium Marnium & Haeredes Ioannis Aubrii; 1604.

Kepler. Dioptrice..., Augustae Vindelicorum: Davidis Franci; 1611.

Kepler. Optics: Paralipomena to Witelo & optical part of astronomy. (trans. and editor Donahue WH). Santa Fe: Green Lion Press; 2000.

Lindberg DC. Theories of vision from Al-Kindi to Kepler. Chicago/London: University of Chicago Press; 1976.

Martinet S. Rôle du problème de la lumière dans la construction de la science cartésienne. XVIIe siècle. 1982;136:285–309.

Maull N. Cartesian optics and the geometrization of nature. In: Gaukroger S, editor. Descartes: philosophy, mathematics and physics. Totowa: Barnes and Nobles Books; 1980. p. 23–40.

Meschini FA. Neurofisiologia cartesiana. Firenze: Leo S. Olschki; 1998.

Nardi A. La luce e la favola del Mondo, Descartes 1629–1633, Annali dell'Instituto di Filosofia, 3. Florence: Olschki; 1981. p. 103–45.

Regius. Physiologia, sive Cognitio sanitatis.... Aeg. Roman: Utrecht; 1641.

Regius. Fundamenta Physices. Amstelodami: L. Elzevirium; 1646.

Regius. Philosophia naturalis. Amstelodami: L. Elzevirium; 1654.

Risner. Opticae thesaurus. Alhazeni Arabis libri septem nuncprimum editi. Eiusdem liber De crepusculis & nubium acensionibus. Item Vitellionis Thuringopoloni libri X.... Basileae: per Episcopios; 1572.

Roux S. La nature de la lumière selon Descartes. In: Biet C, Jullien V, editors. Le Siècle de la Lumière. Bordeaux: Presses Universitaires de l'ENS de Fontenay Saint-Cloud; 1997. p. 49–66.

Sabra AI. Sensation and inference in Alhazen's theory of visual perception. In: Machamer PM, Turnbull RG, editors. Studies in perception: interrelations in the history of philosophy and science. Columbus: Ohio State University Press; 1978. p. 160–85.

Sabra AI. Theories of light from Descartes to Newton. Cambridge/New-York/Melbourne: Cambridge University Press; 1981.

Schuster J. Descartes-agonistes. Physico-mathematics, method & corpuscular-mechanism 1618–1633. Dordrecht/New-York: Springer; 2013.

Simon G. Structures de pensée et objets du savoir chez Kepler, Service de reproduction des thèses, Université Lille III; 1979.

Smith AM. Descartes' theory of light and refraction: a discourse on method, Transactions of the American Philosophical Society. Phila Am Philos Soc. 1987;77(3).

Steno N. Discours sur l'anatomie du cerveau.... Paris: Robert de Ninville; 1669.

Steno N. A dissertation on the anatomy of the brain.... (trans. and editor Garrison-Morton (1733)) Edv. Gotfredsen. Copenhague: A. Busck; 1950.

Sutton J. Philosophy and memory traces: Descartes to connectionnism. Cambridge: Cambridge University Press; 1998.

Chapter 11
Anatomy, Mechanism and Anthropology: Nicolas Steno's Reading of *L'Homme*

Raphaële Andrault

Abstract Nicolas Steno's criticism of *L'Homme* played a major role in the early reception of Cartesianism: from the late 1660s, the *Discourse on the Anatomy of the Brain* has never ceased being used in order to discredit Descartes's philosophy. And yet, the anatomical works of Nicolas Steno are themselves informed by Cartesian method. This paradox has led to the depiction of Steno either as a repentant Cartesian or a non-Cartesian mechanist. In this paper, I clarify such problematic labels by studying the different kinds of relationships between anthropology and anatomy that *L'Homme* may have used to justify. In particular, I show how Descartes' clock analogy was used to defend two different conceptions of the articulation between anatomical observations and functional hypotheses respectively in La Forge and in Steno.

11.1 Introduction

Cartesian anatomy is often the object of two contradictory judgments.[1] On the one hand, the anatomical considerations of *L'Homme* are regarded as fanciful and disconnected from what can be observed in anatomical theaters. On the other hand, this treatise is supposed to have promoted the description of organs as the unique cause

[1] I use the following abbreviations: DESCARTES: Ariew = *Philosophical Essays and Correspondence*, ed. R. Ariew (Indianapolis/Cambridge: Hackett, 2000). AT = *Œuvres de Descartes*, ed. C. Adam and P. Tannery, new presentation by B. Rochot and P. Costabel (Paris: Vrin-CNRS, 1964–1974). *Discours* = Nicolas Sténon, *Discours sur l'anatomie du cerveau* (Paris: Robert de Ninville, 1669). *Epistolae* = N. Steno, *Epistolae et epistolae ad eum datae, quas cum prooemiis ac notis Germanice scriptis edidit*, ed. G. Scherz (Copenhagen: A. Busck 1952). Gaukroger = Descartes, *The World and Other Writings*, trans. and ed. by Stephen Gaukroger (Cambridge: Cambridge University Press, 2004). Maquet = *Nicolaus Steno: Biography and Original Papers of a 17th Century Scientist*, ed. T. Kardel and P. Maquet (Berlin: Springer, 2013).

R. Andrault (✉)
IHRIM, CNRS, UMR 5317. ENS de Lyon, 15 Parvis René Descartes, BP 7000, 69342 Lyon Cedex 07, France
e-mail: raphaele.andrault@ens-lyon.fr

D. Antoine-Mahut, S. Gaukroger (eds.), Descartes' *Treatise on Man* and its Reception, Studies in History and Philosophy of Science 43,
DOI 10.1007/978-3-319-46989-8_11

of the functions of the human body. In other words, Descartes is depicted both as someone who would have dealt negligently with the anatomical science of the human body, and as someone who would have given an explanatory value to anatomy itself. One can read these two kinds of claims for instance in Kurt Sprengel's *Versuch einer Pragmatischen Geschichte der Arzneikunde* published in 1801: the resourceful hypotheses of Descartes allowed scientists to exclude occult qualities and to focus attention on the structure of the parts of the human body, while at the same time it must be recognized that Descartes' theory undermined the sense of observation.[2] One can find also this twofold judgement in Jacques Rogers' *Les sciences de la vie dans la pensée française du XVIIIe siècle* first published in 1963. Roger sharply criticizes the Cartesian use of unverifiable hypotheses while considering that Descartes's conception of life stimulated anatomical research: since for Descartes everything was a matter of shapes and motions, it was essential to discover the shapes of organs.[3] Georges Canguilhem's famous reading of the implications of the 'mechanism' in the life sciences reinforces this twofold judgement. His view was that Cartesian analogies between living bodies and machines allows one to deduce biological functions from anatomical forms, while at the same time imposing a rigid and erroneous conception of the dynamical functioning of living organisms: 'It may thus be said that [Descartes substituted] mechanism for the organism.'[4] Canguilhem then invokes Nicolas Steno's *Discourse on the anatomy of the brain* (1669) to emphasize the distance between Descartes' man and 'the man of the anatomist,' i.e. the man of nature.

As a matter of fact, Steno's reading of *L'Homme*, and, even more so, the early reception of Steno's *Discourse*, maintain a complex relationship with the Cartesian epistemology. A careful reading of Steno's ambiguous judgment on Descartes permits us to clarify both the role played by anatomy in Cartesian anthropology and the medical implications of the mechanist analogy between man and machine. In this chapter, I examine first Steno's reading of *L'Homme* and its early reception. Then, I show in what sense Steno's critique of Descartes' anthropology may be partly based on Descartes' analogy between living bodies and clocks. Finally, by comparing La Forge and Steno, I explain how contradictory claims on the role ascribed to anatomy by Descartes or by the so-called 'mechanists' come from two different interpretations of the machine analogy.

---

[2] K. Sprengel, *Versuch einer Pragmatischen Geschichte der Arzneikunde. Vierter Theil* (Halle: bei Johann Jacob Gebauer, 1801), 379.

[3] J. Roger, *Les sciences de la vie dans la pensée française du XVIIIe siècle. La génération des animaux de Descartes à l'Encyclopédie* (Paris: Armand Colin, 1993), 169, 207.

[4] G. Canguilhem, 'Machine and organism', *Knowledge of life*, ed. P. Marrati and T. Meyers, trans. S. Geroulanos and D. Ginsburg (New York: Fordham University Press, 2008), 75–97, 86.

11.2 Steno's Critique of Descartes and Its Early Reception

No sooner had the Latin translation of *L'Homme* been published, than a certain number of letters and accounts of dissections raised questions about the relevance of Descartes' descriptions or the accuracy of Schuyl's diagrams. Such a reception testifies that Descartes' book constituted an important event for those who practiced anatomy in Europa at the time. To be sure, before the publication of *L'Homme*, the anatomy of the brain proposed in Descartes' *Passions de l'âme* was already being discussed by anatomists – for instance in the third version of the *Institutiones anatomicae* by Thomas Bartholin.[5] But during 1662 a number of new discussions were triggered. In August 1662, Nicolas Steno (Niels Stensen) mentions the publication of *De Homine*, where 'there are some not inelegant figures.' He conceded that 'they have proceeded from a clever brain,' but he doubted 'whether such images can be seen in any brain.'[6] In March 1663, Steno reiterated his judgment. He seemed first to take Descartes's anatomy of the brain seriously, since he was ready to consider the 'pineal gland' in the middle of the brain as the seat of the soul. His own dissections disabused him all the same. Mentioning his dissections of a head of horse, he explained:

> The size of the pineal gland was conspicuous enough but its colour was blackish externally, internally grey marked by many dark sports so that everyone may say that black bile does not leave intact even the seat of the soul. [...] Certainly, the more I open brains, either of other animals or of birds of various kinds, the less the structure of the brain of animals thought out by the noble *Descartes*, most ingenious and otherwise very appropriate to the explanation of animal actions, seems to fit animals.[7]

During his Parisian stay in 1665, the public dissections that Steno performed nearly every day, either at the Faculté de médecine or in Thévenot's salon, disseminated his negative judgment regarding the Cartesian account of the anatomy of the brain. Jean Chapelain, a staunch Gassendist, declared:

> Stensen the Dane has performed the most marvelous experiments ever in this field. He has even forced the obstinate and dogmatic Cartesians to admit the error of their leader with regard to the gland of the brain and its function [...], on which he based all the operations of the reasonable soul.[8]

---

[5] Thomas Bartholin, *Anatomia ex Caspari Bartholini parentis Institutionibus, omniumque recentiorum & propriis observationibus tertium ad sanguinis circulationem reformata, cum iconibus novis accuratissimis* (Lugdunum Batavorum: Franciscum Hackius 1651), 336–337. Beside, Descartes' conception of the movements of the heart, and more broadly, Descartes' physiology, were already discussed in the late 1630 (on the basis of the *Discours de la méthode*, 1637). See on this point Annie Bitbol-Hespériès, 'Cartesian Physiology', in S. Gaukroger, J. Schuster and J. Sutton (eds), *Descartes' Natural Philosophy* (London/New York: Routledge, 2000), 374.

[6] Steno to Bartholin, Leiden, 26 August 1662, in *Epistolae*, 163; trans. in Maquet, 433.

[7] Steno to Bartholin, Leiden, 5 March 1663, in *Epistolae*, 172; trans. in Maquet, 445.

[8] Letter to Huet, 6 April 1665, in Chapelain, *Lettres*, ed. T. de Laroque, Paris: Imprimerie Nationale 1883, II, p. 393, note 3; trans. in *Nicolaus Steno's Lecture on the anatomy of the brain*, ed. and trans. by G. Scherz (Hafniae: A. Busck, 1965), p. 70.

Other accounts show that Steno's dissecting skill was then famous among Parisian scientific circles. For instance, the French scholar André Graindorge claimed that public dissections performed by Steno were 'all the rage.'[9] It partly explains the success of the *Discours sur l'anatomie du cerveau* that Steno pronounced in 1665, in the salon of the polymath Melchisédec Thévenot, who published the short text in 1669.[10] If the *Discours* does not deal exclusively with Cartesian anatomy, it does quote the French edition of *L'Homme* extensively, and calls the 'pineal gland' the 'most famous anatomical question of this century.' Steno rejects Descartes' hypotheses by showing that the gland is not at the entry of the concavities of the brain, that it cannot move from side to side without breaking apart, that it is not surrounded by arteries and, lastly, that it is not located where the animal spirits, i.e. those subtle particles stemmed from the blood and responsible for the sensory-motor actions, are supposed to come from.[11] All these refutations are fundamental to the assessment of Descartes' anthropology.[12] For instance, all the explanations that Descartes gave to psychophysical functions such as sensory perception, voluntary motion or attentiveness are directly or indirectly contingent upon the various inclinations of the gland H, i.e. the pineal gland.[13] If this tiny gland

---

[9] Graindorge to Huet, 5 mai 1665, in L. Tolmer, *Pierre-Daniel Huet, humaniste physicien* (Bayeux: Colas 1949), 330; trans. by Ole Peter Grell, 'Between Anatomy and Religion: The Conversions to Catholicism of the Two Danish Anatomists Nicolas Steno and Jacob Winsløw,' in O. P. Grell and A. Cunningham (eds.), *Medicine and Religion in Enlightenment Europe* (Aldeshot: Ashgate 2007), 205–221, 213: 'This afternoon we saw the eye of a horse. To tell you the truth, compared with him [Steno] we are only apprentices. [...] He is always dissecting. He has an unbelievable patience and through practice he has gained a unique expertise.'

[10] On the circumstances, the audience of the *Discourse* and its publication, see our introduction in *Discours* 2009.

[11] *L'Homme*, AT IX, 179; trans. in Gaukroger, 152: 'Consider also that gland H. is composed of very soft matter which is not joined to or part of the substance of the brain but attached only to certain little arteries whose membranes are somewhat relaxed and pliant, and that it is kept in balance as it were by the flow of blood which the heat of the heart drives in its direction; so that very little is required to make it incline or lean, whether a little or a great deal, whether to this side or to that, and so to make the spirits that issue from it proceed to particular regions of the brain rather than others.' See also Steno, *Discours*, 15–16, where Steno quotes five statements that he finds decisive in *L'Homme*, and 20–21, where he dismisses those statements; trans. in Maquet, 512–513.

[12] See *Discours*, 21; trans. in Maquet, 513: 'The hypothesis of the arteries gathered around the gland and rising to the great channel is a matter of no little consequence for the system of Mr. Descartes since the separation of the spirits and their movement depends on it. However, if you believe your eyes, you will find that it is only a collection of veins.'

[13] See for instance *L'Homme*, AT XI, 183–184; trans. in Gaukroger, 155: 'And when a soul has been put in this machine, this will allow it to sense various objects by means of the same organs, disposed in the same way, and without anything at all changing except the position of the gland [...] Now suppose that gland leans a little further forward, in such a way that points *n* and *o* on its surface are at the places marked *i* and *k*, and that as a consequence it is from them that the spirits entering and issue: the soul would sense what is at *n* and what is at *o* by means of the same hands without them being changed in any way.'

cannot be inclined freely side to side, the entire cerebral physiology that Descartes propounded in the *Passions de l'âme* and in *L'Homme* is challenged. It is thus not surprising that in the *Ethics* Spinoza uses Steno's arguments to dismiss the union of soul and body conceived by Descartes, adding, to his refutation of the idea of an interaction between something corporeal and something incorporeal, that 'this gland is not to be found located in the middle of the brain in such a way that it can be driven about so easily and in so many ways, nor do all nerves extend as far as the cavities of the brain.'[14] Such a polemical use of Steno's anatomical refutation will remain constant. A decade after the *Discourse*, Steno himself, in a letter to Leibniz, will mention the anatomical 'error' of Descartes on the organization of the muscles as a reason not to agree with Cartesian metaphysics:

> I considered the system of Mr Descartes as infallible [...]. I chose a leg of a little rabbit which I had dissected a short time before. The first muscle which I tested revealed to me the first step of the structure of the muscle which so far nobody had known and which demolished the whole system of Mr. Descartes. [...] [If those gentlemen] have deceived themselves in material things which are accessible to the senses, what certitude can they give me against a similar deception if they deal with God and the soul.[15]

Later on, Leibniz will use such a judgment to dismiss Descartes' authority in matter of experimental science.[16] In this way, the dissections performed by Steno were used to discredit Cartesian anthropology as a whole.

11.3 The Man of Descartes and the Man of the Anatomists

What, however, are the real implications of the criticism that the man depicted by Descartes in *L'Homme* does not correspond to the man observed by the anatomists? Descartes himself pointed out the hypothetic status of his 'man.' His aim was not to give a complete physiology of the human body. The *Treatise* aimed rather at showing that it is possible to account for the behavior of a living body without mentioning any occult quality, vegetable power or sensitive soul. In order to do so, Descartes propounded plausible bodily causes of observed effects and functions. By definition, these causes put forward by Descartes are parts and movements that are not

[14] *Ethics*, part. 5, preface, trans. S. Shirley, in Spinoza, *Complete Works*, ed. M. L. Morgan (Indianapolis/Cambridge: Hackett 2002), 365. On the interest of Spinoza for the public dissections of the brain performed by Steno in Leiden, see Pina Totaro,"Ho certi amici in Ollandia': Stensen and Spinoza – science verso faith,' in K. Ascani, H. Kermit, and G. Skytte (eds.), *Niccolo Stenone. Anatomista, geologo, vescovo* (Rome: L'Erma 2002), 27–38. For a more detailed account on this use of Steno's anatomy in Spinoza, see Raphaële Andrault, *La vie selon la raison. Physiologie et métaphysique chez Spinoza et Leibniz* (Paris: Champion, 2014), 308.

[15] Steno to Leibniz, 1677, in *Epistolae*, II, 367–368; trans. in Maquet, 94.

[16] Leibniz to Nicaise, 1692, in *Die Philosophischen Schriften von Gottfried Wilhem Leibniz*, ed. Gerhardt (Hildesheim/New York: Georg Olms 1978), vol. IV, 348.

described in handbooks of anatomy; they are rather invisible components, internal fluids or unobservable movements of the anatomical parts which are shown in anatomical theaters. For instance, regarding the explanations of cerebral functions, Descartes remarks:

> [The] functions that we are concerned with here do not depend at all on the external shape of the visible parts which the anatomists distinguish in the substance of the brain and in its concavities, but solely on three factors, namely, the spirits that come from the heart, the pores of the brain through which they pass, and the way in which the spirits are distributed in these pores.[17]

To be sure, anatomy plays an important role in Descartes' anthropology. First, the explanations of the different functions of the human body should take into account the thin consensus view on the anatomical composition of the human body: we can assume that the hypotheses propounded by Descartes cannot contradict anatomical knowledge.[18] Second, and more generally, anatomy illustrates the possibility of accounting for complex functions, such as nutrition, voluntary motion or even cognitive attention, through *explanantia* that are as simple and as corporeal as mechanical processes observed in machines:

> [The] ignorance of anatomy and mechanics has contributed to [this belief that the soul is the principle behind all our movements], for in considering only the exterior of the human body, we never imagined that it had enough organs or springs in it to move itself in all the different ways in which we see it move.[19]

According to Descartes, anatomy suggests a model of economical explanation, where the causes are homogeneous to the effects while being simpler than them.[20] Third, occasionally, in the *Discours de la méthode*, the anatomical structure seems to constitute by itself the cause of physiological functions. It is the case for the circulation of the blood, a central function of the animal body that seems to depend only on the different shapes of the parts of the heart and blood vessels:

> [This movement of the blood] which I have just been explaining follows just as necessarily from the mere disposition of the organs that can be seen in the heart by the naked eye, and from the heart that can be felt with the fingers, and from the nature of blood, which can be known through observation, as does the movement of a clock from the force, placement, and shape of its counterweights and wheels.[21]

---

[17] *L'Homme*, AT XI, 166; trans. in Gaukroger, 140.

[18] See *L'Homme*, AT XI, 121, trans. in Gaukroger, 99–100.

[19] *La description du corps humain*, AT XI, 224; trans. in Gaukroger, 170.

[20] On the principle of economy in the explanation, see *L'Homme*, AT XI, 201 (nature acts always by the most easy and simple ways); see also *Principes de la philosophie*, IXb, 319–320: the *explanans* must be simpler than the *explanandum* and conceived on the model of the things one can feel.

[21] *Discours de la méthode*, AT VI, 49–50; trans. in Ariew, 69.

But it remains the case that anatomy does not constitute by itself the unique source of the explanations proposed by Descartes.[22] To put it briefly, one should not confuse the way Descartes presents his physiological explanations (the *modus exponendi*), which in the case of the heart perfectly illustrates the possible deduction from anatomical structures to biological functions, with the *modus inveniendi* that Descartes really adopted in order to elaborate his physiology: most of the time the empirical data provided directly by anatomical observations correspond rather in Descartes to phenomenal effects than to underlying causes.[23]

It is precisely what La Forge's remarks on *L'Homme* suggest in a passage where he comments on Descartes' statement according to which the functions of the brain do not depend on 'the external shape of the visible parts which the anatomists distinguish in the substance of the brain:'

> There is nothing truer than what Monsieur Descartes claims here, since we see that it is impossible to explain, nor to account for any functions of the brain, by the sole conformation of its sensible parts; what clearly demonstrates that [such a conformation] is not sufficient, and that it is necessary to seek another cause, which goes beyond what we feel.[24]

For La Forge, it is necessary to resort to conjectures on the invisible parts of the human body to explain how it functions. In this matter, hypothetical-deductive reasoning is legitimate and does not rule out the relevance of the physiological explanations propounded.[25] Both the remarks on *L'Homme* and his own *Traité de l'esprit humain*, published soon after, suggest that La Forge granted a very limited role to anatomy. First, as La Forge held, the anatomy of dead bodies dissected in anatomical theaters does not necessarily correspond to the exact configuration of internal

[22] For instance Descartes mentions the 'common experience of surgeons' to prove the circulation of the blood: 'He [Harvey] proves this very effectively from the common experience of surgeons, who, on binding an arm moderatly tightly above the spot where they open the vein, cause the blood to flow out in even greater abundance than if they had not bound the arm at all.' Beside this point, a lot of readings, experiments and analogies underlie Descartes' descriptions and explanations. On the anatomical knowledge of Descartes, see Annie Bitbol-Hesperies, 'Cartesian Physiology', in *Descartes' Natural Philosophy* (London/New York: Routledge, 2000), eds. S. Gaukroger, J. Schuster and J. Sutton, 349–382.

[23] We cannot develop this point here. For a discussion of the idea of 'anatomical deduction' on which Canguilhem based his critique of Descartes' mechanism, see our book *La raison des corps. Mécanisme et sciences médicales* (1664–1716), Paris, Vrin, 2016, chap. 1.

[24] La Forge, 'Remarques sur le *Traité de l'Homme*', in *L'homme* (Paris: Charles Angot, 1664), 287: 'il n'y a rien de plus vray que ce que dit icy Monsieur Descartes, puis que nous voyons que l'on ne peut expliquer, ny rendre raison, d'aucune des fonctions du cerveau, par la seule conformation de ses parties sensibles; Ce qui monstre clairement qu'elle n'est pas suffisante, & qu'il en faut chercher une autre cause, qui ne tombe pas sous les sens.'

[25] See also Descartes, *Principes de la philosophie*, part. IV, art. 201, AT IXb, 324; trans. in Ariew, 269: 'But it seems to me to be doing great wrong to human reason if we do not consider that knowledge goes beyond what we see.'

organs and fluids in a living body.[26] Second, cerebral actions that Descartes explained in his *Treatise* are allegedly caused by invisible parts such as animal-spirits and pores, which precisely go beyond the limits of anatomical observations. Consequently, Descartes' physiology cannot be refuted simply by invoking anatomical observations that contradict it: Descartes' automaton depicted in *L'homme* is not supposed to resemble the corpses observed in anatomical theaters and described in anatomical handbooks, since a certain number of parts and springs of this automaton by definition escape the power of observation that limits anatomical knowledge. The idea that Descartes' man is not the man observed by the anatomists could thus be read in a favorable light.

Steno's reading of *L'Homme* agrees with La Forge's view on one point: Descartes' machine described in *L'Homme* does not pretend correspond to the man described by anatomists. But such a common assumption leads Steno to a completely conflicting view regarding the relevance of Descartes' physiology and the role of anatomy: Steno regarded *L'Homme* as an interesting modelling only, not as the plausible and useful reconstruction of the most hidden parts of the human body:

> As far as Mr Descartes is concerned, he knew too well the shortcomings of the description that we have of man to explain his true structure. Therefore, he does not undertake to do that in his *Traité de l'homme* but he explains to us a machine that would do everything men are able to do. [...]
>
> Mr Descartes thus must not be condemned if his system of the brain is not strictly in agreement with experience.[27]

Steno gives a very specific role to Descartes's treatise. According to him, *L'Homme* is not uninteresting in itself, but it should not be read as a medical book and should not be seen as a solid basis for the emergent modern human physiology:

> I should have been content to admire [this treatise] with some other people as the description of a nice machine all of his invention, if I had not met many people who take it quite differently and who want to present it as an exact report of that which is deeply concealed in the recesses of the human body. Since these people do not agree with the very evident demonstrations of Mr. Sylvius, who has often showed that the description of Mr. Descartes

[26] *Treatise on the Human Mind*, ed. and trans. by Desmond Clarke (Dordrecht: Springer 1997), 153: 'Thirdly to their objection [*i. e.* the objections made by anatomists like Steno against Descartes' anatomy of the brain] that this gland cannot move, I reply that if they can convince us that all the parts of a living animal brain are as compacted as those of the head of a dead calf, their objection may be acceptable and we would possibly agree with it. But there is no reason to believe that is the case while the animal is alive...' Such a statement may be based on Descartes, *Description du corps humain*, AT XI, 224; trans. in Gaukroger, 170 : 'And we have been confirmed in this error in judging that dead bodies have the same organs as living ones, for they lack nothing but the soul...'

[27] *Discours*, 13; trans. in Maquet, 511: 'The excellence of his mind which principally appears in his *Traité de l'homme* covers the errors of his hypotheses. We see that very skilled anatomists such as Vesalius and others could not avoid making similar errors. If these great gentlemen who spent most of their lives in dissections have been forgiven their errors, why would you be less indulgent to Mr. Descartes who has spent his time very happily on other speculations?'

> is not in agreement with the dissection of the bodies which it describes, I must, without reporting here all this system, point out some places where, I am sure, if they want, they will see clearly and acknowledge a big difference between the machine which Mr. Descartes has imagined and that which we see when we make the anatomy of human bodies.[28]

The anthropological interest of *L'Homme* seems thus rather limited. On this point, Steno and La Forge differ radically. Yet, they both defended their views by using the clock analogy.

11.4 Steno and La Forge: Two Opposite Understandings of the Clock Analogy

When Steno points out the epistemological interest of Descartes' undertaking, he describes it as a 'mechanical' explanation:

> Nobody else [other than Descartes] has explained mechanically all the actions of man and principally those of the brain. The others describe man himself. Mr Descartes speaks only of a machine, which however, lets us see the insufficiency of what the others teach and lets us know a method of looking for the functions of the other parts of the body as evidently as he demonstrates the parts of the machine of his man, which nobody has done before him.[29]

Nowhere in Steno's texts there is a true definition of 'mechanical' or 'mechanism.' Moreover, it is well-known that there was no consensus about the meaning of this notion among his contemporaries.[30] Sometimes, 'mechanical' meant strictly an explanation based on the shapes and movements of small corpuscles, or, to put it in Boyle's words, an explanation based on 'motions and other affections of the minute particle of matters' that 'are obvious and very powerful in Mechanical Engines.'[31] Sometimes 'mechanism' implies only the thesis that 'whatever comes about in matter arises from the prior state of matter, according to the laws of change,'[32] if we

[28] *Discours*, 14; trans. in Maquet, 511.

[29] *Discours*, 13; trans. in Maquet, 511.

[30] See for instance A. Gabbey, 'What was 'Mechanical' about *The Mechanical Philosophy* ?', in C. P. Palmerino and J. M. M. H. Thijssen (eds.), *The reception of the Galilean Science of Motion in Seventeenth-Century Europe* (Dordrecht: Kluwer Academic Publishers, 2004), 11–23, and S. Roux and D. Garber, 'introduction', in S. Roux and D. Garber (eds.), *The Mechanization of Natural Philosophy* (Dordrecht/Heidelberg/New York/London: Springer, 2013), xi.

[31] R. Boyle, 'Some Specimens of an Attempt to make Chymical experiments useful to Illustrate the Notions of the Corpuscular Philosophy', in *The works*, eds. Michael Hunter and Edward B. Davis (London: Pickering & Chatto, 1999–2000), 2, 87.

[32] Leibniz, *The Leibniz-Stahl Controversy* (New Haven: Yale University Press, 2016), ed. and trans. F. Duchesneau and J. Smith, Animadversiones, §2, where Leibniz adds: 'And this is what is meant, or ought to be meant, by those who say that all things in bodies can be explained mechanically.' See also: 'Leibniz's Exception to Exception XXI' ('While I state that all things happen in the body mechanically, I do not thereby dwell on the exquisite figures of the pores, but in this instance I grant a greater part to motions than to figures.')

stick to Leibniz's own mechanism. In this last case, a mechanical explanation does not necessarily identify itself with a corpuscular explanation. Leibniz regarded with harshness the corpuscular explanations proposed by Descartes in *L'Homme*, but he propounded all the same a natural philosophy that he described himself as 'mechanical' or true to the 'mechanism.' This natural philosophy is based on local motions, but did not seek to explain growth and nutrition by putting forward the various shapes of tiny corpuscles or the fitting between the shape of the particles of the secreted fluid and the shape of the pores of the secreting gland according to the modelling of the sieve.

In accordance with Descartes' epistemological principles, Steno's natural science never refers to occult entities or invisible powers that would be irreducible to the local motions of corporeal parts.[33] But Steno does not present this postulate as 'mechanical.' In addition, if we stick to Steno's *Discours*, 'mechanical' refers very generally to the evidence and clarity of the theories that presuppose that the human bodies is as decomposable as a machine: 'Descartes … lets us know a method of looking for the functions of the other parts of the body as evidently as he demonstrates the parts of the machine of his man, which nobody has done before him'.[34] In the anatomical vocabulary of Steno, the word 'demonstrate' refers to the visual and public display of a bodily part during the process of dissecting or experimenting. Hence, the machine provides an analytical model according to which the complex behavior of a machine is explained by means of the arrangement of its observable components. Steno uses himself the analogy between the brain and a machine to indicate the necessity to base every functional explanation on the exhaustive description of bodily parts under consideration:

> There are two ways only to arrive at the knowledge of a machine, one that the master who made it discloses us its artifice, the other to take it to pieces to the last spring and to examine all these separately and together […]. The brain being indeed a machine, we must not hope to find its artifice through other ways than those which are used to find the artifice of the other machines. It thus remains to do what we would do for any other machine; I mean to dismantle it piece by piece and to consider what these can do separately and together.[35]

And later on in the *Discours sur l'anatomie du cerveau*:

> I did not say anything so far of the functions of the parts, nor of the actions called animal because it is impossible to explain the movements occurring through a machine if the artifice of its parts is not known.[36]

The comparison between an organ and a machine conceived on the model of a clock indicates that the description of parts is a necessary condition for understanding the functioning of the whole.

Negatively, this analogy means that one should suspend all hypotheses about the explanation of an action if one cannot accurately describe the real figures and situ-

---

[33] See Steno, *De Solido intra Solidum naturaliter contento dissertationis Prodromus* (Florentiae: ex Typographia sub signo Stellae, 1669), 10–11.

[34] *Discours*, 13; trans. in Maquet, 511.

[35] *Discours*, 32; trans. in Maquet, 516.

[36] *Discours*, 53; trans. in Maquet, 521.

ations of the components of the machine performing this action. That is why, regarding the brain, Steno recommends as a first matter to consider 'the description of the parts, in which one must determine what is true and certain to be able to distinguish that from propositions which are false or uncertain.'[37] Plausible structures of the small parts of the human body are not considered as a sufficient basis for functional explanations. Steno takes up Descartes' arguments in the *Discours de la méthode* according to which there are several possible ways of producing a single effect, but to better oppose Descartes' reasoning in *L'Homme*:

> These are the true means of knowing the artifice of a machine and, however most people have believed that they had better guessed it than it was easy to see it by examining it closely with one's senses. They were content with observing its movements and, on these observations alone, they have built systems which they presented as truths when they have believed that they were able thus to explain all the effects which had come to their knowledge. They did not consider that one thing can be explained in different ways and that only the senses can assure us that the idea which we have formed about it is consistent with nature.[38]

Both authors agree on the fact that experiments are needed for knowing which cause, among all the possible causes, has really produced the observed effect.[39] But for Steno, it means that one should wait for a more advanced state of knowledge before propounding functional hypotheses: assertions that are not firmly based on observations are regarded by Steno as useless and deceptive. In this respect, Cartesian physiology would just be a premature theory – not a useful modelling and program.

Positively, the machine analogy mentioned by Steno implies making anatomy a true experimental science, and not only a point of departure on which one can build a conjectural physiology. But in order to make anatomy such a solid and essential science, it is however necessary to reform it, and notably to develop technical innovations and new experimental procedures.[40] For instance, Steno recommends inventing a circular saw that could rotate on a fixed axis, or a liquor that would dissolve

---

[37] *Ibid.*, p. 518.

[38] Maquet, 516.

[39] *Discours de la méthode*, VI, AT V, 65, trans. in Ariew, 76: 'But I must also admit that the power of nature is so ample and so vast, and these principles are so simple and so general, that I notice hardly any particular effect without at once knowing that it can be deduced in many different ways from them, and that ordinarily my greatest difficulty is to find in which of these ways it depends on them. For, to this end, I know of no other expedient at all except to search once more for some experiments which are such that their outcomes are not the same, if it is in one of these ways rather than in another that one ought to explain the outcome.'

[40] Steno considers also institutional reforms and a modification of the taxonomy, see *Discours*, 35; trans. Maquet, 517: 'The boundaries between these two professions have been so poorly marked that true knowledge of the machine of the human body, which was most necessary, is neglected as not being in the province of anatomy nor of that of the physician nor of the surgeon. I say that to make researches which would teach us the truth requires a man entirely, a man who has nothing else to do. Even the one who makes profession of anatomy is not suited to that since he is compelled to carry out public demonstrations that prevent him from engaging in this application …'

bones, in order to observe the brain more easily[41]: at the time being, the methods of dissecting, and particularly the violence required to remove the brain from the cranial cavity, damaged the brain and altered the organization of the cerebral parts.

The decisive role that Steno ascribed to anatomy by using the machine analogy thus implies a certain understanding of the word 'anatomy.' In particular, 'anatomy' requires dissections, experimentations, vivisections, compared anatomy, pathological anatomy and embryological anatomy.[42] Such a broad meaning of the term is not peculiar to Steno. In the early modern period, anatomy often included local excisions or techniques such as infiltration of wax or colored ink in the vessels. The division between anatomy and physiology did not correspond to the division between, on one hand, a descriptive science of observable structures (anatomy), and, on the other hand, a science of the functions of living beings (physiology).[43] The study of what we would call today a 'function' (respiration, reproduction and so on) is not detailed in physiological treatises, but in anatomical accounts: it is in books named 'anatomy of…' that functions or uses of organs are first studied.[44]

Hence, the machine analogy used by Steno in order to promote anatomy does not suggest that observing the structures of the main parts of the body would be sufficient for understanding the functioning of the whole: (1) according to Steno, it is necessary – not sufficient – to observe the structures of the parts to understand precisely the way the functions are performed; (2) the functions are assimilated with the observable effects of the machine, or at least the effects that the anatomist seeks to explain – not with the *terminus ad quem* of the anatomical demonstrations[45]; (3) the observations that Steno considers essential to the explanation of the whole are not limited to the direct observation of its most coarse components; they include more complex procedures and experiments, more subtle divisions of the parts, several public demonstrations according to different methods of dissection, targeted excisions and even deductions from pathological and compared anatomy. The components of the machine that Steno deems necessary to know experimentally are not necessarily solid parts visible to the naked eye: they may be also fluids or more subtle components.

---

[41] See for instance *Discours*, 45; trans. in Maquet, 519–520.

[42] See for instance *Discours*, 54; trans. in Maquet, 521.

[43] See on this point Andrew Cunningham, 'The pen and the sword: recovering the disciplinary identity of physiology and anatomy before 1800. Old Physiology – the Pen', *Studies in History and Philosophy of Biological and Biomedical Sciences*, 33 (2002), 631–665; 'Old Anatomy – the Sword', *Studies in History and Philosophy of Biological and Biomedical Sciences*, 34 (2003), 51–76. Physiology corresponds to a systematic discourse on the whole animal or human body, while anatomy corresponds to the experimental knowledge on such or such aspect of this body.

[44] For a distinction between function (*fonctio*) and use (*usus*), see Bartholin, *Anatomia reformata*, 2–3. To put it briefly, 'function' is more general (respiration is a function, generation too), while 'use' means only what an organ or such structural aspect of this organ do, or help to do (for instance a motion).

[45] For instance Steno discovered the existence of the canal parotid in investigating on the saliva and the way the glands produced the saliva. The first element in the *modus inveniendi* was thus the apparent function and not the anatomical structure.

Strikingly enough, La Forge's *Remarques*, in which he seeks to defend Descartes' 'man', are underlain by a totally different conception of anatomy. The machine analogy is also used in order to indicate the scientific place that one should give to the direct observation of the bodily components, but in an exact opposite way. For La Forge, the different wheels of the clock are perceptible by our senses, when the inner parts of the human bodies are not. As a result, it may suffice to observe the shape and situation of the wheels in order to understand how the clock indicates the hours, when, in the case of the human body, we are compelled to resort to rational hypotheses about the structure of invisible parts:

> I believe that one will not deny that, if, by the sole inspection of the conformation of every perceptible part, one could conceive how everything that is observed in the human body is accomplished, in the same manner as one clearly understand in which manner every movement of a clock is accomplished when one has examined the shape and the situations of all its wheels, one would never have guessed that there were in the human body a quantity of things that are not perceived in any way by our senses.[46]

There 'is a lot of things in the human Body that our senses do not perceive in any way'.[47] For those reasons, it is not only irrelevant to dismiss Descartes's anthropology with anatomical observations, but it is also inappropriate to pretend explain more generally the functioning of the human body by means of anatomy. La Forge emphasizes the clear separation between the invisible causes invoked by philosophers and the exterior figures observable by anatomists: according to him there is no more connection between the figures observed by anatomists and the functions executed by the human body than between the exterior structure of the clock and its capacity to indicate the hours. And indeed, the apparent structure that it is observable can be changed in various manners without altering the function of the clock:

> The author [Descartes] does not say simply that these functions do not depend on the figure of the parts and on those of the cavities of the brain; otherwise he would contradict himself, since our body, being regarded as deprived from a reasonable soul that would be united to it, is nothing more than an Automaton, of which every movement depend on the conformation of its parts. But he claims that these [functions] do not depend on the figure that may

[46] 'Remarques de Louis de La Forge', in *L'homme de René Descartes et un Traité de la formation du fœtus du même auteur, avec les remarques de Louis de la Forge* (Paris: Charles Angot, 1664), 171–408, 215: 'Je crois que l'on ne niera pas non plus, que si par la seule inspection de la conformation de toutes les parties sensibles, nous pouvions concevoir comment se fait tout ce qui se remarque dans le Corps humain, de la même façon que l'on comprend clairement de quelle manière se font tous les mouvements d'une horloge quand on a examiné la figure et la situation de toutes ses roues, on n'aurait jamais supposé qu'il y a quantité de choses dans le Corps de l'homme que les sens n'aperçoivent en aucune façon.'

[47] *Ibid.*, 216: 'Ce n'est pourtant pas que [ces hypothèses] soient fausses à cause que les sens n'en découvrent rien ; nous serions bien ignorants si nous devions douter de tout ce que nous ne voyons point.'

> be observed by the senses. As if a clockmaker said that the power according to which a watch indicates the time does not come from its outwards shape, all the more so because it may be changed in thousand ways without stopping to produce the same effect.[48]

In sum, for Steno the analogy between the body and a machine suggests an analytical model for explaining the actions of the human body: it is legitimate to consider the human body as a machine than one can and should dismantle if one wants to explain the way it functions. For La Forge, it is impossible to explain the functioning of the human body by the mere observation of its components as it is the case for the clock, since most often the key components of the human body are precisely not observable. In both cases however, for La Forge and Steno, the image of the machine is not used in order to model some definite physiological process, but, far more generally, to demonstrate or invalidate the relevance of the autopsy of a human body for the explanation of its living functioning.[49] The analogy allows one to promote a certain kind of connection between the different branches of medical sciences or between the different parts of natural philosophy.

11.5 Conclusion: Mechanism and Anatomy as Polysemic Labellings

Those two uses of the comparison between the clock and the human body in La Forge and Steno imply two opposite understandings of the notion of visibility and of the role of observations. By means of the machine analogy, La Forge points out the opposition between what is inside the body and remains invisible and what is outside and observable. The underlying idea is that the inside of human body will always escape our powers of observation. Accordingly, anatomy cannot display the hidden causes of the observed effects; it can only display some intermediary effects.[50] On the contrary, according to Steno's understanding of the machine

[48] *Ibid.*, 287: 'L'Autheur ne dit pas simplement, que ces fonctions ne dependent point de la figure des parties & de celles des cavités du cerveau, autrement il se contrediroit; car nostre corps, estant consideré comme n'ayant point d'Ame raisonnable qui luy soit unie, n'est rien autre chose qu'un Automate, de qui tous les mouvemens dependent de la conformation de ses parties: Mais il declare qu'elles ne dependent pas de la figure exterieure qui peut tomber sous les sens; Comme si un horloger disoit que ce n'est. pas de la forme exterieur d'une monstre, que vient le pouvoir qu'elle a de monstrer les heures, d'autant qu'elle peut être changée en mille façons, sans qu'elle cesse d'avoir le mesme effet.'

[49] About this distinction, see S. Roux, 'Quelles machines pour quels animaux? Jacques Rohault, Claude Perrault, Giovanni Alfonso Borelli', in A. Gaillard, J.-Y. Goffi, B. Roukhomovsky and S. Roux (eds.), *L'automate. Machine, métaphore, modèle, merveille* (Pessac: Presses universitaires de Bordeaux, 2013), 69–113.

[50] See C. Salomon-Bayet, *L'institution de la science et l'expérience du vivant* (Paris: Flammarion, 1978), 180–181 : 'la dissection ne donne pas la raison des faits de surface, elle donne une autre série de faits constatés à un autre niveau.'

analogy, the inside is always likely to be rendered observable by the new techniques and the dexterity of the observers: regarding bodily structures, the inside and the outside are relative notions. The aim of dissections is precisely to push back the limits of the experimental analysis.[51] For him there is not an insurmountable separation between the main organs observed by the anatomist at first sight and the unseen internal mechanisms according to which these organs move and function together. Hence, it is not wrong to say that the machine analogy promotes an epistemological model according to which the intelligibility of corporeal phenomena is contingent upon their visibility. As Guenancia put it, 'the machine is by excellence the example of the integral visibility to which science is supposed to reduce all natural phenomena.'[52] But the conceptions of this visibility may vary widely. This visibility may just be a model of intelligibility, and in this case one may then conceive of small unobservable components, like Descartes did in *L'Homme*. Or this visibility may be an experimental requirement that implies to really observe the components put forward, like Steno recommended. Those two opposite understandings of the intelligibility of the human body and of the machine analogy may be seen however as two readings of only one Cartesian claim: the claim that the different functions of the human body 'follow naturally from the disposition of its organs alone.'[53] Indeed, on this last point, La Forge and Steno would agree. But this agreement indicates nothing specific about the way we have access to the disposition of the organs (experimentally or by conjectures?), or about what we can call an organ (just the main organs, the most subtle parts and fluids or the unobservable corpuscles mentioned by Descartes?).

These two concurrent readings of Descartes precisely gave rise to two different understandings of the specificity of mechanism in respect to anatomy. The first understanding, following La Forge, identifies medical mechanism with the hypothetical reconstitution of hidden structures and movements.[54] The other understanding associates mechanism with an analytical method that experimentally brings back the functioning of a whole to the description of its various components.[55]

---

[51] Steno mentions sometimes the *'analysis sensibus'*, see the *Elementorum Myologiae Specimen seu Musculi Descriptio Geometrica* (Florentiae: ex Typographia sub signo Stellae, 1667), 4.

[52] P. Guenancia, 'La signification de la technique dans le *Discours de la méthode*', in H. Méchoulan (ed.), *Problématique et réception du* Discours de la méthode *et des* Essais (Paris: Vrin, 1988), 213–223, 215. See also C.Wilson, *The Invisible World: Early Modern Philosophy and the Invention of the Microscope* (Princeton : Princeton University Press, 1995), 113.

[53] Descartes, *L'homme*, AT XI, 202.

[54] Claude Perrault adopts for instance this kind of epistemological model. See *Œuvres de physique et de mécanique de Mrs. C. & P. Perrault, de l'académie royale des sciences et de l'académie française* (Amsterdam: chez J.-F. Bernard, 1727), vol. IV, 513: the dissection displays only the outside of the organs; conjectures and reflections are needed to go further (orig. 'La dissection, qui présente à l'œil la composition et la structure artificieuse de toutes les parties des organes, n'en fait voir, pour ainsi dire, que le dehors. Pour être instruit autant qu'il est possible, de ce qui se fait dans les organes, il faut entrer plus avant, et passer outre, si l'on peut, par l'entremise des conjectures et des réflexions que les différents phénomènes peuvent fournir').

[55] It is the kind of mechanism that epistemologists identify today in R. Cummins' understanding of the notion of 'function' (see 'Functional Analysis', *Journal of Philosophy*, 1975, 72, 741–764).

It seems to us that these two conceptions of what defines Cartesian mechanism and its relationship to anatomy are often not distinguished. This would explain the contradictory claims about the place of anatomy in Descartes' natural philosophy that we mentioned in introduction.

Accordingly, Steno's *Discourse* may be seen as a mechanist manifesto only if one interprets 'mechanist' as 'analytical' and finds decisive the very occurrence of the machine analogy. But this machine analogy was very common and sometimes used to defend opposite claims. In the late seventeenth century, there was neither agreement on the exact definition and role of anatomy, nor on the exact implications of the general comparison between the human body and a clock. Accordingly, it does not seem relevant to relate the so-called biological 'mechanism' with a certain epistemological priority granted to anatomy understood as the mere description of bodily structures.

Bibliography

Andrault R. Mathématiser l'anatomie: la myologie de Stensen. Early Sci Med. 2010;15(4–5):505–36.

Andrault R. La vie selon la raison. Physiologie et métaphysique chez Spinoza et Leibniz. Paris: Champion; 2014.

Andrault R. La raison des corps. Mécanisme et sciences médicales. Paris: Vrin; 2016.

Ariew R. Descartes and the first cartesians. Oxford: Oxford University Press; 2014.

Bartholin T. Anatomia ex Caspari Bartholini parentis Institutionibus, omniumque recentiorum et propriis observationibus, tertium ad sanguinis circulationem reformata, cum iconibus novis accuratissimis. Lugdunum Batavorum: Franciscum Hackium; 1651.

Bartholin T. Anatome ex omnium veterum recentiorumque observationibus imprimis Institutionibus b. m. parentis Harvejanam, et vasa lymphatica quartum renovata. Lugdunum Batavorum: ex Officina Hackiana; 1673.

Bitbol-Hespériès A. Cartesian physiology. In: Gaukroger S, Schuster J, Sutton J, editors. Descartes' natural philosophy. New York: Routledge; 2000.

Brown H. Scientific organizations in seventeenth-century France (1620–1680). Baltimore: William & Wilkins Cie; 1934.

Canguilhem G. Knowledge of life. In: Marrati P, Meyers T, editors. (trans: Geroulanos S). New York: Fordham University Press; 2008.

Chapelain J. Lettres. In: Laroque T, editor. Paris: Imprimerie Nationale; 1883.

Cummins R. Functional analysis. J Philos. 1975;72:741–64.

Cunningham A. The pen and the sword: recovering the disciplinary identity of physiology and anatomy before 1800. Old physiology – the pen. Stud Hist Philos Biol Biomed Sci. 2002;33:631–65.

Cunningham A. Old anatomy – the sword. Stud Hist Philos Biol Biomed Sci. 2003;34:51–76.

Descartes R. De Homine, figuris et latinitate donatus a Florentio Schuyl. Lugdunum Batavorum: Petrum Leffen & Franciscum Moyardum; 1662.

Descartes R. L'homme de René Descartes et un Traité de la formation du fœtus du même auteur, avec les remarques de Louis de la Forge. Paris: Charles Angot; 1664.

Descartes R. Œuvres. New presentation by B. Rochot and P. Costabel. In: Adam C, Tannery P, editors. Paris: Vrin-CNRS; 1964–1974.

Descartes R. Philosophical essays and correspondence. In: Ariew R, editor. Indianapolis/Cambridge: Hackett; 2000.

Descartes. The world and other writings. (trans. and editor Gaukroger S). Cambridge: Cambridge University Press; 2004.

De Houdeville, editor. Journal des savants, I (1665–6), Amsterdam: Pierre Le Grand; 1685.
Gabbey A. What was 'mechanical' about *The mechanical philosophy* ? In: Palmerino CP, Thijssen JMMH, editors. The reception of the Galilean science of motion in seventeenth-century. Dordrecht: Kluwer Academic Publishers; 2004. p. 11–23.
Gaukroger S, Schuster J, Sutton J, editors. Descartes' natural philosophy. New York: Routledge; 2000.
Gaukroger S. The resources of a mechanist physiology and the problem of goal-directed processes. In: Gaukroger S, Schuster J, Sutton J, editors. Descartes' natural philosophy. New York: Routledge; 2000. p. 383–400.
Grell OP. Between anatomy and religion: the conversions to Catholicism of the two Danish anatomists Nicolas Steno and Jacob Winsløw. In: Grell OP, Cunningham A, editors. Medicine and religion in enlightenment Europe. Aldeshot: Ashgate; 2007. p. 205–21.
Guenancia P. La signification de la technique dans le *Discours de la méthode*. In: Méchoulan H, editor. *Problématique et réception du* Discours de la méthode *et des* Essais. Paris: Vrin; 1988. p. 213–23.
Huygens C. Œuvres complètes. In: Société hollandaise des Sciences, editor. Den Haag: Martinus Nijhoff; 1880–1950.
La Forge, Louis de. Traité de l'esprit de l'homme et de ses facultés et foncitons et de son union avec le corps suivant les principes de René Descartes, Paris: Theodor Girard, 1666.
La Forge Louis de. Treatise on the human mind. (editor and trans. Clarke D). Dordrecht: Springer; 1997.
Leibniz GW. Sämtliche Schriften und Briefe. Berlin: Akademie-Verlag; 1923.
Perrault C. Œuvres de physique et de mécanique de Mrs. C. & P. Perrault, de l'académie royale des sciences et de l'académie française. Amsterdam: chez J.-F. Bernard; 1727.
Roger J. Les sciences de la vie dans la pensée française du XVIIIe siècle. La génération des animaux de Descartes à l'Encyclopédie. Paris: Armand Colin; 1993.
Roux S. Was there a Cartesian experimentalism in 1660s France? In: Dobre M, Nyden T, editors. Cartesian empiricisms. Dordrecht: Springer; 2013a. p. 47–88.
Roux S. Quelles machines pour quels animaux? Jacques Rohault, Claude Perrault, Giovanni Alfonso Borelli. In: Gaillard A, Goffi J-Y, Roukhomovsky B, Roux S, editors. L'automate. Machine, métaphore, modèle, erveille. Pessac: Presses universitaires de Bordeaux; 2013b.
Roux S, Garber D. Introduction. In: Roux S, Garber D, editors. The mechanization of natural philosophy. Dordrecht: Springer; 2013.
Salomon-Bayet C. L'institution de la science et l'expérience du vivant. Paris: Flammarion; 1978.
Schiller J, Théodoridès J. Sténon et les milieux scientifiques parisiens. In: Scherz G, editor. Steno and brain research in the seventeenth century. Oxford: Pergamon Press; 1968.
Spinoza B. Complete works. In: Morgan ML.(trans: Shirley S). Indianapolis/Cambridge: Hackett; 2002.
Sprengel K. Versuch einer Pragmatischen Geschichte der Arzneikunde. Vierter Theil. Halle: bei Johann Jacob Gebauer; 1801.
Steno N. Elementorum myologiae specimen seu Musculi descriptio geometrica. Florentiae: ex Typographia sub signo Stellae; 1667.
Steno N. Discours sur l'anatomie du cerveau. Paris: Robert de Ninville; 1669a.
Steno N. De Solido intra Solidum naturaliter contento dissertationis Prodromus. Florentiae: ex Typographia sub signo Stellae; 1669b.
Steno N. Opera theologica. In: Larsen K., Scherz G, editors. Hafniae: A. Busck; 1944.
Steno N. Epistolae et epistolae ad eum datae, quas cum prooemia ac notis germanice scriptis. In: Scherz G, editor. Hafniae: A. Busck; 1952.
Steno N. Lecture on the anatomy of the brain. In: Scherz G, editor. Hafniae: A. Busck; 1965.
Steno N. Chaos-manuscript, Copenhagen 1659. (complete edition editor and trans. Ziggelaar A). Copenhagen: Munksgaard; 1997.
Steno N. Discours sur l'anatomie du cerveau. In: Andrault R, editor. Paris: Classiques Garnier; 2009.
Steno N. Nicolaus Steno: biography and original papers of a 17th century scientist. In: Kardel T, Maquet P, editors. Berlin: Springer; 2013.

Tolmer LJAL. Pierre-Daniel Huet, humaniste physicien. Bayeux: Colas; 1949.

Totaro P. 'Ho certi amici in Ollandia': Stensen and Spinoza – science verso faith. In: Ascani K, Kermit H, Skytte G, editors. Niccolo Stenone. Anatomista, geologo, vescovo. Rome: L'Erma; 2002. p. 27–38.

Wilson C. The invisible world: early modern philosophy and the invention of the microscope. Princeton University Press: Princeton; 1995.

Chapter 12
The Art of Cartesianism: The Illustrations of Clerselier's Edition of Descartes's *Traité de l'homme* (1664)

Steven Nadler

Abstract One of the more difficult tasks that Clerselier faced in bringing out his 1664 edition of the *Traité de l'homme* was securing the illustrations, eventually composed by La Forge and Gutschoven. After considering the chronology of this frustrating process, which is interesting in its own right, I will examine the illustrations themselves, comparing them with Schuyl's illustrations for his 1662 Latin edition, and especially in the light of what Clerselier says were the intended purpose of such illustrations.

By early 1662, Claude Clerselier's patience was wearing thin. The Parisian lawyer was in possession of a manuscript copy of a treatise by the century's greatest philosopher, and several years of effort to get it into print still had not paid off. The text itself was ready to go to press, but some essential artwork he had commissioned to illustrate the work was long overdue. To make matters worse, a Latin translation of the treatise would soon appear with a foreign publisher. Clerselier was concerned that this translation was based on a poor, unreliable copy of the original, and so he tried to postpone the book's appearance with a polite letter. But a failure to communicate due to the Dutch translator's insensitivity to the niceties of the French language had led to the impression that Clerserlier was, in fact, endorsing the project.[1]

Clerselier was convinced that his edition of the treatise, in the original French, would be better than the Latin version: more authentic, based on what he claims to

[1] See Clerselier's chronology leading up to the publication of his edition of the *Traité de l'homme* in his preface to the 1664 volume: *L'Homme de René Descartes et un traitté de la formation du foetus du mesme autheur, avec les Remarques de Louis de la Forge* (Paris: Charles Angot, 1664; henceforth, "Preface").

S. Nadler (✉)
Department of Philosophy, University of Wisconsin-Madison,
5185 Helen White Hall, 600 North Park St, Madison, WI 53706, USA
e-mail: smnadler@wisc.edu

D. Antoine-Mahut, S. Gaukroger (eds.), Descartes' *Treatise on Man* and its Reception, Studies in History and Philosophy of Science 43,
DOI 10.1007/978-3-319-46989-8_12

be the philosopher's autograph manuscript,[2] and the most valuable and marketable to a broad European audience.

If only he had those illustrations.

12.1 The Background

In late 1653 or early 1654, Clerselier acquired many of the papers left behind by his late friend, René Descartes, who, after living for many years in the Netherlands, had died in Sweden in 1650.[3] As Descartes's de facto literary executor, Clerselier's first act was to bring out an edition of much of the philosopher's correspondence. Starting in 1657, and over the next ten years, he would produce three volumes representing four decades of letters to and from Descartes. He did exercise a little editorial discretion in his selection of letters, and often had to do serious reconstructive work from old and mutilated scraps of paper. Moreover, in his preface to the second volume of the *Lettres de Mr. Descartes* (1659), Clerselier tells the reader that, unlike the first volume, which contained a mix of letters in Latin and French, the contents of this one are entirely in French. Various booksellers had complained that the first volume did not sell well, just because of the large number of letters it contained in Latin, and "plusiers personnes qui n'ont point de commerce avec cette langue ne l'avoient pas l'acheté."[4]

Now, while working on the third volume of letters—it would not appear until 1667, in part because of the time required for him to draw the relevant illustrations ("parce qu'il m'a fallu beaucoup de temps pour en tracer les figures"[5])—Clerselier was also putting in order two of the unpublished philosophical treatises Descartes had left behind (and that may or may not have been found among the inventories of

[2] In fact, there is reason to believe that, despite what he says, Clerselier did *not* have an autograph original of the *Traité de l'homme*; see Sylvain Matton, "Un témoignage oublié sur le manuscrit du *Traité de l'homme* de Descartes", *Bulletin Cartésien* XXXVI, in *Archives de Philosophie* 71 (2008): 148–9. Theo Verbeek has suggested to me that what Clerselier possessed was a copy that was no less removed from an original than the copy used by Florent Schuyl for his 1662 Latin translation. For the provenance of Descartes's posthumous papers, including those found in the "Leiden suitcase" and those listed in the Stockholm inventory, see the "Introduction" by Verbeek and Erik-Jan Bos in *The Correspondence of René Descartes, 1643*, eds. Theo Verbeek, Erik-Jan Bos, and Jeroen van de Ven (Utrecht: Zeno Institute, 2003), xvi-xxiii. See also Franco Aurelio Meschini, "Filologia e scienza. Note per un'edizione critica de *L'Homme* di Descartes", in Franco Aurelio Meschini, *Le opere dei filosofi e degli scienziati. Filosofia e scienza tra testo, libro e biblioteche* (Florence: Olschki, 2011), 165–204; and the review of Meschini's book by Matthijs van Otegem, *Bulletin cartésien* XLII, 3.1.65. Meschini and Van Otegem seem to leave open the possibility that Clerselier did have an autograph original.

[3] For the dramatic but unconfirmed story of their delivery in Paris, see Adrien Baillet, *La Vie de Monsieur Descartes*, 2 vols. (Paris: Daniel Horthemels, 1691), II.428.

[4] *Lettres de Mr. Descartes* (Paris: Charles Angot, 1659), xii.

[5] *Lettres de Mr. Descartes* (Paris: Charles Angot, 1667), i.

Descartes's effects ca. 1653).[6] These were in fact, as Clerselier was the first to recognize, closely related parts of Descartes's earliest philosophical project. Taken together, they constitute the first and second sections of *Le Monde*, an ambitious work composed in the late 1620s and early 1630s in which Descartes lays out, for the first time, the fundamental principles of his philosophical system and uses them, with great audacity, to explain (as he says to his friend, the Minim priest Marin Mersenne) "tous les Phainomenes sublunaires."[7] Clerselier would begin with what he identified as the second part of *Le Monde*, the *Traité de l'homme*, an essay on human anatomy that presents "toutes [les] principales fonctions [de l'homme]" as if his body were just a machine, without assuming the presence and activity of a soul.[8] However, from certain letter-labels in the manuscript that he possessed, Clerselier could tell that this relatively short work was missing its illustrations. He knew, then, that before he could publish the treatise he would have to find a way to supply these; which is precisely where, like the final volume of letters, things seem to have gotten held up.

Clerselier, despite working on the "figures" for the correspondence, did not believe himself to have the artistic skills for this more specialized and detailed work; he claims that he did not feel "assez fort pour les inventer de moy-mesme."[9] Thus, he was quite thrilled—he calls it "une faveur du Ciel"—when, in 1657, he learned from Louis and Daniel Elzevirs, of the Dutch publishing firm which that year had brought out the first volume of Descartes's letters, that a man named Huyberts had already done some illustrations for the treatise.[10] Clerselier feared that Huyberts had worked from an unreliable copy of the manuscript, however, and so he sent a fair copy of what he claimed to be the original to the Elzevirs to forward to Huyberts, in order that he might refine his figures accordingly. Huyberts wrote back to Clerselier promising to deliver the illustrations "quand il leur auroit donné la meilleure forme qu'il auroit pû."[11] Unfortunately, this seems to have been the end of the trail. Clerselier never heard back from Huyberts—perhaps, Clerselier suspects, because

[6] It is not entirely clear *how* Clerselier came into possession of the *Traité de l'homme*. It may have been among the items conveyed to him by his brother in-law, Descartes's friend, and French ambassador to Sweden Pierre Chanut (who in turn got it from either (a) the Leiden or Stockholm inventories of Descartes's effects, although it is not listed in the latter [see AT X.5–12]; or (b) some other source, as the document published by Matton ["Un témoignage oublié"] implies). Or, as Verbeek and Bos suggest (*The Correspondence of René Descartes*, xxi–xxii), the manuscript may have been given to Clerselier by Tobias Andreae. Or perhaps he received several copies of the manuscript, from these different sources.

[7] To Mersenne, 8 October 1629, AT I.23.

[8] To Mersenne, Nov. or Dec. 1632, AT I.263. Clerselier would not publish the first part, the *Traité de la lumière*, until 1677, when he brought out both parts of *Le Monde* together in a single volume.

[9] "Preface", v-vi (there is no pagination in the original text, but these are the page numbers that would be there).

[10] If Huyberts did in fact produce illustrations for *L'Homme*, he must have had access to one of several manuscript copies of the treatise that were apparently circulating in the Netherlands among a small coterie of Cartesian devotés.

[11] "Preface", xvii.

the latter had become rather ill and maybe even died—and he soon gave up hope on this lead.

In April of 1659, Clerselier made a second attempt to secure illustrations. He believed himself to have a natural candidate in Henricus Regius (Henri le Roy), a professor of medicine at the University of Utrecht and an unorthodox (and somewhat troublesome) disciple of Descartes. Given Regius's Cartesian credentials and his medical background, Clerselier could say that "je ne connoissois alors personne, que je crusse plus capable d'executer ce dessein."[12] Regius had had an unpleasant falling out with Descartes some years earlier—Descartes did not feel that Regius was representing his philosophy in a proper and responsible way—and there were bitter recriminations on both sides. Regius was also not on good terms with Clerselier, since he apparently resented Clerselier's including Descartes's letters to him in the first volume of correspondence without permission, while Clerselier had taken Regius to task for removing in the second edition of his *Fundamenta physices* the praise of Descartes that he had included in the first edition. But a collaboration on the *Traité de l'homme* seemed to Clerselier like a good opportunity to mend fences and for Regius to publicly reconfirm his "ancienne amitié" with the late Descartes and re-enter the good graces of his followers.

Regius, however, declined the offer. As Clerselier tells it, Regius was afraid that, were he to work on the illustrations for the *Traité de l'homme*, people would suspect that he had long been familiar with the treatise, and thus that his own work in physics and human anatomy was nothing but a plagiarization of Descartes. Regius' relationship with Descartes had been poisoned in part by disagreement over who should get credit for what, and he was again worried about being perceived as the pupil and not as the master. (Clerselier, for his part, was skeptical that Regius could, all on his own, really have come up with exactly the same ideas—and the same words!—as Descartes, and he claimed to have the letters to prove who really learned from whom.[13])

With Regius's rejection of the commission, it was back to square one.

Clerselier at this point—probably the summer of 1659, just after the disappointment with Regius—decided to make a more public plea for help with the illustrations. In the final paragraph of his preface to the second volume of Descartes's letters that was published later that year, he wrote that among the items from Descartes "qui m'ait esté mis entre les mains par celuy qui a esté le Depositaire de tous les biens de son Esprit"—that is, by Pierre Chanut, Descartes's friend, France's ambassador to Sweden and Clerselier's brother in-law[14]—is a work titled *L'homme*, an "ouvrage tout à fait curieux auquel il eût esté à souhaiter pour sa derniere perfection, que son Autheur y eût pû mettre la main luy-mesme." Clerselier notes that he is now trying to put this work into the best form possible. But because, among other

[12] "Preface", vi.

[13] "Preface", vii–viii.

[14] Whether Clerselier did in fact receive the manuscript from Chanut (or exclusively from Chanut) is open to question; see note 6 above.

things, it lacks the necessary illustrations, "j'invite tous les Sçavans de me vouloir aider à les suppléer." He continues this solicitation as follows:

> Si quelque obligeante personne, jalouse de la reputation de Mr Descartes, & de la sienne propre, vouloit s'offrir à ce glorieux travail, je le prie de vouloir m'en donner avis. Fût-il Estranger, pourvû qu'il me donne de seûres & de fidelles adresses, je luy feray mettre entre les mains tout ce qui sera necessaire; & ne stipuleray point d'autre condition avec luy, sinon que ce Traitté ne sera point imprimé en nostre Langue dans les Païs Etrangers, qu'il ne l'ait premierement esté en France.

At around the same time, Clerselier, persuaded that Descartes himself must have created some illustrations—or even just sketches—for the treatise to match up with the labels in the manuscript, sent a young emissary to the Netherlands to see what he could discover. Pierre Guisony did not find any figures by Descartes during his travels, but in Flanders he did come into contact with Gerard van Gutschoven (1615–1668), a professor at the University of Louvain. As a physician, anatomist, and mathematician—and as someone who, during a visit to Holland in the late 1630s, got to know Descartes personally and admired his philosophy—Gutschoven seemed perfectly well qualified to take on this task. Best of all, he was willing to do so.

Unfortunately, Clerselier did not have any information on how to send Gutschoven the manuscript. After a year went by with still no word from Louvain, Clerselier was ready to write this off as yet another dead-end. However, in mid-1660, he received a visit in Paris from one Monsieur de Nonancourt, a Flemish gentleman who was acquainted with Gutschoven and who brought word that the Louvain professor was indeed still interested in doing the illustrations, and he would begin just as soon as he received a copy of the manuscript.

By great coincidence, on that very same day (or so he says) Clerselier also received a letter out of the blue from a man, previously unknown to him, named Louis de la Forge, a medical doctor in Saumur, France, who was a devotee of Cartesian philosophy. La Forge had learned of the *Traité de l'homme* project from Clerselier's preface to the second volume of Descartes letters, and was in fact quite eager to contribute to the edition and do the illustrations, as well as provide some commentary on the text.

Clerselier now faced an embarrassment of riches. He had two highly qualified people willing and exceptionally able to provide the book with the requisite figures. On the one hand, he did not want to insult Gutschoven by now rescinding the invitation; on the other hand, he did not want to pass up the opportunity being offered by La Forge, "une personne de nostre Nation, dont la profession respondoit à la connoissance que demondoit dans un homme le travail auquel il s'offroit."[15] Clerselier decided to make a kind of competition of it and have them both do the work, independently of each other. In fact, he says he would let neither know that someone else was preparing illustrations as well, lest they become "paresseux et negligens" and unwilling to put their best effort into a job in which there was a risk that they would

[15] "Preface", xi.

not derive all the glory.[16] He gave each of them a copy of the manuscript, and his plan was that at each point in the book where an image was needed he would use whoever's illustration was better.

All of this was neatly arranged by the fall of 1660 at the latest.[17] At around the same time, Clerselier also learned from Descartes's friend Alphonse Pollot about the Latin translation of the *Traité de l'homme* being prepared in Leiden by Florent Schuyl (1619–1669), a naturalist and professor of philosophy at the university there. Pollot also told Clerselier about some illustrations ("quelques figures") for the treatise by Descartes himself that Schuyl had in his possession. Clerselier thus wrote to Schuyl and asked if he would send him Descartes's images, a copy of the manuscript he was using for the translation, and any illustrations he may have himself done for the Latin edition. "Si je les trouvois justes", Clerselier added in his letter, he would use them in his own edition.[18]

Schuyl was happy to accommodate Clerselier's request and forwarded everything to Paris: Descartes's drawings—Schuyl refers to them as "two figures crudely drawn by the hand of Monsieur Descartes"[19]—his own drawings for the text (or prints thereof), and a copy of the manuscript he was using. Clerselier was careful not to show any of Schuyl's illustrations to either La Forge or Gutschoven, so that they might do their best work on their own. (Even after receiving a copy of Schuyl's book in 1662, Clerselier says he did not bring it to the attention of his two anatomist-artists, in order to safeguard their claims of originality.)[20]

Less than a year after receiving La Forge's letter—and so probably by mid-1661—Clerselier received from the Saumur physician not only the promised illustrations and an extensive, detailed commentary on the treatise, but also a draft of La Forge's own *Traité de l'esprit de l'homme*, in which La Forge supplements Descartes's study of the human body with an account, on Cartesian principles, of the human mind and its relationship to the body.[21] However, Clerselier did not yet have Gutschoven's figures. A clean copy of the treatise itself was almost certainly ready to go to the printer by this time. It is possible that as late as the spring of 1662, Clerselier still did not have the other set of illustrations promised by his perpetually

[16] "Preface", xi.

[17] We have a letter from Clerselier to La Forge in December 1660 (it appears as the final entry in Clerselier's third volume of Descartes's correspondence, published in 1667), in response to at least one letter from La Forge, and so their arrangement must have been concluded before that.

[18] "Preface", iv.

[19] See the final paragraph of Schuyl's "Ad Lectorem", in *Renatus Des Cartes De homine figuris et latinatate donatus a Florentio Schuyl* (Leiden: Franciscus Moyardus, 1662; henceforth, *De homine*), n.p. : "Additis duabus figuris a Des Cartes rudi Minerva exaratis".

[20] "Preface", v.

[21] La Forge was, in fact, completing Descartes's project, since the discussion of the human body in the *Traité de l'homme* was, as Descartes says in the opening paragraph of the treatise, supposed to be followed by a description of "l'âme aussi à part" and then an account of "comment ces deux natures doivent être jointes et unies." In 1661, however, Clerselier felt that there was enough work to be done on *L'Homme* without taking on the project of publishing La Forge's treatise as well. Moreover, he felt that La Forge's work was not quite ready, that it needed a bit more work ("il luy faut encore quelques coups de lime pour la mettre à sa perfection" ["Preface", xii]).

tardy Dutch collaborator.[22] From Clerselier's account of this whole process in his preface to the *Traité de l'homme*, however, we can infer that Gutschoven's figures did arrive sometime *before* the publication of Schuyl's Latin translation in later 1662.[23] Still, it would be two more years before Clerselier could finally bring out his own, French edition of Descartes's *Traité de l'homme* (henceforth, *L'Homme*), and he would have to wait another thirteen years before seeing his edition of its companion treatise, the *Traité de la lumière*, in print.

12.2 Clerselier's Selection

In the end, then, Clerselier had five sets of illustrations to consider for his edition of *L'Homme*. First, there were the two "crudely drawn" figures by Descartes that he is supposed to have received from Schuyl. Then there were Schuyl's own figures from the Latin translation, which Clerselier apparently had permission to use in his French version if he so wanted. He also had the images provided by La Forge and those finally rendered by Gutschoven. Moreover, as we shall see, Clerselier claims that at some point before publication he came across yet another, previously unknown original figure from Descartes.

Before considering the images themselves and assessing their value as illustrations of Descartes's text, let us first look at what purpose Clerselier believed such illustrations should serve. That is, what criteria did Clerselier have in mind when deciding which figures to include in his edition?

The project of *L'Homme* is to describe the human body (or an imaginary fascimile thereof) on its own—that is, a body unrelated to a human soul or any other immaterial animating power, such as an Aristotelian substantial form. In keeping with the strategic rhetorical posture adopted in the first part of *Le Monde* (the *Traité de la lumière*)—in which the origin, elements, and processes of the world that Descartes describes are presented as "une Fable"[24]—the body under consideration in *L'Homme* is only a hypothetical body of a hypothetical being, although the ruse is rather thin and Descartes often calls attention to this body's resemblance to the actual human body. The hypothetical body is a kind of machine:

> Je suppose que le corps n'est autre chose qu'une statue ou machine de terre, que Dieu forme tout exprès, pour la rendre la plus semblable à nous qu'il est possible … elle imite toutes celles de nos fonctions qui peuvent être imaginées procéder de la matière, et ne dépendre que de la disposition des organes.[25]

---

[22] Clerselier says in his preface to the 1664 edition that if not for Gutschoven's delay, Descartes's *Traité de l'homme* could have been published two years earlier.

[23] Clerselier had Gutschoven's illustrations before Schuyl's book appeared in 1662, because he says that he when he received "le présent que Monsieur Schuyl m'avoit fait de son livre", presumably just after the book's publication, he already possessed "toutes les figures de ce Traité, que chacun de ces Messieurs avoit faites" ("Preface", v).

[24] AT XI.31.

[25] AT XI.119–20.

Clerselier was clearly aware of the general difficulty of Descartes's text. He admits that the density, even occasional opacity of Descartes's discussion might cause trouble for readers trying to grasp the finer details of his account of human anatomy, and especially the explanations of muscular motion, sensation, digestion, memory, and other phenomena that take place in the body. The text, Clerselier says, "est serré, et … dit beaucoup de choses en peu de paroles."[26]

The basic structure of Descartes's "homme"—including the circulatory system, the arrangement of nerves and muscles, and the constitution of the brain and its pores—may itself be rather simple; as Descartes says, "la Nature agit tousjours par les moyens qui sont les plus faciles de tous, et les plus simples."[27] But the text still demands serious effort to be followed and understood. It especially requires some work on the part of the reader's imagination, and not only because Descartes makes frequent reference to labeled elements of illustrations that are assumed to have originally accompanied the treatise. One needs to be able to picture what exactly Descartes is describing when, for example, he says that such and such a muscle has a number of interior valves that move back and forth over some opening in a nerve canal. Like the figures embedded in the demonstrations of Euclid's *Elements of Geometry*, the illustrations of *L'Homme* were not intended by Descartes to prove or confirm anything, but they were at least supposed to facilitate an understanding of the work's scientific content. As one scholar notes, "the text of Descartes, which was written entirely around the pictures, is often incomprehensible without them."[28]

Clerselier was concerned by this. He notes in his preface that what Descartes says in one article (which is about the muscles responsible for moving the eyelids) is "fort concis, et assez difficile à entendre sans figure, comme chacun le peut éprouver, s'il se veut donner la peine de lire le texte sans jetter les yeux dessus."[29] He is thus quite clear about what he expects from the illustrations. He says that they should be able to "servir à faire … entendre le texte" and "en rendre l'intelligence facile."[30] The figures are to function as aids for the reader in visualizing, literally—in seeing on paper—the bodily elements and processes that Descartes is describing and to help him understand how everything is supposed to work. The point of "les figures dans ce Livre" is "pour faire comprendre quelle est en cela la pensée de Monsieur Descartes"[31] and "seulement d'expliquer … ce que Monsieur Descartes avance dans son Livre, où il ne parle le plus souvent que de choses qui ne tombent point sous les senses, lesquelles il a fallu rendre sensibles, pour faire qu'elles devin-

---

[26] "Preface", xv.

[27] AT XI.201.

[28] Claus Zittel, "Conflicting Pictures: Illustrating Descartes' *Traité de l'homme*", in Sven Dupré and Christoph Lüthy, eds., *Silent Messengers: The Circulation of Material Objects of Knowledge in the Early Modern Low Countries* (Münster: LIT Verlag, 2011), 217–60, p. 221.

[29] "Preface", xx-xxi.

[30] "Preface", xiii-xiv.

[31] "Preface", xx.

sent plus intelligibles."[32] The illustrations must allow even the non-specialist reader to "deviner sa [Descartes's] pensée."[33]

By the same measure, Clerselier does not intend the illustrations to provide insight as to what these elements and processes might actually look like were one to peer inside the body. Some of the things that Descartes talks about are not visible in fact; others are not even visible in principle, and so do not have a "look" that can be captured on paper. Other than the gross organs, musculature, connective tissues, and circulatory channels, many of the items in Descartes's anatomical picture are much too small to be seen by the naked eye, while some are presumably so minute that they would not be accessible even to microscopic view, given the state of the art of microscopy at the time. And a good number of the functional components of Descartes's *homme* are simply invisible—for example, the animal spirits. (Still other features of Descartes's explanations cannot possibly be represented in a single image: how, for example, can motion be depicted in one figure?) Conversely, many of the bare diagrammatic illustrations that Clerselier ended up using in the text do not look at all like anything that would be found in an actual human body.

In other words, what guided Clerselier in his selection of figures was their adequacy as illuminating schematic representations, not any kind of naturalistic verisimilude. Claus Zittel is thus quite right to insist that "the criterion of 'faithfulness to the object'—a confirmation of the image's truthfulness through personal observations—therefore had to be ruled out" as a criterion for assessing the adequacy of the images.[34] As Clerselier himself says, this is not an anatomy book.

> Si les figures ne ressemblent pas au Naturel, il ne s'en faut pas etonner, puisque l'on n'a pas eu dessein de faire un livre d'Anatomie, qui fist voir exactement comment les parties du Corps humain sont faites, et le raport ou proportion qu'elles ont entr'elles.[35]

He adds that nothing is easier than drawing pictures that naturalistically resemble the way things actually look. It is a much more difficult task to come up with illustrations that facilitate understanding of objects and mechanisms "qui ne tombent point sous les senses." The book needed to be accessible to lay readers into whose hands it might fall and who might not be willing or able to take the trouble to "examiner les choses de si près"; thus, Clerselier has, with the figures, attempted to tailor the treatise "à tout le monde.".[36]

With illustrations of this kind, however, what the reader sees on paper provides only the first step in the process of understanding. Because the figures are schematic diagrams and not naturalistic representations, they need to be aided by the reader's own imagination. Together with the text, they stimulate and guide the reader's mind to fill in the missing details. This is especially true when Descartes discusses various kinematic features of the circulatory, nervous, and sensory systems. The images do

[32] "Preface", xxv.

[33] "Preface", xviii.

[34] "Conflicting Pictures", 221.

[35] "Preface", xxv.

[36] "Preface", xvii-xviii.

not actually show the animal spirits inflating and deflating muscles; and when Descartes talks about how the pineal gland is moved this way or that because the spirits flow out of it toward this or that pore in the brain lining, all one can do is try to visualize what he has in mind. Clerselier seems to be aware of this, in so far as he notes that the illustrations by Gutschoven and La Forge still leave some things "à l'imagination à supléer."[37]

Clerselier suggests that at times the illustrations not only help the reader by supplementing the text and allowing him to visualize (with his senses and his imagination) what Descartes is talking about, but even do a better job than Descartes at explaining something, especially when the text is too short on words ("fort concis").[38] And if one image will not do the trick in clarifying a particular matter and making Descartes's meaning clear, Clerselier will include two or three, by different hands, so that, as he says, what was not seen through one might be seen through another. Thus, in one instance, related to Descartes's account of the motion of parts of the body, "le plus importante action que l'Autheur ait eu à décrire, et à expliquer", Clerselier saw fit to publish the relevant figures from Gutschoven, La Forge *and* Descartes himself. Referring to the three figures related to the explanation of the motion of the eye by its surrounding muscles (Figs. 12.1, 12.2 and 12.3), Clerselier says that "j'ay jugé necessaire de mettre celle que chacun a inventée; à cause qu'estant toutes trois differentes, ce qu'on ne pourra comprendre par l'une, sera peut-estre suplée par l'autre."[39] As we shall see, on at least one other occasion, Clerselier informs us, the illustrator—La Forge—went beyond the call of duty and actually corrected Descartes.[40]

12.3 The Illustrations

With the criteria of selection now at hand, let us consider the various options that Clerselier had and what he ended up doing with them.

12.3.1 Descartes's Figures

First, there were the two illustrations by Descartes that Clerselier says he received from Schuyl. The fate of these are a bit of a mystery. Schuyl included only *one* figure by Descartes in his Latin translation (Fig. 12.4); presumably it was one of the

[37] "Preface", xxv. Clerselier notes that the reader's imagination is especially required both when it comes to filling in the details that cannot be captured in woodcut printing and for supplying the motions that are to take place in the body, "lequel on ne sçauroit representer" (xxvi).

[38] "Preface", xx.

[39] "Preface", xiv-xv.

[40] "Preface", xix-xx.

Fig. 12.1 Gérard van Gutschoven's illustration of the muscles around the eye; from Clerselier's edition, *L'Homme* (1664), p. 16

drawings that he forwarded to Clerselier. But Clerselier did not include either it or its missing companion in his edition. From Schuyl's claim that these two images were poorly drawn ("duabus figuris a Des Cartes rudi Minerva exaratis"), we can infer that Clerselier found them unsuitable for his own purposes.

However, Clerselier says that independently he did come across another illustration by Descartes, which he assumed was meant for *L'Homme*. What he found, he

Fig. 12.2 Louis de la Forge's illustration of the muscles around the eye (*top*); *L'Homme*, p. 18

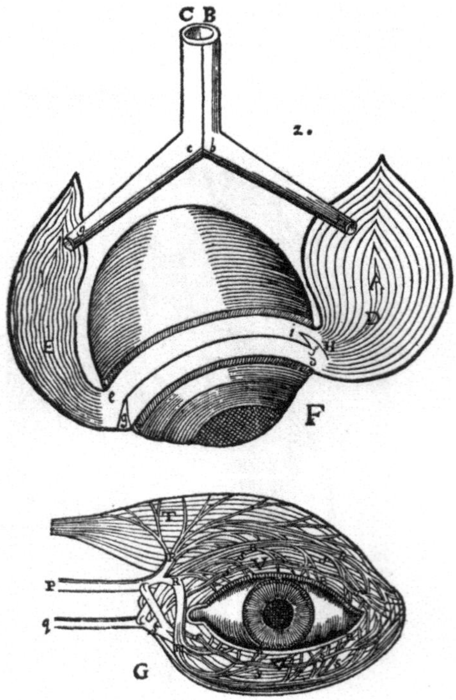

Fig. 12.3 Descartes's illustration (re-drawn by Clerselier) of the muscles around the eye; *L'Homme*, p. 17

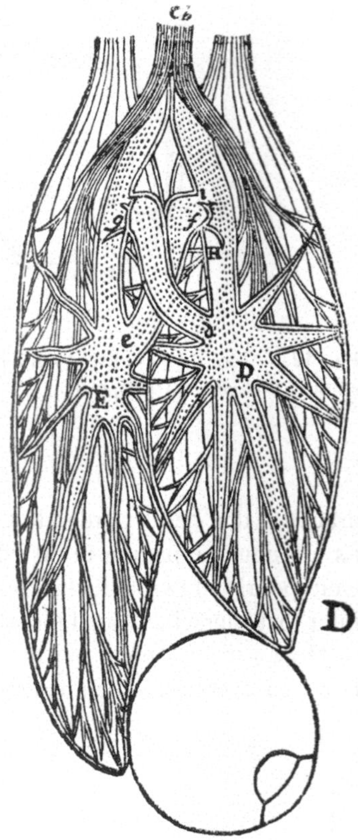

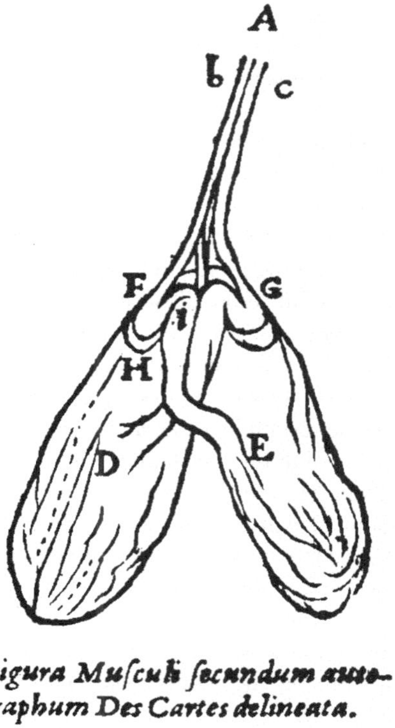

Fig. 12.4 An original drawing by Descartes of the eye muscles; from Florent Schuyl's edition, *De homine* (1662), p. 25

says, is a rough draft on a small, torn, disfigured piece of paper, something that was in such bad shape that someone else might easily have overlooked it:

> Dans un brouillon, que tout autre que moy auroit jetté au feu, tant il est petit, déchiré, et défiguré, j'ay trouvé un essay qu'avoit autrefois griffoné M. Descartes, lors qu'il tentoit les moyens de s'imaginer une figure qui pust répondre et satisfaire à ce qu'il avoit dans l'esprit.[41]

To judge from the sketch that Schuyl included in his edition, as well as some other figures hand-drawn by Descartes,[42] the philosopher seems not to have had great artistic skills, or at least not to have taken great care in such matters. Thus, the "brouillon" that Clerselier found needed to be redrawn and better rendered not just because of its poor physical condition and its low artistic quality (he calls it "mal desseiné"[43]), but also in order to serve as a proper illustration for the text. Clerselier did this himself as best he could. "La figure ... au bas de laquelle il y a un D, est une copie de ce brouillon de Monsieur Descartes, don't j'ay parlé cy-dessus, que j'ay

[41] "Preface", xv.

[42] See the insert following AT XI.634.

[43] "Preface", xviii.

tirée le mieux que j'ay pû."[44] Among the things he had to do to improve Descartes's drawing was clarify one of its details so it better fit the text. Apparently, Descartes's original image seemed to have three folds ("replis") on each valve in the nerve extending into the muscles around the eye and through which the animal spirits would flow, whereas the text speaks of only twofolds. After comparing the drawing with the text, however, Clerselier realized that what he was taking for a third fold was in fact only a small hook that serves to facilitate the transit of the spirits from one muscle to the next by holding down the valve to which it is attached.[45] Having improved the image to make all this clear, Clerselier then included it in his edition as one of the three illustrations of the eye muscles (Fig. 12.3). As we shall see, while Clerselier did correct Descartes's image for consistency with the text, what he did *not* do was correct it for scientific accuracy.

What is striking, however, is the degree to which this finished illustration can be seen to resemble, in a general enough way, the rough figure by Descartes that Schuyl included in his translation (Fig. 12.4), minus the eyeball but including many of the labels. This suggests two possibilities: either Schuyl and Clerselier each had a different sketch by Descartes of the same thing (the eye's muscles), or the sketch that was redrawn by Clerselier and included in his edition was not in fact some "brouillon" that he found on his own ("j'ay trouvé") but just the drawing that he received from Schuyl or saw in his book. What may seem to count against this second possibility, however, is the fact that the Descartes drawing published by Schuyl does not exhibit the ambiguous valve flaps that Clerselier claims had him confused. Moreover, Clerselier says that he still has Descartes's "brouillon" in his possession, and is willing to show it to "ceux qui en auront la curiosité."[46]

Another issue is the remarkable similarities in design between Clerselier's polished "D" (for "Descartes") image and Gutschoven's figure for the same text. These suggest that Clerselier might have referred to Gutschoven's rendering for guidance in his improvement of Descartes's drawing. It is hard to believe that, relying on the complicated text alone, Clerselier and Gutschoven would have come up with such similar figures. Perhaps, then, Clerselier took Descartes's rough sketch of the eye muscles that he found on his own (or saw in Schuyl's edition) and drew up a finer and more detailed version based on Gutschoven's illustration.[47] Why would he bother with an improved copy of Descartes's drawing if he had Gutschoven's meticulous illustration of the text in question? After all, he concedes that, with respect to

[44] "Preface", xviii.

[45] "Preface", xviii.

[46] "Preface", xviii.

[47] I am grateful to Annie Bitbol-Hespériès for sharing her insights on this with me through correspondence. She is skeptical about the discovered "brouillon", especially given that what Clerselier says about this alleged autograph drawing ("je le garde pour le faire voir à ceux qui en auront la curiosité" ["Preface", xviii]) is just like what he says about the autograph original of *L'Homme* he claims to have possessed ("l'original que j'ay, et que je feray voir quand on voudra" ["Preface", iii]), a claim about which many scholars are dubious (see note 2 above). Her view is that when Clerselier re-drew Descartes's drawing, he was influenced both by the crude figure by Descartes published by Schuyl and by Gutschoven's rendering.

these eye muscles, "je trouve que Messieurs de Gutschoven et de la Forge [les] ont mieux rencontré que Monsieur Descartes mesme, et que la disposition qu'ils ont donnée à la valvule et à ses deux replis, est plus conforme au texte, et le jeu de leur valvule plus aisé à comprendre."[48] Perhaps Clerselier wanted to be able to say that his edition, like Schuyl's, contains one of Descartes's own original figures.

12.3.2 Schuyl's Figures

Schuyl's Latin translation contains a large number of discrete, original illustrations, fifty-seven in all (some repeated throughout the text). Clerselier had access to all of them, both because Schuyl sent at least some of them to him before publishing his edition and because the book was published in 1662 while Clerselier was still at work on the French edition. Schuyl's illustrations include relatively schematic, bare-bone illustrations—literally, in some cases, as Schuyl depicts the human skeletal structure—of the relevant text (Fig. 12.5); and more elaborate and naturalistic representations (Figs. 12.6 and 12.7). Some of these latter images—I hesitate to call them 'figures', given their degree of artistic flair—include elements (such as landscapes, architectural ruins, even weapons of war and punishment) that are, strictly speaking, extraneous to their supposed role in elucidating the scientific and philosophical content of the text (Fig. 12.8).[49] Some of the illustrations even combine a realistic rendering with a schematic figure (Fig. 12.9).

Clerselier was impressed by Schuyl's artistic ability and the care he took in creating such detailed images. Compared to the figures he would end up putting in his own edition, he found Schuyl's illustrations expertly produced and aesthetically superior. He says that Schuyl's figures, which were done as copper-plate engravings, "emportent sans doute de beaucoup sur celles que j'ay fait mettre à ce Livre, si l'on a simplement égard à la graveure et à l'impression."[50] However, Clerselier ended up using none of Schuyl's material, at least directly. He finds "quelques défauts" in Schuyl's figures.[51] Some of these, it seems, are a matter of substantive content, in so far as Schuyl has used a faulty manuscript copy of the original work (or so Clerselier claims). But more important from Clerselier's point of view is the quality of the illustrations themselves. They simply do not suit his purposes. Schuyl's illustrations are "pour la pluspart ... moins intelligibles que les autres, et

[48] "Preface", xix.

[49] Rebecca M. Wilkin has argued, on the other hand, that these items in fact play a role in the overall philosophical message conveyed by Schuyl's illustrations, although she claims that, with their realism and worldliness, they function as a kind of *memento mori* and reminder of the transience of things human and thereby undermine Descartes's dualist project; see "Figuring the Dead Descartes: Claude Clerselier's *Homme de René Descartes* (1664)", *Representations* 83 (2003): 38–66.

[50] "Preface", ii.

[51] "Preface", iv.

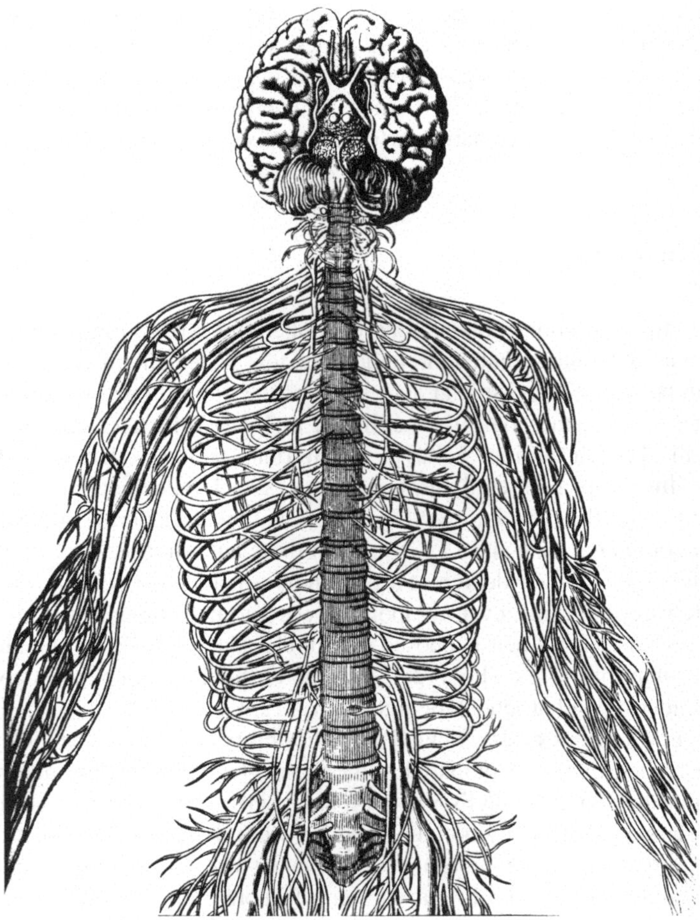

Fig. 12.5 Schuyl, *De homine*, fol. 110, fig. II

moins propres à l'intelligence du texte."[52] As naturalistic renditions, they are nicely done and well drawn, but they are too opaque, too fussy to do the work that Clerselier expects his book's illustrations to do. This will become particularly clear when we compare them with the figures that he received from Gutschoven and La Forge.

12.3.3 Gutschoven and La Forge

Clerselier received a large number of illustrations from his Dutch and French collaborators. We do not know how many figures altogether were submitted, but we know (from Clerselier's preface) that it was more than actually appear in the 1664

[52] "Preface", ii.

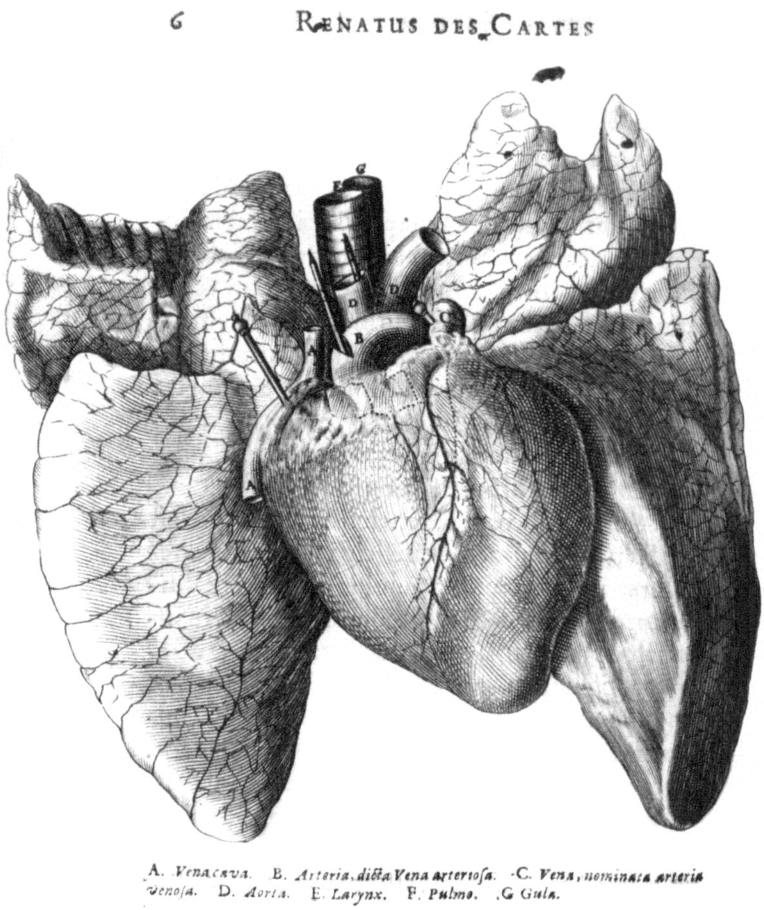

Fig. 12.6 Schuyl, *De homine*, p. 6

edition. He decided against including a figure either when it was unnecessary for clarifying a part of the text which was sufficiently clear in itself and easy enough for the reader to visualize on his own, or when it would have led to a needless duplication of illustrations (an exception being the instance mentioned above, where he decided he had good reason to include all three illustrations—by Gutschoven, La Forge, and Descartes—for the text discussing the eye muscles). In the end, Clerselier published forty figures altogether: six by La Forge, his own copy of the drawing by Descartes, and the rest by Gutschoven. If he had figures by La Forge and Gutschoven on the same text and there was no reason to use both, he generally went with the one by Gutschoven. Here is how Clerselier describes his selection process, and

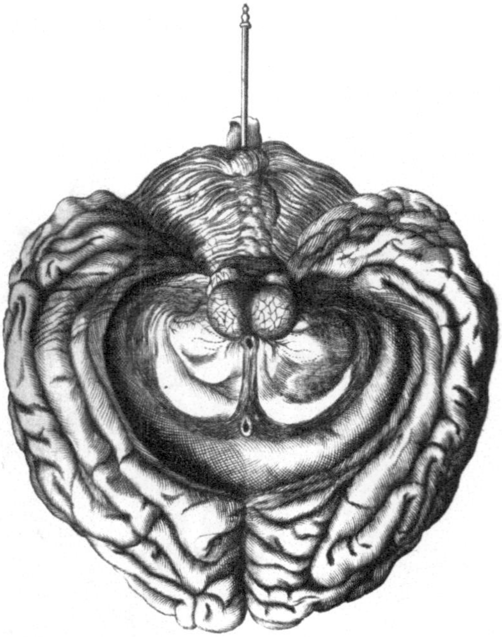

Fig. 12.7 Schuyl, *De homine*, fol. 118, fig. LIV

Fig. 12.8 Schuyl, *De homine*, p. 98

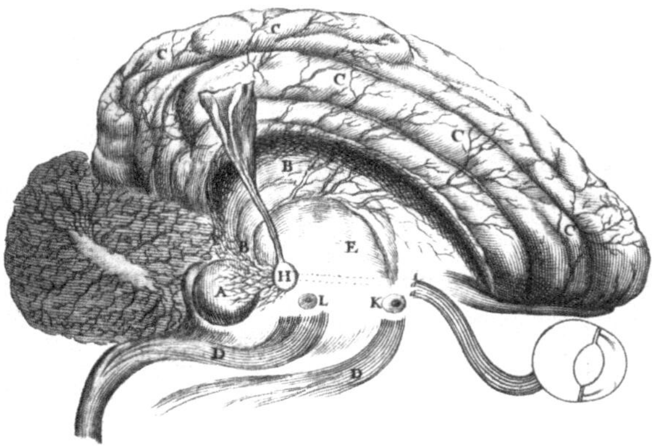

Fig. 12.9 Schuyl, *De homine*, p. 73

especially his decision to give priority to Gutschoven's illustrations over those of La Forge, primarily on the basis of their higher artistic quality:

> Comme la pluspart des Figures que ces deux Messieurs avoient tracées chacun à part, estoient semblables, ou que la difference qu'il y avoit entr'elles n'estoit pas essentielle, et ne regardoit que la disposition exterieure du corps de la figure, j'ay pensé qu'il estoit inutile de faire voir deux fois une mesme chose, et me suis contenté de me servir pour la pluspart des figures de M. de Gutschoven, qui estoient mieux dessinées que les autres; Mais pour celles où la difference estoit notable, et qui pouvoient servir à des usages particuliers, comme sont celles des muscles et du cerveau, je les ay mises des deux façons.[53]

Clerselier then notes that when he did end up reproducing figures by both of the illustrators, he labeled the one by Gutschoven with a 'G' and the one by La Forge with an 'F'. There are in fact only two instances where Clerselier included figures by both Gutschoven and La Forge for the same textual item: in the case of the eye muscles (Figs. 12.1 and 12.2), and in the case of the interior of the brain with the pineal gland (Figs. 12.10 and 12.11). It is odd, then, that Clerselier still labels quite a few of the solo figures by Gutschoven with a 'G', when there is no corresponding figure by La Forge.

In contrast with his assessment of Schuyl's figures, Clerselier was quite pleased with the illustrations that he received from Gutschoven and La Forge. First, he appreciated the degree to which both illustrators were willing to heed his critical recommendations and make changes to their figures. They sent drafts of their images to Clerselier, who offered "quelques petits advis" on how to make revisions so that some of the illustrations might do a better job of illuminating the text. On at least one occasion, Clerselier himself modified Gutschoven's figure, changing its orientation from a frontal view to a side view, in order to "faire mieux voir" what Descartes

[53] "Preface", xiv.

Fig. 12.10 La Forge, from *L'Homme*, p. 62

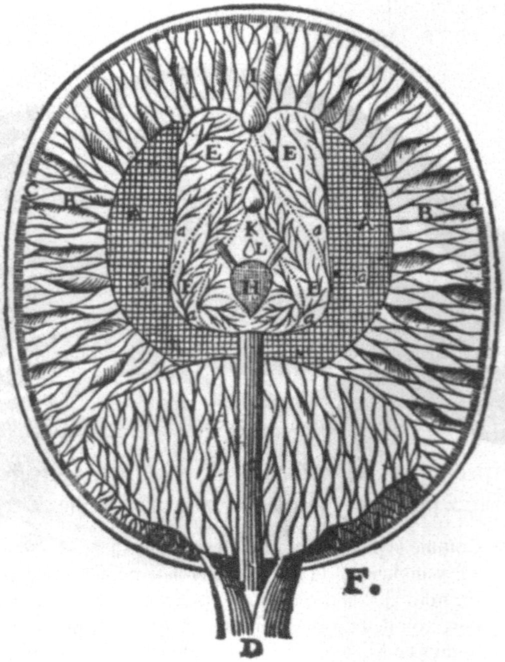

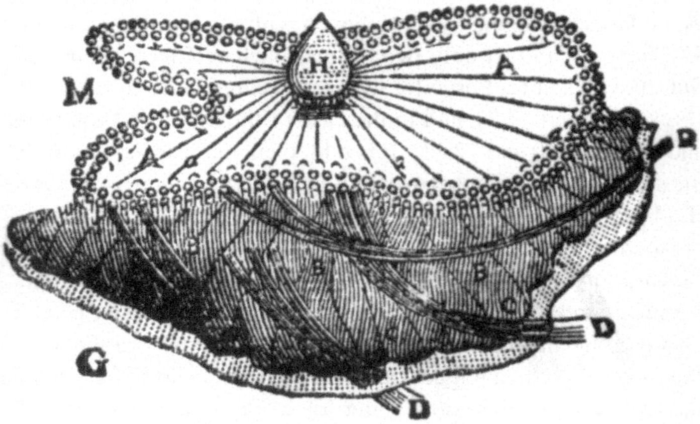

Fig. 12.11 Gutschoven, from *L'Homme*, p. 91

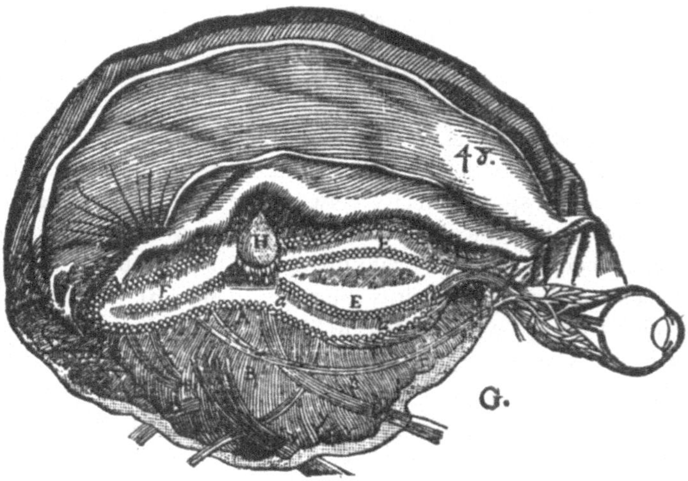

Fig. 12.12 Gutschoven, from *L'Homme*, p. 63

was talking about when he discusses the arterial pathways taken by the parts of the blood that circulate to and from the brain (Fig. 12.12).[54]

As we have seen, Clerselier considered Gutschoven the better artist, and for this reason most of the figures in the 1664 edition are his. However, Clerselier also notes that La Forge was "plus hardy", bolder or more daring. What he means by this is that La Forge did not uncritically follow Descartes's lead, and that he tailored his illustrations not only to fit the text and do the explicatory work that Clerselier expected, but also, on occasion, to make sure that what was shown in the figure was scientifically correct. "Il n'a point fait difficulté de s'éloigner icy de la pensée de l'Autheur, et de substituer la sienne en sa place."[55] La Forge would then, in the extensive commentary that he composed for Clerselier's book—and "par de bonnes et vives raisons" (in Clerselier's judgment)—justify any ways in which his figures departed from Descartes's text. In short, La Forge was a good and faithful Cartesian. His goal in interpreting and completing Descartes's project, as much in his illustrations as in his commentary, and then in his own *Traité de l'esprit de l'homme*—and as a philosopher and a physician—was not simply to proceed in exactly the way Descartes would have (although this was an important consideration), but to improve that system and its constituent parts by moving everything toward greater internal consistency and, no less important, toward greater correspondence with the latest

[54] In fact, the resulting image appears to be a side-angle view of the figure produced by Schuyl for the same text, suggesting that either Clerselier or Gutschoven referred to Schuyl's work for guidance; compare the figure in *De homine*, p. 14 with the figure in *L'Homme*, p. 9.

[55] "Preface", xix.

scientific and medical discoveries.[56] La Forge's fealty was as much to reason and experience as to Descartes.

For example, as Clerselier notes, La Forge's illustration of the eye muscles places the valves between the muscle tendons rather than in the nerves leading into the muscles, which is how Descartes had it in his drawing (and which Clerselier kept in his rendering of that "brouillon"). Similarly, La Forge's image shows the nerves discharging the animal spirits directly between the fibers of the eye muscles to inflate them, whereas Descartes says in his text (and shows in his illustration) that the nerve canals themselves extend into the muscle tissue and the muscles inflate when the canals are filled by the spirits (Figs. 12.2 and 12.3).[57] However, as I have noted, Clerselier wanted the illustrations to "faire comprendre quelle est … la pensée de Monsieur Descartes" on various matters; thus, in addition to La Forge's relatively simple, artistically inferior but scientifically correct figure, he felt he should also include his enhanced copy of Descartes's useful but, he agrees, incorrect illustration (as well as Gutschoven's detailed figure for the same passage).[58]

In many instances, and without looking at Clerselier's labels, it is not too difficult to tell which figures are by Gutschoven's hand and which are by La Forge once we know each illustrator's style. Clerselier was quite right to say that Gutschoven is the better artist. His pictures are, for the most part, denser and richer in detail, capturing more information with finer lines, whereas La Forge seems to opt for clarity and ease of reading. Compare, for example, the illustrations of the eye muscles examined above (Figs. 12.1 and 12.2), or their respective images of the brain and the pineal gland (Figs. 12.10, 12.11, and 12.12). As we see in the latter case, La Forge often goes for a two-dimensional cut-away view, whereas Gutschoven tends to give his illustrations some depth. There is a certain flatness to La Forge's figures, and they can, in fact, seem quite primitive (Figs. 12.13 and 12.14). This may be why Clerselier adopted the policy of going with the images by Gutschoven when he had figures from both illustrators on the same passage and no compelling reason to display both. (On the other hand, I would not claim that the difference between a Gutschoven and a La Forge is *always* so clear cut [see, e.g., Figs. 12.15 and 12.16].)

[56] Here is how Clerselier's describes La Forge's various contributions: "Il n'y a point de difficultez qu'il n'ait resolües, point de scruples qu'il n'ait levez, point d'obscuritez qu'il n'ait éclaircies" ("Preface", xxiii-xxiv).

[57] "Preface", xx. La Forge explains his departure from Descartes on this point in his commentary on *L'Homme* (228). His commentary (as well as Clerselier's preface) is also accessible in a modern edition: see *L'Homme de René Descartes et Un Traité de la Formation du Foetus du Mesme Autheur, Avec les Remarques de Louys de la Forge* (Paris: Fayard, 1999; henceforth, *L'Homme de René Descartes*), 230. Interestingly, the crude and barely understandable drawing by Descartes that Schuyl used (fig. 4), unlike the one reproduced by Clerselier, seems to show the nerves discharging the animal spirits into the muscle, as La Forge's illustration does.

[58] As far as I can tell, this is the only instance in which one of the illustrations "corrects" Descartes's text. Thus, Zittel's claim may be a bit of an exaggeration when he says that Gutschoven and La Forge, as trained medical doctors, made an effort "to bring Descartes's book up to date with the latest scientific knowledge. As a consequence, their main objective in drawing the images was often not the exact reconstruction of what Descartes might have intended, but the reflection of current scientific knowledge" ("Conflicting Pictures", 221).

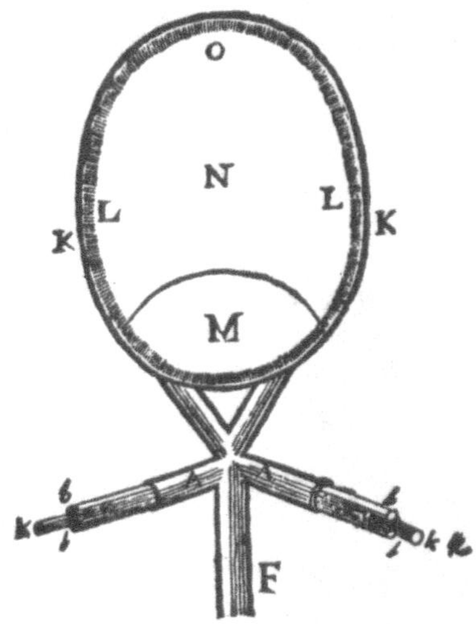

Fig. 12.13 La Forge, from *L'Homme*, p. 15

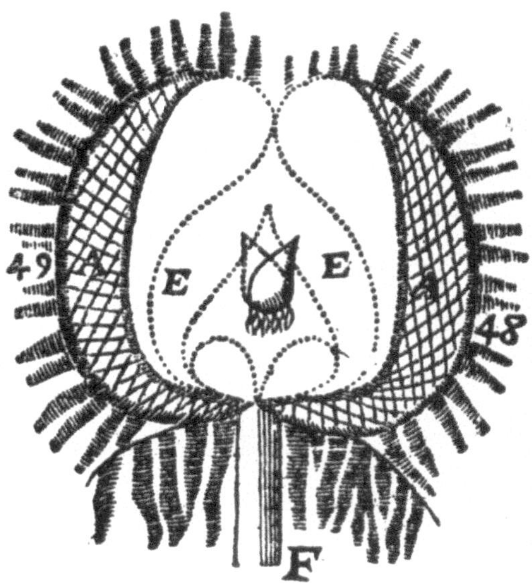

Fig. 12.14 La Forge, from *L'Homme*, p. 64

La Forge himself is conscious of the occasional disparity between their figures. He notes in his commentary that his own spare, structural illustration of the skull and the nerves descending from the brain (Fig. 12.13), while adequate for displaying what La Forge wanted it to show relative to the text, is in fact not quite at the level of Gutschoven's illustration. "La premiere partie de la figure de Monsieur de Gutschoven est plus exacte que la mienne."[59] The latter, with its richer detail,

[59] *L'Homme de René Descartes*, 227.

Fig. 12.15 La Forge, from *L'Homme*, p. 24

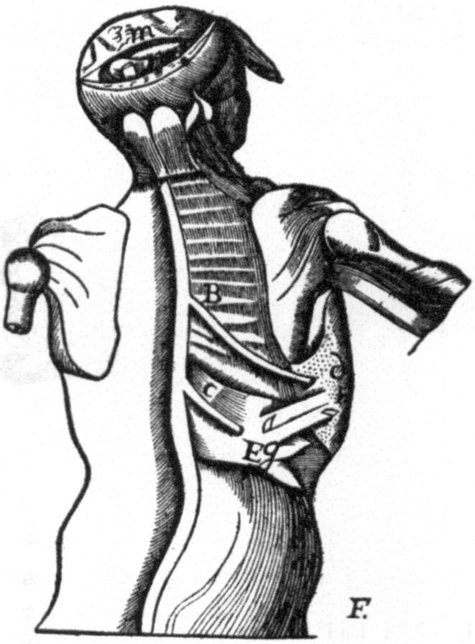

Fig. 12.16 Gutschoven, from *L'Homme*, p. 27

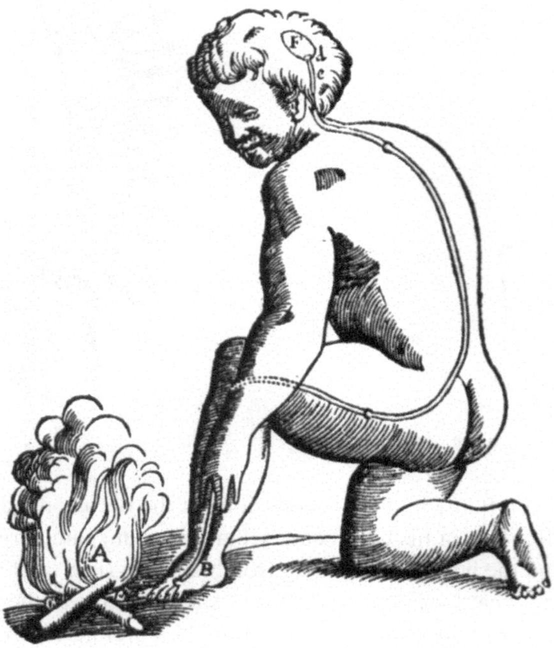

succeeds in displaying a number of things that La Forge's own figure, he concedes, leaves only to the imagination. As for most of the others, however, he insists that they are "toutes semblables, et ne disent que la mesme chose."

The more telling and interesting comparison among the illustrations, of course, is between those by Gutschoven and La Forge in the 1664 French edition, on the one hand, and, on the other hand, the ones by Schuyl in the 1662 Latin edition.

12.4 Gutschoven and La Forge vs. Schuyl

To begin with, a natural question arises with respect to the illustrations by Gutschoven and La Forge when looked at beside those of Schuyl. Clerselier claims in his preface that his two illustrators worked totally independently of each other and without consulting Schuyl's work. However, one might suspect that Gutschoven, at least, did in fact have a look at Schuyl's figures. For example, Gutschoven's schematic depiction of the internal structure of the eye is remarkably similar to the one rendered by Schuyl (Figs. 12.17 and 12.18). One wonders whether Clerselier sent copies of Schuyl's images to him, or maybe Clerselier is not to be trusted when he says that he had Gutschoven's figures in hand before the publication of Schuyl's book. Fortunately, we need not suspect Clerselier of violating the letter (if not the spirit) of his challenge to Gutschoven to come up with illustrations all on his own, and especially without seeing those of Schuyl. For the more obvious explanation of the similarity between Gutschoven's eye and Schuyl's eye is that they both took their lead from the same source, namely, Frans van Schooten, the illustrator of Descartes's *La Dioptrique*, which was published as one of the essays accompanying the *Discours*

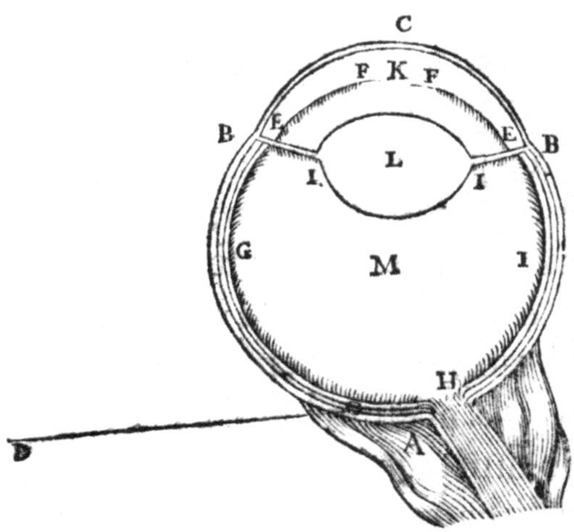

Fig. 12.17 Schuyl, *De homine*, p. 44

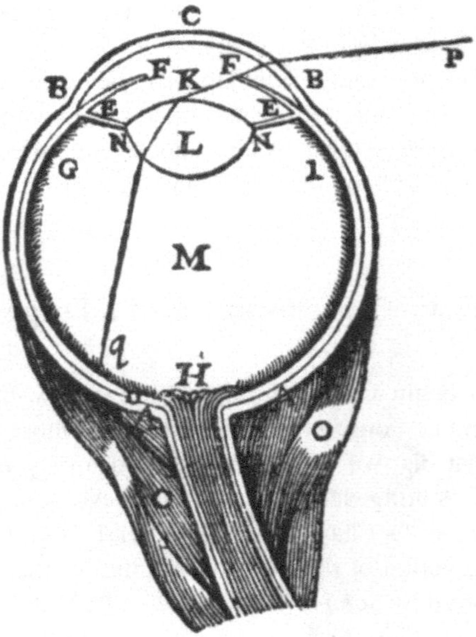

Fig. 12.18 Gutschoven, from *L'Homme*, p. 38

de la méthode in 1637. Van Schooten's diagram of the eye is clearly the model for both of the later artists (Fig. 12.19).[60] Moreover, it is obvious that Gutschoven was consulting Van Schooten's figures for some of his other illustrations as well (see Figs. 12.20 and 12.21).[61]

There is little mystery as to why, given his stated goals for the illustrations in his edition, Clerselier rejected the figures by Schuyl. While often intriguing to look at, they do not have the rational clarity of the figures by Gutschoven and La Forge. Schuyl's figures do not really succeed as diagrams that are supposed to supplement and illuminate a scientific text. They are busy, with unnecessary detail and extraneous features, and thus hard to read. Even when there is no landscape setting, the renderings offer too naturalistic a representation to be intellectually useful. Clerselier is quite right to say that they are "moins intelligibles que les autres, et moins propres à l'intelligence du texte."[62] In this sense, and contrary to Clerselier's express purpose, the illustrations in Schuyl's edition do indeed look like the illustrations of an anatomy textbook (see, for example, Fig. 12.6). This is especially clear when we consider his figures along with those in a contemporary anatomy book—Caspar Bauhin's *Theatrum anatomicum* (first published in 1605)—that, as Annie Bitbol-

[60] AT VI.106.

[61] Van Schooten's image is at AT VI.135–6. On this borrowing from Van Schooten, see Annie Bitbol-Hespériès, "Introduction" to René Descartes, *Le Monde, L'Homme*, eds. A. Bitbol-Hespériès and Jean-Pierre Verdet (Paris: Editions du Seuil, 1996), iii–liii (p. xlviii).

[62] "Preface", ii.

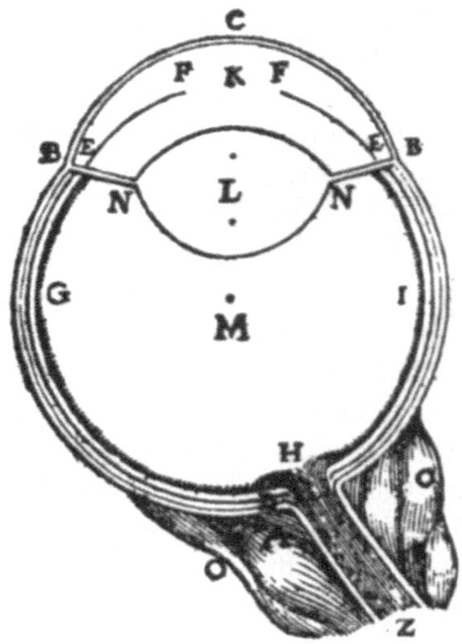

Fig. 12.19 Frans van Schooten, from Descartes, *La Dioptrique*, AT VI, 106

Fig. 12.20 Gutschoven, from *L'Homme*, p. 50

Fig. 12.21 Frans van Schooten, from Descartes, *La Dioptrique*, p. 135

Hespériès has shown, Descartes consulted while composing *L'Homme* (see Figs. 12.5, 12.6, 12.22 and 12.23).[63] Compare, for example, Schuyl's illustration of the brain with the pineal gland in the center (Fig. 12.7) with Gutschoven's rendering of the same thing (Fig. 12.11). Schuyl's picture looks like a brain; Gutschoven's, not at all. But Gutschoven's image helps the reader see what Descartes is talking about much better than Schuyl's.

Part of the "problem" for Schuyl may be the respective media in which the illustrators were working. Schuyl's illustrations are copper-plate engravings, and this technique, popular for art prints of the period (especially reproductions of paintings and sculpture), affords the artist richer graphic opportunities. An engraving allows for finer detail than the older and artistically more primitive technique of woodblock printing, which is what was used for the figures by Gutschoven and La Forge. In fact, Clerselier alerts the reader to the fact that on at least one occasion, regarding the diverse, criss-crossing paths that can be taken by the animal spirits as they descend from the brain cavity through the nerves, the woodblock technique was not up to the task of providing the requisite fine detail, and so the reader will have to use his imagination to fill in what is missing. "[L]e bois qu'on a employé pour tailler les figures ne pouvoit pas souffrir cetter delicatesse … et l'imagination fera le reste."[64] But it is precisely the fact that he took advantage of engraving's opportunity for greater detail that worked against Schuyl, at least in the eyes of Clerselier.[65]

(The interesting question here is why Clerselier opted for the more old-fashioned woodblock prints for his illustrations rather than engravings. One possible answer is economic—a book using woodblock prints would be significantly less expensive to produce than one using copper-plate engravings; it could also therefore be priced more affordably for a larger audience. Moreover, bear in mind that Clerselier was, by 1664, anxious to see his edition of *L'Homme* finally published; and woodblock

[63] Bitbol-Hespériès, "Introduction", xxiv–xxvii; and *Le Principe de vie chez Descartes* (Paris: J. Vrin, 1990).

[64] "Preface", xxvi.

[65] Interesting and compelling analyses of the illustrations for *L'Homme*, especially comparing Gutschoven and La Forge vs. Schuyl, are offered by Zittel ("Conflicting Pictures") and Wilkin ("Figuring the Dead Descartes"). In fact, Zittel and Wilkin disagree on a central issue. According to Wilkin, Schuyl's worldly illustrations subvert Descartes's scientific project, and especially the mind-body dualism, by underscoring human mortality and the transience of bodily things, whereas the more abstract and spare figures by Gutschoven and La Forge figures "perpetuate the philosopher's project of postponing death by diverting the reader's attention from the corruptible nature of the body" (45). Zittel, on the other hand, insists that Gutschoven's and La Forge's illustrations go beyond Descartes's own project by prioritizing "the reductionist notion of man as a machine", and represent Clerselier's ideal (and that of other latter-day Cartesians) more than Descartes's. Zittel insists that "from the point of view of Descartes's supposed intentions, … the arguments in favour of Schuyl's visual language would definitely be stronger" (230).

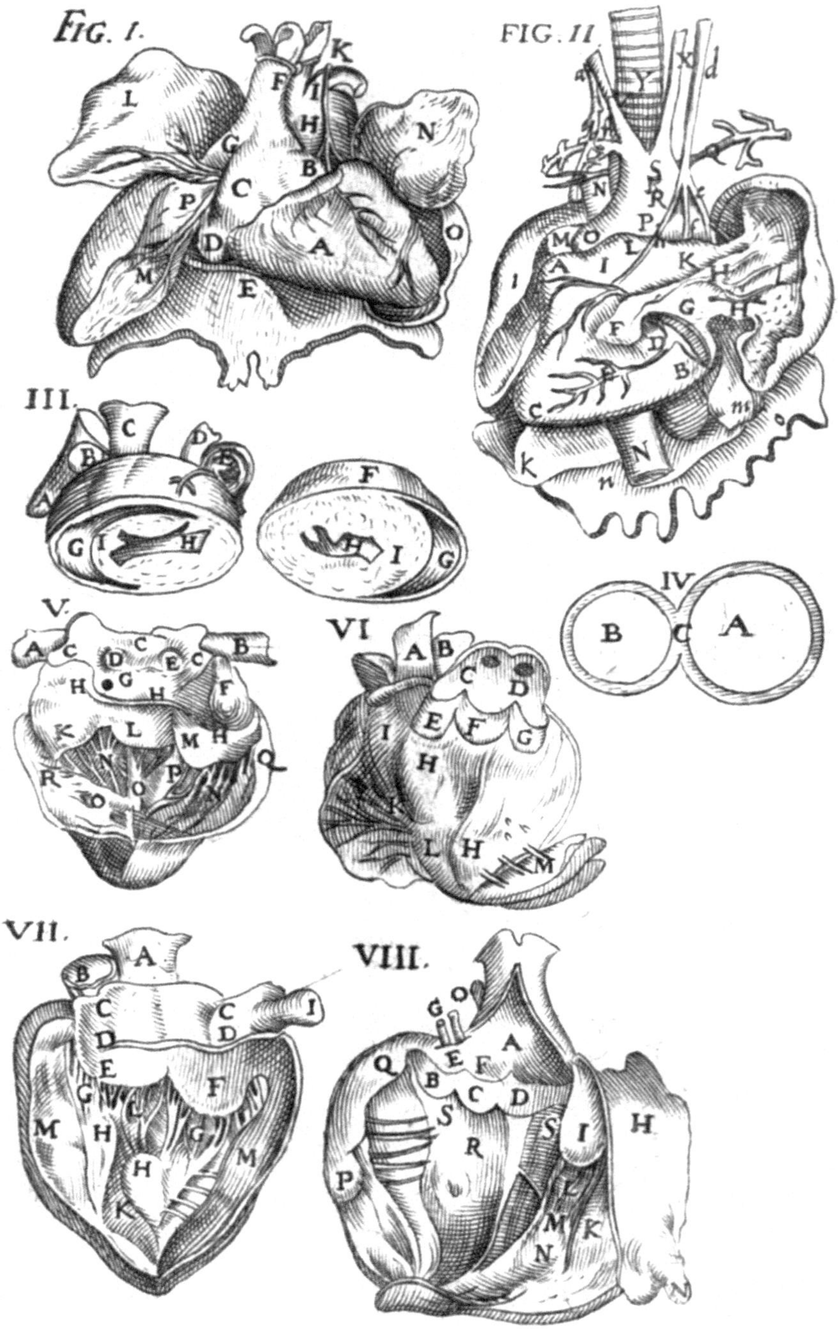

Fig. 12.22 From Caspar Bauhin, *Theatrum anatomicum* (heart)

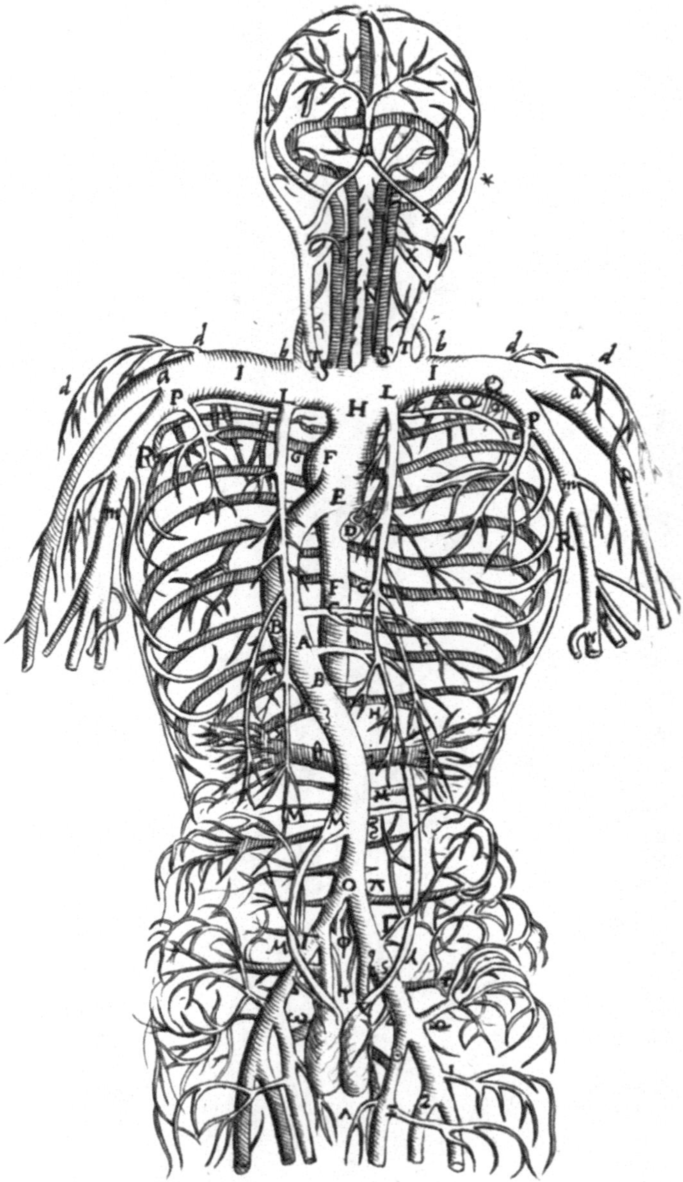

Fig. 12.23 From Bauhin, *Theatrum anatomicum*, (skeleton)

prints could be produced much more quickly than the extremely time-consuming method of engraving, which would have delayed publication even longer.[66])

*

The illustrators for Clerselier's 1664 edition of Descartes's *Traité de l'homme* faced a difficult task, as Clerselier himself notes. They had to come up with figures that would illuminate for a wide range of readers a dense, often terse and elliptical text of human anatomy according to the principles of the new mechanistic science. Some of the images are more successful than others. In the end, however, we have no choice but to admire their handiwork. Gutschoven and La Forge may not have been great artists, but they understood the charge they were given and the challenges they faced. They more than satisfied the expectations of their client. Clerselier, for one, believed that they had succeeded admirably. So did a reviewer in the very first issue of the *Journal des sçavans* (5 January 1665), who had this to say about the book: "M. Des Cartes avoit laissé ce traité dans une si grande confusion, qu'il ne seroit pas intelligible si M. Clercelier ne l'avoit mis en ordre, & si Mess. de la Forge et Guscoven ne l'avoient esclaircy pas des figures."[67,68]

[66] For a discussion of the relative advantages and disadvantages of woodblock-print illustrations vs. engravings in early modern texts, see Sachiko Kusukawa, *Picturing the Book of Nature: Image, Text, and Argument in Sixteenth-Century Human Anatomy and Medical Botany* (Chicago: University of Chicago Press, 2012), pp. 29–34; and "Illustrating Nature", in Marina Frasca-Spada and Nick Jardine, eds., *Books and the Sciences in History* (Cambridge: Cambridge University Press, 2000), pp. 90–113.

[67] *Journal des sçavans*, 5 January 1665, 9–11 (11). Accessible online: (http://gallicalabs.bnf.fr/ark:/12148/bpt6k56523g/f13.image)

[68] My work on this essay was made possible by a senior fellowship at the Institute for Research in the Humanities at the University of Wisconsin-Madison, and greatly facilitated by a Residency at the American Academy in Rome (April 2015). My thanks to Margaret Maida for her bibliographic suggestions. I am also indebted to Robin Rider, Special Collections, Memorial Library, Univexrsity of Wisconsin-Madison, and to Micaela Sullivan-Fowler and Joanna Baisch, Rare Books and Special Collections, Ebling Library, University of Wisconsin-Madison, for their help.

The images from Schuyl's edition of *De Homine* are courtesy of the Daniel and Eleanor Albert Collection, Department of Special Collections, Memorial Library, University of Wisconsin-Madison.

The images from Bauhin's *Theatrum Anatomicum* are courtesy of Ebling Library, Rare Books & Special Collections, University of Wisconsin-Madison.

Part III
L'Homme and Early-Modern Anthropology

Chapter 13
A *Treatise of Human Nature*, a *Treatise of the World*?

Claude Gautier

Abstract My contention is to understand in what manner Hume's Treatise of Human Nature should be considered: is it a Traité de l'homme or is it a major contribution to modern historical anthropology? I shall emphasize the fact that from Descartes to Hume the problem of what "human nature" is contrasts very strongly. Such contrasts are to be based (1) on the very specific dimension of Hume's scepticism; (2) on the fact that human nature should be now considered as basically historical.

My subject does not bear upon the textual and historiographical aspects of the publication of the *Treatise of Man*; nor does it bear upon the doctrinal exegesis characteristic of readings of Descartes, or of those which relate the *Treatise* to the state of scientific knowledge – physical, medical, etc – of the time.

I will endeavour to put in perspective a movement that appears to be important when considering the philosophical situation of Hume vis-a-vis the philosophy of the Classical Age. In a certain way, I wish to revisit some of the ruptures introduced by *A Treatise of Human Nature*, and it seems to me that the *Treatise of Man* constitutes, from this point of view, a quite remarkable term of comparison. I am seeking to understand, through particular comparisons, how some of the contours are drawn of what we could call an 'anthropology' of Hume, based on his empiricist scepticism.

C. Gautier (✉)
Triangle-CNRS, UMR 5206, Département des sciences humaines, ENS de Lyon,
15 Parvis René Descartes, BP 7000, 69342 Lyon Cédex 07, France
e-mail: claude.gautier@ens-lyon.fr

D. Antoine-Mahut, S. Gaukroger (eds.), Descartes' *Treatise on Man* and its Reception, Studies in History and Philosophy of Science 43,
DOI 10.1007/978-3-319-46989-8_13

13.1 The Science of Man as First Philosophy

Unlike Descartes – but this is also true for the philosophers of the Classical Age up to Leibniz – Hume didn't practice science, and this trait, as we might suspect, is not at all incidental for it is the expression of a new situation in philosophy. The great philosophers of the Classical Age were all scientists and their scientific practices had a theoretical companion in a certain concept of philosophy as a totality of knowledge, that is to say as a project of the integral systemisation of knowledge.

However with his *Treatise*, Hume takes account of the henceforth irreversible distinction between moral philosophy and natural philosophy. This split doubtless expresses a quite clear acknowledgement, that of the development of natural philosophy which had freed itself from all need of and dependence upon a metaphysical foundation. The first of these philosophies set itself out positively as an ensemble of results supported by tested methods.

In a certain manner, *A Treatise of Human Nature* inherits from this break between science and philosophy. And in this way, Hume is without doubt the modern writer who, more than others, admits that philosophy is no longer the total system of sciences ordered under its sovereignty.

If the model of accomplished science remains that of Newton, it is in a very special way that Hume recognises this, and places it, following his theorised scepticism, as a science of phenomena, a science which only reaches phenomena by the so-called 'real' – that is to say mental *data*. However, to say that the physical is a science of phenomena, is to say that its theoretical scientific status stems from a preliminary determination relying on the status of 'appearances', and in that way it falls under the legislation of moral philosophy since, for Hume, appearances are psychic *data* which stem from the science of man.

It is in this sense, probably, that the radical affirmation of Book 1 on *Understanding* which states 'Human Nature is the only science of man'[1] can be understood. The sole objective of science is man and mathematics is only a science as far as we understand it as something human, that is to say dependent on the structure of the human mind.

So the meaning to be given to this first characteristic of the philosophy must not be mistaken. Introducing experimental method to moral subjects as indicated by the secondary title of *A Treatise* is another way to emphasise that the science of man is total and that exact sciences cannot be assigned their theoretical status until the science of man has been constructed.

But the science of man which is in play here should not be understood as an extension to a new object, which would be 'man', of the method of a science which has proved itself elsewhere – for example in the study of matter or of nature; it should rather be understood as a *First Science* starting from which it becomes possible, consciously, to practice sciences.

---

[1]David Hume, 2002: *A Treatise of Human Nature*, Ed. David Fate Norton & Mary J. Norton, Oxford, 2002, Oxford University Press, Col. « Oxford Philosophical Texts », I.4.7.:177.

> 'Tis evident, that all the sciences have a relation, greater or less, to human nature; and that however wide any of them may seem to run from it, they still return back by one passage or another. Even Mathematics, Natural Philosophy, and Natural Religion, are in some measure dependent on the science of MAN; since they lie under the cognizance of men, and are judg'd of by their powers and faculties[2].

As a First Philosophy, it thus appears that this science of man is, above all, a philosophy of the mind, that is to say a philosophy capable of making rulings on the 'knowledge', the 'powers' and the 'faculties' of human understanding that it knows. If it is thus First, it is so in a critical fashion or in order to reply to the question of the conditions of possibility, which comes back precisely to say what these 'faculties' are, or, which amounts to the same thing, what they can do.

It is within this framework that the question of knowing whether *A Treatise of Human Nature* can be understood as a *Treatise of Man* becomes pertinent.

13.2 Sensibility as a Substantial Zone of Man

The work is entitled *A Treatise of Human Nature*. But what is the status of such a title? Since the start of the seventeenth century, *Treatises of Man* appear, that is to say texts which specifically aim their work at the study of the human being. Hobbes wrote two major works of which the first part is *Human Nature*, and then *De Homine*; Descartes composed a projected *De l'homme* and the alternate title of the first part of *Passions de l'âme* [*'Des passions en général'*] is *'et par occasion, de toute la nature de l'homme'*. Thus, since the dawn of the Classical Age, there were texts devoted to the construction of the science of man.

Even in the seventeenth century, man is not constituted as an autonomous object of science. Thus, the *Méditations métaphysiques* of Descartes are not a *Traité de l'homme*, which is not an object of metaphysics. They are the inauguration of a philosophy of the subject; but to be precise, the subject is not man, it is *res cogitans*, that is to say the soul as distinct from the body.

It is only in effect as a substantial union of the soul and the body, that is to say outside the founding region of metaphysics, that man is envisaged as a being. This signifies that the knowledge of man is deduced, or derived, from an earlier knowledge, founded in reason or experience. It is not first insofar as is suggested by the critical gesture of Hume in his *Treatise*.

Thus, for Hobbes, the total system of science is tripartite: body, man, citizen. The knowledge of man supposes preliminary constitution of the physical, since man is a body. Similarly for Descartes, the science of man comes from the branches of philosophy's tree, that is to say the metaphysical roots and the physical trunk are supposed to have already been created. Man is thus simply one of the objects of science in general and not that which conditions the possibility of it.

---

[2] David Hume, *A Treatise of Human Nature*, Op. cit.: 4; Emphasises Hume.

The reasons for which *Human Nature* is not treated explicitly as such before the seventeenth century, are complex. In the logic of Porphyry, man is defined as a « rational animal ». His place, in the global structure of the being, is perfectly assigned to him. However, this definition is refused by the great philosophical works of the seventeenth century.

Thus Descartes, in the *Méditation seconde*, returns one last time to his naïve opinions, those which he held before undertaking doubt of everything:

> Qu'est-ce donc que j'ai cru être ci-devant ? sans difficulté, j'ai pensé que j'étais un homme. Mais qu'est-ce qu'un homme ? Dirai-je que c'est un animal raisonnable ? Non certes: car il faudrait par après rechercher ce que c'est qu'animal, & ce que c'est que raisonnable, & ainsi d'une seule question, nous tomberions insensiblement en une infinité d'autres, plus difficiles et embarrassées [...][3].

The significance of this 'embarrassment' is not psychological but properly metaphysical: we cannot discover what we are starting from a supposedly valid knowledge of that which is [animality, reason] because we can only know that which is starting from a knowledge of that which is certain. And that knowledge first supposes that we are certain of what we are. Descartes doesn't first find himself as a reasonable animal, but as a *res* that thinks. The same refusal is found in Locke, this time on the empiricist flank: what is bad in the scholastic definition, is that it proceeds from a poorly made abstraction. For if we know that a dog talks, and therefore reasons, we still don't call it a man, however the dog too is a rational animal[4].

The reasons for which classical thought refused to define man as a 'rational animal' are profoundly anchored in its most intimate logic. First of all, the birth of mathematical and experimental physics forced thought to abandon the ontology of gender and its differences. The real was no longer structured like a joint to hold things together, but as a collection of corpuscular phenomena, co-ordinated according to a mathematical logic, no longer by a syllogistic logic. This signifies that the relationship between the particular and the universal is no longer of the order of participation or of belonging, but of occurrence and of the law.

Next, this mutation of scientific rationality brought about a mutation of philosophical rationalism. The foundation of mathematical and experimental physics rests on dualism, that is to say upon the radical separation of extension and of thought. The study of nature can only be geometricized when it is reduced to quantity, which supposes that everything stemming from quality must be placed in perception, that is in the mind of the thinking subject. And it is only within this dualist theoretical structure that the question of human nature can be posed anew.

For it can be seen, on the one hand, that man is not a being of nature, caught up in nature, like other beings. Nature could only be invented in the seventeenth century by the constitutive activity of a subject which had begun to extract itself from it. It is for these reasons that it was no longer satisfactory that the definition of man

---

[3] René Descartes, *Méditations métaphysiques*, Méditation Seconde, in *Œuvres de Descartes*, Adam & Tannery *eds*, Paris, Vrin Librairie Philosophique, Volume IX: 20.

[4] John Locke 1998: *An Essay Concerning Human Understanding*, Penguin Classics, Book III « Of Words », Section VI « Of the Names of Substances »: 580-s.

was as a 'rational animal', that is to say as a living corporal substance that also participated in reason.

So it is not as a man but as a *res cogitans* that the new philosophy discovered the subject for which there is nature. But strictly speaking, man is neither a rational animal nor a *res cogitans*. There is in fact mediation between the soul and the extension, a zone that is very difficult to understand, that stems neither from pure thought nor from matter, but which is within thought, and has something to do with matter; this is the domain of qualities, that is sensibility and sentiment. The object of my pure thought is order, that is to say measure, thus of quantity. The object of the science of nature is geometry, quantity again. But the world in which we live is that of quality, of sensation and the affect. And it is this world that is the world of man, precisely since it is not a case of the soul as a rational substance but as a thinking substance, subject however – paradoxically – to be passively affected by something which to it is not only exterior, but radically heterogeneous if not contradictory – the body.

That which makes of man a man is therefore neither his animality nor his rationality. And when it is said that he is a union of the two, a substantial union of body and soul, this still does not bring us back to a close or analogous concept, to the scholastic concept of the 'rational animal'. Precisely because the place of that union, the element of this substantial zone of man, is not reason but *sensibility*, imagination, and the affects, that is to say all that which the elder philosophers placed on the animal side[5]. What makes a man a man, from the seventeenth century on, are the passions.

Such are doubtless some of the reasons that explain the appearance from the beginning of the seventeenth century of something like the treatises of man; and the same reasons explain that these treatises could not use man as their starting point but had inevitably to think of man by starting from physics and physiology, thus making of man one of the objects of science. For the way in which man appeared was a problematic one – neither animal, nor rational, nor rational animal – which supposes as its own condition of possibility the split, which is constitutive of modernity, between matter and mind, soul and body. Man could thus only be thought of as the aporetic junction point of these two orders of reality which had already been posed as adequate objects of science: the metaphysics of the soul, and the physical mathematics of geometricized bodies.

[5] For Aristotle, the soul had levels: it is nutritive, has sensibility, is imaginative, and finally rational: it is only at this last level that it is human. With Descartes, the soul is one and undividable, and sensibility, imagination and reason are only modes of its essential attribute which is thought. And it only envisaged as human as much as it has sensibility and is affective. Therefore this is not a simple change in the doctrine of man, but a real and total turnaround.

13.3 Is a Treatise of Human Nature a Treatise of Man?

With the *Treatise* of Hume, we are in a wholly other theoretical universe. The science of man does not rest upon the physical, and ignores physiology consciously and resolutely. Manifestly, the science of man is not a branch of science. This can be seen simply by the structure of the work, the order of the three books, *Of the Understanding*[6], *Of the Passions*[7], and *Of Morals*[8]. However this structure is unlike the works of the time. These generally treat understanding or the faculties of knowledge apart.

From *Regulæ* to the *Critique*, from *On the Improvement of the Understanding* of Spinoza to *An Essay Concerning Human Understanding* of Locke, knowledge was the object of a specific theory; it is a separate philosophical object. It is only afterwards that we pass from the soul to man; and there is a fundamental disconnection in the discursive regime, since between the theory of knowledge and the theory of passion, there must be the physical. It is the same thing for morality: it is deduced from metaphysics, that is to say that it is thought, starting with Descartes, as a question which engages the articulation – possible or impossible – between the rational dimension of the soul and the insertion of the soul in the passivity of the body.

It is these complex articulations that Hume ignores. Human nature is no longer an object that must be known between the theory of knowledge and morality; human nature encompasses it all. It is the same human nature whether it works in understanding, in the passions, in morality, in politics, and in aesthetics.

The discursive regime is here unified. According to what logic? Is there a reason to this order, and if so what type of order is it? Firstly, it has to be remarked that there is no disconnection in the *Treatise*. The same categories, very few in number, command the humean discourse in its totality.

The fact of changing from Book to Book does not change the object for us. Thus passing from understanding to passions, is not to pass from the soul to the man, and passing from passions to morality, is not to pass from descriptive and explicative to normative and to axiology. The passions are in fact at the same time affects and values in themselves, and reciprocally, Hume's moral discourse means to be analytical. As for the elements of knowledge, they are themselves thought in terms of affects, since the departure point for Book 1 is the distinction, internal to perception, between the impression and the idea, the which distinction is only one of intensity.

The order of the work is neither an order of reasons nor an order of matters. It is an order which moves from the most simple to the most complex, without it being their rationality which qualifies the simple and the complex. That which appears simple in fact, is not the clarity of rational evidence, but the power of an impression which cannot be analysed. It is therefore an intensity felt as such. Starting from the most simple, Hume composes the totality.

---

[6] *A Treatise of Human Nature*, Op. cit.: 7-s.

[7] *A Treatise of Human Nature*, Op. cit.: 181-s.

[8] *A Treatise of Human Nature*, Op. cit.: 293-s.

Because the most simple, the sensitive impression, is the smallest quantity of being, that is to say of *data*, which analysis can reach. But it is as such absolutely separated from all the rest, and first of all from other impressions: blue is the ultimate *data* of the consciousness, but it is only blue. However we don't see blue, we see things which are blue. So the law of composition of elements must thus be understood in everything.

Put in another way, it is necessary to understand how the mind passes from the chaos of impressions to the world. The *Treatise* is therefore the book of the creation of the world. The world as object firstly, then the world of intersubjectivity, and finally social, political and cultural world. So Hume has to restitute the processes of generation of the world in its totality, starting from the sensitive origin, given that this origin is not a first moment in the time of the species, or of individual history.

Hume's proposition is thus anthropological in a wholly other sense in relation to the proto-anthropology that was developing in his time: how the human mind comes to forge nature around itself – and not what is the place of man *in* nature. All that is given up to philosophical analysis, is the pure *data* of perception. That which is experienced, is a nature as an ensemble structured universally and necessarily by phenomena, themselves experienced as extra-mental realities.

In this sense, the concrete, that is to say the experience of common consciousness, is the existence of a nature in which we 'subjects' insert ourselves, and the impression or perception is an abstraction. But it is just this which poses Hume's philosophical problem. How to constitute that which appears to us as concrete, immediately given – the world, nature, society, other people, the self itself – whilst the rigour of analysis imposes upon us that the only *data* is perception, which seems 'abstract', though it alone is immediate reality. It is thus a case of understanding how that which the mind constructs, a nature existing outside and independently of us, can present itself masked, can exhibit itself as *data* while hiding the process of its construction.

In Book 1 the approach is also an approach of increasing complexification. It goes from the impression to the thing, from inside to outside. The analogy with Descartes is striking. Descartes also begins by a sceptical moment by which the outside is evacuated, and he comes across the first truth, the *cogito*, which defines itself as that which cannot be doubted, as that from which is discovered a first external existence, that of God. Descartes thence deduces nature and the existence of bodies. But with Hume if one starts from inside the mind (perceptions), the start point is not the subject, since the subject is analysed at the end of the path (Part IV, Section 6). The interiority of the mind is not personal, the subject is not the first *datum*, no more than it is the metaphysical condition of the *data*: it is constructed.

And we can only understand the law of its construction after having understood the law of construction of the thing *as* external to the mind: the First Part of this Book analyses ideas, that is to say the fundamental *data* of the mind, as content of 'consciousness'. The Second Part analyses space and time, but especially space as 'form' or 'condition' of exteriority. The Third Part analyses causality, or rather the nature of causal reasoning such as it is, that is to say independently of the existence of things supposed to be causes and effects of one other. In this part, we have here the first moment of the filling-up of the world, since by causal inference I move past the *data* of immediate perception. If the soil is wet and it is not raining, I 'know' that it has rained. Causality enlarges the world, or to reuse the consecrated expression of Hume, 'it peoples it'. By this the chaos of possibilities arranges itself in series.

Finally, the Fourth Part treats our certitude in the existence of things independent from our perception. Amongst the 'things' that we believe transcend, is found the *self*, 'fictional idea' for Hume, who finishes thus the doctrine of understanding by the deconstruction of that which Descartes posed as the first principle of philosophy.

One might sometimes query whether Hume, who wrote *A Treatise* at La Flèche, in homage to Descartes, wanted to produce a kind of ironic parody of *Méditations*. These start with doubt, discover the *cogito*, and finish by finding again the certitude of nature and the objectivity of material bodies.

Book I of *A Treatise* starts from certitude – we have in Hume a sceptic who doesn't doubt but who affirms. It begins with sensible perception, finishes by deconstructing the *cogito*, and concludes by placing radically in doubt everything which has been reached during the long enquiry of Book I. Whether sought after or not, the rhetorical chiasmus between *A Treatise* and *Méditations* is striking.

13.4 In What Sense Does *A Treatise of Human Nature* Speak of Man?

I would like, to conclude, to return briefly to the sense of this comparison between the two treatises. If man is truly at the heart of the humean approach, it is as 'nature' for, in a sceptical gesture that is classical although much more radical, to consider the 'nature' of man is a manner of reassigning man as an integral part of nature, starting from that of animals, if not plants. It is doubtless useful to recall that in *A Treatise of Human Nature,* just as in *An Enquiry Concerning Human Understanding*, it is concerning 'reason' that the Humean demonstration proposes to remind us of the profound unity of animal nature and human nature. It is precisely there where the scholastic definition of the 'rational animal' or the classical definition of the substantial and cognitive 'unity' of the *res cogitans* and of the *res extensa* making of the *aisthesis* the object itself of a specific knowledge, it is there, then, that Hume returns to unity. A unity which encompasses differences, certainly, but differences of degrees, not of nature.

To consider nature then it must be admitted that animal reason and human reason share a common background which is that of habits and customs, that of dispositions and corrected forms of spontaneities – in a word mores. If it is perhaps inexact to say that Hume breaks with the constitutive dualism of classical thought of man, it remains that he gives multiple perspectives which relativize it. This relativisation is readable on two levels at least: that of the liaison between 'nature' and 'history', and that of the liaison between 'reflexivity' and 'sensibility'.

In the first case, the liaison allows understanding that to talk of a nature of man is simply to register it in a history which is that of the constitution of his habits and that of his modifications; 'nature', thus, is an operant form of a continued adjustment, ceaselessly changing, of an organism and an environment. And to take account of this, there is no need, as in the *Treatises of Man* to enter the domain of the physical or the physiological; there is no need to support hypotheses; it suffices to limit oneself to taking account of events: nature is then grasped as that which unfolds in the effects.

In the second case, the liaison allows understanding that the active form of these continued adjustments is given in dispositions and that these are the acting forms of the liaison between 'reason' *and* 'sentiments', between 'reason' *and* 'affects'. Disposition is precisely that which is offered as a practical reason which doesn't necessitate, once again, in order to be identified and restituted as such, an investigation in the domain of natural philosophy. It is only another way of considering the historicity of nature in certain of its effects from the time when it is placed at the centre of the gaze which, once again, unifies the animal and man, the *life* that manifests their conduct.

If therefore, to finish, man is truly grasped as a nature, it is precisely because this nature is thought as an active modality of the liaison between spontaneity of tendencies and correction of effects. And on this point, man is not to be separated ontologically *from* animal. Fundamentally, can it not be supposed that from *Traité de l'homme* to *A Treatise of Human Nature* is the movement itself of the re-inclusion of man in nature which operates? Inclusion which can only arrive, if we follow Hume, through the express condition of not holding to distinctions of 'forms', but, through the bias of a requalification of the powers of the imagination which make up understanding, of comprehending that which stems from reason is precisely our common nature to be affected.

However, to understand this, or at least, give reasons for it, does not justify leaving moral philosophy behind. On the contrary, such an enterprise belongs precisely to moral philosophy and it is in this way that it can pretend to be, without contradiction, and remain, in Hume's system, a *first* philosophy.

Chapter 14
What the Body Can Do: A Comparative Reading of Descartes' *Treatise on Man* and Spinoza's Physical Interlude

Julie Henry

Abstract The means of exposition and the content of Descartes' Traité de l'homme and Spinoza's Physical Interlude are quite dissimilar. One is a long treatise meticulously exploring a number of functions of a machine exactly resembling to vital and sensitive functions of a human body. The other looks like a short digression about physical aptitudes of bodies which aren't specifically mentioned as living bodies. These two texts have yet an approach in common: proposing a physical explanation of body's functions, taking notice of what a body can by itself, independently of any animation or deliberate set in motion. I propose in this paper to explain the dissimilarity between these two texts by their taking root in different philosophical plans. Spinoza's Physical Interlude takes place in a book which has ethical aims; it induces Spinoza to regard physical aptitudes of human bodies as conditions of the possibility of an ethics progression.

14.1 Introduction

The two texts of Descartes' *Treatise on Man* and Spinoza's Physical Interlude are starkly dissimilar in their form as in their content: on the one hand, a long treatise that thoroughly explores the various functions of a machine entirely comparable to the vital and sensory functions of the human body; and on the other hand, something that may seem like a short digression, situated between Propositions XIII and XIV of Part II of the *Ethics*, composed of several axioms, lemmas and propositions that allude to the physical abilities of bodies, in which at no point is it made clear to us

J. Henry (✉)
Département des Sciences Humaines et sociales, Centre de lutte contre le cancer Léon Bérard, 28 rue Laënnec, 69008 Lyon, France
e-mail: julie.henry@ens-lyon.fr

D. Antoine-Mahut, S. Gaukroger (eds.), Descartes' *Treatise on Man* and its Reception, Studies in History and Philosophy of Science 43,
DOI 10.1007/978-3-319-46989-8_14

that it refers to living bodies. The term "human body [*corpus humanum*]" subsequently appears without segue in the final postulates of this interlude.[1]

It may therefore seem pointless, if not artificial, to provide a comparative reading of these two texts. And yet, they share a fundamentally common approach: bringing attention to *what the body can do* regardless of any animation by a soul, like any movement determined by a decree of the mind. Both texts therefore involve questioning what the body can do from itself and by itself. It is then in this context of the functions of the body that I would like to offer a brief comparison of these two texts, reflecting on how, from a seemingly common method and approach (to place the body once again in nature and offer a physical explanation for its functions), Spinoza goes as far as significantly diverging from a Cartesian approach due to the fundamentally different aim he sets himself.[2]

14.2 Neither Wondrous Divine Nor Specific Action of the Mind: The Body by Itself

Descartes' *Treatise on Man* and what is sometimes referred to as Spinoza's Physical Interlude firstly share a common requirement in explaining the functions of the body: offering strictly physical explanations, which therefore require no recourse to any occult faculty. The clearest Cartesian formulation of this is well known and takes place at the very end of the *Treatise on Man*:

> I desire, I say, that you should consider that these functions follow in this machine simply from the disposition of the organs as wholly naturally as the movements of a clock or other automaton follow from the disposition of its counterweights and wheels. To explain these functions, then, it is not necessary to conceive of any vegetative or sensitive soul, or any other principle of movement or life, other than its blood and its spirits which are agitated by the heat of the fire that burns continuously in its heart, and which is of the same nature as those fires that occur in inanimate bodies.[3]

The affirmation that the fire that burns in the heart and the fire in *inanimate* bodies are of the same nature speaks of Descartes's refusal to conceive of *animated* bodies as the being through a soul that is the principle of life. This is apparent for example, in the body of the text, in the strictly physical explanation of the movement of spirits, according to the laws of nature that are common to all bodies:

[1] Starting from Postulate I: "The human body is composed of a number of individual parts, of diverse nature, each one of which is in itself extremely complex." These postulates come after the note of Lemma VII, which addresses the question of increasingly complex bodies.

[2] Namely, a fundamentally ethical aim. As John Toland subsequently notes, in an admittedly polemical intervention but while rightly identifying the aim of Spinoza's philosophy, "the *Ethics* (the title to which [Spinoza] has reduced his whole philosophy) is his veritable system, and we only find his true sentiments regarding philosophy in this book and in his letters" (§ 11 of *Letter IV to Serena*).

[3] AT XI, 202.

> these spirits, being like a wind or a very fine flame, must flow promptly from one muscle to another as soon as they find a passage, even though they are propelled by no other power than the inclination that they have to continue their motion in accord with the laws of nature.[4]

In the same vein, Spinoza mentions the human body within the Physical Interlude, in which bodies[5] are distinguished from one another "in respect of motion and rest" (Lemma I) in strictly physical terms.[6] It is true that this lemma clearly distinguishes the Spinozist conception from the Cartesian conception of bodies, asserting that they do not constitute separate substances, but it remains that human bodies are covered by Spinoza in the propositions of the small Physical Interlude, immediately following the study of complex bodies and compositions of complex bodies, without any mention of the different laws by which these bodies, as living bodies in general, would act. There is therefore increasing complexity, which gives rise to different natures, but these do not involve conceiving specific physical principles that isolate complex bodies from the common laws of nature. It is moreover significant that, according to the Spinozist perspective, human bodies also differ between themselves in respect of motion and rest at a base level (before distinguishing themselves *also* by an ability to be affected in various ways at once). We thus discover, in both cases, a lack of reference to a principle of life and motion outside the body in order to explain the functions of the human body.

Not having recourse to a principle of life and motion other than corporeal nature also involves straying from the tradition to praise God's work in the admiration given to the functions of creatures. This rupture is manifest in the Cartesian text, and is even assumed from the very first page, when Descartes attests that God has put into the machine all pieces (we would speak of parts for a living body) needed "to make it walk, eat, breathe, and imitate all those functions we have which can be imagined to proceed from matter and to depend solely on the disposition of our organs."[7] There is thus, in this passage and in the insistence on what may happen *on the sole disposition of organs*, no mention of the "wondrous" character of the work completed by God or of the "admiration" it is likely to arouse in us, despite this being *de rigueur* in medical treatises. It is therefore a strictly mechanical explanation that does not serve as a pretext to praise God.

---

[4] AT XI, 137.

[5] Namely, all types of the extended substance characteristic, and therefore both living and inert bodies as well as human and animal bodies. Here we see a replacing of human bodies in the physical laws of movement to which Descartes could subscribe, the slight (yet meaningful) difference being that the complexity of human bodies in Spinoza's thought gives them, amid common ontological and physical principles, *specifically* human abilities, thus without the need to conceive of their union with a soul to make the bodies *of men*.

[6] Similarly, Spinoza writes in the fourth part of the *Ethics* that "the action of striking" is considered "physically [*physice*]", in which case we "look to the fact that a man raises his arm, clenches his fist, and moves his whole arm violently downwards". This is explained by the sole structure [*fabrica*] of the human body, which is in itself neither good nor bad.

[7] AT XI, 119.

However, Spinoza will retain this point in his use of the term *fabrica*, comparable to Descartes' use of the French term "*fabrique*" (composition). Thus, on page 132 of the *Treatise on Man*, Descartes writes that he wants to "tell you about the composition of the nerves and the muscles, and to show you how, from the sole fact that the spirits in the brain are ready to enter into certain of the nerves, they have the power to move certain bodily parts at the same instant."[8] We can see in the use of this term "*fabrique*" a reference to Vesalius' *De humani corporis fabrica*; there is, however, a difference of scale here, in that the structure of the body refers to, according to Vesalius, the admirable work of God and the work which he demonstrated by associating nerves with muscles. Meanwhile Descartes, while speaking of the *structure* or *composition of* bodies, adds in the same sentence that this singlehandedly aids in understanding the movements of muscles; there is no mention in this context of the admiration that should rightly produce divine work. The term *fabrica* does not appear in Spinoza's Physical Interlude, but it is indeed also in this perspective when he employs it at the end of Part I of the *Ethics* and early in the third part. In Appendix I of the *Ethics*, Spinoza points out that when men "survey the frame of the human body [*corporis humani fabricam*], they are amazed," and that, instead of seeking "to understand natural phenomena as an intelligent being" (that is to say, by explaining the functions from a strictly physical point of view), they just "gaze at them like a fool". We thus find here the will to offer a mechanistic explanation of the functions of the body, as distinguished from those striving to praise through these functions the admirable work of God. The second occurrence of *fabrica* in the *Ethics* confirms this perspective. In the proposition of the *Ethics* III, 2, Spinoza states that nobody has so far determined "what the body can accomplish solely by the laws of nature", precisely because "no one hitherto has gained such an accurate knowledge of the bodily mechanism [*corporis fabricam*], that he can explain all its functions".[9] Just as in Descartes' *Treatise on Man*, the composition of the body is related to its functions, and with the strictly physical explanation of these functions, irrespective of any reference to God's wonderful work. This is perhaps seen to a greater extent in the Cartesian text, where what the body can do is subject to its power, in a positive and affirmative way, which will prove crucial to Spinoza's ethical perspective, one in which he will specifically diverge from Cartesian conceptions despite this common and required point of departure.

[8] The original French reads as follows: "parler de la fabrique des nerfs et des muscles, et montrer comment, de cela seul que les esprits qui sont dans le cerveau se présentent pour entrer dans quelques nerfs, ils ont la force de mouvoir au même instant quelque membre."

[9] The remainder of the sentence states that nobody can even explain, for lack of having sufficient knowledge of the structure of bodies, what the body cannot do unless it is determined by the mind. Exactly this will be the subject of the continuation of this note, in which Spinoza notably makes use of the example of animals (whose actions can "far transcend human sagacity") and sleepwalkers (who, in their dreams, do many things that they would not dare do when awake). Spinoza thus associates in one motion a separation of the functions of the body from admirable divine work and a separation of actions of the body from decrees of the mind that would render them solely possible.

Finally, bringing attention to what the body can do by itself amounts to explaining its functions independently of any recourse to any sort of intervention of the mind. This is the significance of the plan initially outlined at the beginning of the *Treatise on Man*: "And I must describe for you first the body on its own; and then the soul, again on its own; and finally I must show you how these two natures would have to be joined and united".[10] Explaining the functions of the body can therefore be carried out before any intervention of the soul. Descartes, on this point, puts himself in opposition to the theories developed by Aristotle in the second book of *On the Soul* and in the first book of *On the Parts of Animals*, two treatises according to which the soul is the principle of nutrient and sensory faculties and linked to the motion of the body, and which are largely repeated in subsequent medical treatises. This is confirmed by the fact that, when the soul is mentioned in the *Treatise on Man*, it is restricted to its rational dimension. We can thus read on page 143 that "when God unites a rational soul to this machine, as I intend to explain later on [the treatise effectively ends when Descartes is about to address the rational soul, and therefore even before the union can be mentioned], He will place its principal seat in the brain". This rational soul therefore has no function in connection with the life of the body, and it is moreover this distinction that will require Descartes to consider the union as a third element in its own right, faced with rendering commensurate a physical body and a rational soul.

Spinoza, albeit for other reasons,[11] follows a similar initial plan in the second part of the *Ethics*: thus, within the section entitled "On the nature and origin of the mind", he announces from the annotation to Proposition XIII that "no one will be able to grasp [the nature of the union between mind and body] adequately or distinctly, unless he first has adequate knowledge of the nature of our body", which is the reason why he "must premise a few propositions concerning the nature of bodies."[12] And the question of what the body can do by the laws of nature, which I mentioned earlier when citing the note of the *Ethics* III, 2, precisely targets the independence of the body and its functioning in relation to any intervention of the mind, as evidenced by the remainder of this first citation:

> No one hitherto has gained such an accurate knowledge of the bodily mechanism, that he can explain all its functions; nor need I call attention to the fact that many actions are observed in the lower animals, which far transcend human sagacity, and that somnambulists do many things in their sleep, which they would not venture to do when awake: these instances are enough to show, that the body can by the sole laws of its nature do many things which the mind wonders at.

---

[10] AT XI, 119.

[11] The mind and body being, according to Spinoza, a single and same thing conceived under two different characteristics (firstly, thought, and secondly, extended substance), there can be no question of firstly addressing the body, then the mind, then giving oneself reason to reflect on their union. On the other hand, the body being the object of the idea that constitutes the mind (according to Proposition XIII of the *Ethics* II), its knowledge is required to be aware of this idea, which justifies the introduction of the Physical Interlude at the beginning of the second part of the *Ethics*, devoted to the study of the mind.

[12] Which, of course, results in the Physical Interlude, following Proposition XIII of the *Ethics* II.

These Spinozist passages therefore fit well with the Cartesian sense of returning the body to physical nature and the parallel physical explanation of its functions, without any mention of an action of the mind or a soul playing a role in either these functions or in the "animation" of the body.

14.3 Spinozist Radicalisation Brought on by a Distinct Aim: From One Philosophical Perspective to Another

Both Descartes and Spinoza insist on the fact that they limit themselves in their remarks to what is useful for their purpose, without going into further details. Thus, when Descartes addresses the structure of the eye on page 152 of the *Treatise on Man*, he states that he "shall do so briefly, omitting many superfluous details which the curiosity of anatomists has uncovered here", when medical treatises traditionally devote many pages. In an even more marked fashion due to the pronounced brevity of the Physical Interlude, Spinoza concludes the note following Lemma VII as follows:

> I should feel bound to explain and demonstrate this point at more length, if I were writing a special treatise on body. But I have already said that such is not my object; I have only touched on the question, because it enables me to prove easily that which I have in view.[13]

These texts are thus both determined philosophical projects, yet they differ. This explains why, from the common background that I have just discussed, this results in two different and sometimes even diverging perspectives.

The first point of divergence concerns the extension given by Spinoza to these mechanistic-type explanations, in which he covers the mind. Indeed, later in the second part of the *Ethics*, Spinoza writes that.

> the mind is a fixed and definite mode of thought, therefore it cannot be the free cause of its actions; in other words, it cannot have an absolute faculty of positive or negative volition; but it must be determined by a cause, which has also been determined by another cause, and this last by another, etc.[14]

In other words, it is not only bodies whose functions can be explained in mechanistic terms; the same applies to the mind and the sequence of its ideas. However, while Descartes addresses in the *Treatise on Man* movements that we might qualify as spontaneous or that one performs without thinking (like breathing, digestion or even walking), insofar as he speaks of the body separately, he does not rule out that

[13] Namely, what leads to blessedness, according to the introductory text of the *Ethics* II: "I now pass on to explaining the results, which must necessarily follow from the essence of God [...]; not, indeed, all of them (for we proved [...] that an infinite number must follow in an infinite number of ways), but only those which are able to lead us, as it were by the hand, to the knowledge of the human mind and its highest blessedness."

[14] *Ethics* II, Prop. 48, proof. In the note of the same proposition, Spinoza adds that by the will he means "the faculty, whereby the mind affirms or denies what is true or false, not the desire, wherewith the mind wishes for or turns away from any given thing."

there are other types of movement, this time related to the will. This is what can be deduced from the passage that explains how, in the case of walking, a single action of the mind can cause one foot to advance after the other: "and this is applicable to everything I said earlier about respiration and similar movements which do not usually depend on an idea; and I say 'usually' because they may sometimes depend on them."[15] These movements linked to the will are discussed as an example in the *Passions of the Soul*.[16] Instead, the note of the *Ethics* II, 3, referenced above, was intended to demonstrate that *no* movement of the body was determined by the will of the mind, as men considered themselves free precisely because, in being aware of their actions, they ignore the causes that influenced them.[17]

Another difference, indeed a direct comparison between the two texts, lies in the fact that, although his text is significantly shorter than Descartes', Spinoza does not limit himself to solely physiological or sensory considerations – or even to considerations linked to the various physical movements of the body. He immediately addresses the broad issues *relative* to these physical abilities of the body, which is where we see his own perspective emerge. Thus, on the one hand, Descartes revisits the list of thematics addressed at the end of the *Treatise on Man* as follows:

> the digestion of food, the beating of the heart and the arteries, the nourishment and growth of the bodily parts, respiration, waking and sleeping; the reception of light, sounds, odours, smells, heat, and other such qualities by the external sense organs; the impression of the ideas of them in the organ of common sense and the imagination, the retention or imprint of these ideas in the memory; the internal movements of the appetites and the passions [read: the bodily movements that cause or accompany the passions, understood as perceptions of the soul]; and finally the external movements of all the bodily parts....[18]

Despite Spinoza's Physical Interlude being considerably less detailed in regard to these functions (most of them are not even mentioned), he does immediately link the increasing complexity of bodies to a growing ability to be affected in various ways, which entails significant ethical issues. We read successively in the note following Lemma VII that, aside from bodies composed solely of the most simple character, there is another type "composed of several individuals of diverse natures," which "can be affected [in a number of ways]...without losing its nature", and then "a third kind of individuals composed of individuals of this second kind", which can

[15] AT XI, 197.

[16] Thus, after having addressed in Article 16 "every movement we make without any input from our will" (among which we find, once again, breathing and walking), in Article 18 Descartes distinguishes between two types of will: "actions of the soul that aim only at something in the soul itself, as when we will [...] apply our mind to some object that isn't material; and actions of the soul that aim at some event in our body, as when we will to walk."

[17] Therefore, for example, in *Letter 58 to Schuller*, dated October 1674, Spinoza takes the example of the movement of a stone that, gifted with thought and invested in its movement, "believe[s] that it is very free and perseveres in motion purely because it wills to". Spinoza then extends the consequences of this example to men and concludes as follows: "this is the famous human freedom everyone brags of having, which consists only in this: that men are conscious of their appetite and ignorant of the causes by which they are determined."

[18] AT XI, 201–202.

in turn "be affected in a still greater number of ways without changing their actuality." This is required, however, in order to take a certain ethical path, consisting of increasing one's perfection without changing form or nature[19]; in other words, to transform while somehow remaining the same.

Thus, the difference between the two texts is partly explained by the inclusion of the Physical Interlude in a philosophical project with an ethical aim: indeed, the physical functions constituting the anthropological conditions of the possibility of taking an ethical path are addressed. It is significant in this regard that Spinoza criticises the Cartesian recourse to the pineal gland (which he considers a "hypothesis, beside which occult qualities are commonplace") not in the Physical Interlude for anatomical or physiological reasons relating to its location or to the seat of the soul, but in the preface to Part V, when discussing human freedom. Spinoza, when he explicitly mentions Descartes, is not targeting his physiological explanations, but what he sees as a false trail from an ethical perspective, in that it maintains our illusions of a free arbitrator and the possibility that we could exercise absolute dominion over our emotions. This is why, in the Physical Interlude, Spinoza conversely insists upon the absence of the action of the mind on the body: Lemma III states in this way that "[a] body in motion or at rest must be determined to motion or rest by another body", and not by an idea or the will of the mind. In understanding this we directly seek an ethical path not in the illusion of free will that has effects on the body, but in the gradual acquisition of new *corporeal* emotional habits.

Hence the need, in Spinoza's view, to insist in the Physical Interlude on differences between individuals at the level of the body itself, without reducing it to an addition of the soul; not in postulating the different laws of nature for the human body, but by envisaging the increasing (physical) aptitudes that are affected in various ways with respect to the increasing complexity of the body's composition. Consequently, this explains that, according to Axiom I following Lemma III, "different bodies may be moved in different modes by one and the same body", a body of equal complexity to that of a human body being the only one able to "move external bodies, and arrange them in a variety of ways" (Postulate VI). This is an ethical *possibility* rendered conceivable by an anthropological condition: the physical complexity specific to human bodies. In this context, the temperament is not addressed in physiological terms of abundance, size, or even agitation of minds, as Descartes does in pages 166–167, but in terms of emotional memory, insofar as we are *today* affected by the encounters we have experienced *in the past*.

The two different usages of the example of language that Descartes and Spinoza employ testify to this point. Thus, at the beginning of *The World*, Descartes uses the comparison of language to make known to his reader that ideas do not resemble the objects from which they proceed in the case of feelings (that today we would call sensations):

[19] Spinoza thus states in the preface of the *Ethics* IV that to go from a lesser to greater perfection is not, for him, to exchange his essence or form for another: "a horse would be as completely destroyed by being changed into a man, as by being changed into an insect." Similarly, a man would be as completely destroyed by being changed into a horse, as by being changed into God…

> if words [...] are sufficient to make us think of things to which they bear no resemblance, why could not Nature also have established some sign which would make us have a sensation of light, even if that sign had in it nothing that resembled this sensation?[20]

Meanwhile Spinoza, shortly after the Physical Interlude, uses this comparison to demonstrate that the way in which we are affected depends on our factual history up until now, as long as we do not make an effort to develop our way of being affected in several ways at once[21]:

> the mind from the thought of one thing, should straightway arrive at the thought of another thing, which has no similarity with the first; for instance, from the thought of the word pomum (an apple), a Roman would straightway arrive at the thought of the fruit apple, which has no similitude with the articulate sound in question [...]. Thus every man will follow this or that train of thought, according *as he has been in the habit* of conjoining and associating the mental images of things in this or that manner.[22]

The Cartesian and Spinozist philosophical projects differ; the uses of the common project that Descartes and Spinoza employ to reflect on what the body can do by itself also differ as a result. This is what allows us to understand how, from a common approach and method that is drawn from Cartesian texts, Spinoza seeks to develop theories to which Descartes would not have subscribed.

14.4 Conclusion

We do not know if Spinoza read the *Treatise on Man* insofar as he does not mention it at any point, not even in his correspondence. It is thus an imaginary dialogue that I propose in this comparative reading of the two texts. However, this text appeared in his library and was published between the drafting of the *Short Treatise* (in which he makes no reference to what the body can do in terms of the power to act and its own abilities, regardless of it being set in motion by God or a determination by a decree of the mind) and the *Ethics*, in which we find the Physical Interlude, on which I have based this comparison between the Cartesian and Spinozist perspectives. We know that Spinoza read and studied in detail the *Principles of Philosophy* and retained some of its grand principles, such as the approach of reflecting on what the body can do by itself, without recourse to occult faculties and without any specific intervention of a soul. Pierre-François Moreau has stated that we could consider the *Principles of Cartesian Philosophy* to be Spinoza's first work, "as a

---

[20] AT XI, 4.

[21] Which is made possible by the complexity of our bodies, composed of very complex bodies, according to the note following Lemma VII.

[22] *Ethics* II, Prop. 18, note; my emphasis. The manner in which we become accustomed to conjoining things is forged throughout our life experience (emotional and mnemonic history which has *in fact* become ours), but we can progressively vary this way of associating ideas between themselves in the context of an ethical future (a singular order of the series of emotions in our mind). We will be thus *differently* and *actively* determined to sequence the images of things in the future.

propaedeutic and a strategic preparation for the explanation of his own philosophy." While Spinoza's philosophy was already taking shape in this book, we can consider it fully developed when he wrote the *Ethics*. Moreover, the insertion of the Physical Interlude in his ethical project demonstrates both the importance of the principles he borrowed from Descartes, such as the physical conditions of the possibility of taking an ethical path, and his own particular aim, which explains the appropriate and sometimes divergent use of the grand Cartesian principles.

Chapter 15
Hobbes and Descartes on Anthropology: Is There a Debt of Hobbesian Anthropology to *L'Homme*?

Arnaud Milanese

Abstract Hobbes didn't know *L'Homme* of Descartes when he outlined his materialism, and controversies that opposed them (on optics and *Meditationes*) did not directly relate to physiology. However, this was probably the area where they were the closest. That's why we here propose to trace the scheme of the Hobbesian physiology and its sources to estimate to what extent they are related or not to Descartes, or whether (which we support here) they seek common sources.

Strictly speaking, the answer to this question is negative. *L'Homme* was published in 1662 and 1664, and Hobbes had already written those of his treatises that deal with anthropological development in 1640 for *Elements of Law*, 1651 for the publication of *Leviathan,* and 1658 for the publication of *De Homine*. Of course, Descartes wrote his treatise in 1633, but obviously Hobbes did not have access to this text. He discovered Descartes through his *Dioptrics*, between 1637 and 1640, and the *Meditationes de prima philosophia* (1640–1). Concerning the optics, Hobbes received a letter by Digby, sent on 4 October 1637, announcing that he would bring back the book of 'Monsieur des Cartes (que Mydorge admire tant)' (Digby came back from Paris in july 1638). So, it seems that Hobbes was interested in Descartes because of Mydorge's admiration for Descartes. But, in correspondence, neither Descartes nor the Mersenne circle are mentioned before the end of March 1640: so it is difficult to know whether Hobbes studied Descartes's *Dioptrics* before the end of 1639 or not. But, on 5 November, Hobbes sent fifty pages to Mersenne: a critical reading of Descartes' *Dioptrics* (this letter is lost). Descartes received a part of this text in January 1641, and answered Mersenne that he would await the remaining pages before reading it. But, immediately afterwards, he wrote

A. Milanese (✉)
ENS de Lyon, CNRS (Triangle – UMR 5206), Lyon, France
e-mail: arnaud.milanese@ens-lyon.fr

D. Antoine-Mahut, S. Gaukroger (eds.), Descartes' *Treatise on Man* and its Reception, Studies in History and Philosophy of Science 43,
DOI 10.1007/978-3-319-46989-8_15

again to Mersenne: after having read two pages (to which he answered in this letter to Mersenne), Descartes considered that he would not read what followed. A new exchange nevertheless occurred (letter from Hobbes, 7 February), and Descartes wrote to Mersenne (4 March) that he did not want any contact with Hobbes, accusing him of plagiarism. On 20 March, Hobbes answered Mersenne: Newcastle and Charles Cavendish could testify that a few years ago Hobbes already developed his views concerning light and images and that there could be no plagiarism on these subject-matters.[1]

So, even if *L'Homme* cannot have played any role in the formation of the Hobbesian anthropology, physiology was one of the matters about which both thinkers could understand each other, and the interest of Hobbes in *Dioptrics* testifies for this fact. According to received opinion, however, there was no dialogue between them: not competent enough to understand *Dioptrics,*[2] Hobbes was involved in a discussion of *Meditations* to which neither Hobbes nor Descartes wanted to take part.[3] If we add to this the fact that Hobbes wrote his objections in the carriage, which took him from Paris to Rouen, we have all the elements leading to the construction of the fiction of a missed appointment. It of course seems difficult—though not all scholars would agree with us—to deny that some formulations of the objections to *Meditations* do not have the same accuracy as those he wrote, almost at the same time, on Thomas Whites's *De Mundo*. However, it does not mean that Hobbes did not read *Meditations* closely, and, above all, that his objections are all wrong. This idea of a missed appointment, due to circumstances, is then fanciful: the result of the Hobbesian objections and of the Cartesian answers is mostly to be explained by a difference of substantive positions, already well established. The philosophical oppositions between Hobbes and Descartes are well known: dualism, innatism and exclusion of politics from philosophy, on one side, materialism, sensualism and a mathematical science of politics, on the other. In any case, these divergences should not have prevented them, as mechanists, agreeing on physiology

[1] Hobbes again urged these points in dedicatory epistle of the *First Draught of the Optiques* (1645–6) and in *Six Lessons*, and added that Walter Warner took him the idea that images are subjective (1656). Anyway the letter to Newcastle on 16 October 1636, in which Hobbes rejected sensible species, and developed a theory of the propagation of light by the milieu (a letter containing a critic of Galileo) already claimed obviously that light and colours are nothing but effects of a motion within the brain.

[2] Concerning Hobbesian criticism of *Dioptrics*, see J. Bernhardt, "La *Dioptrique* de Descartes dans le *Tractatus opticus* II", in *Revue Internationale de Philosophie*, n°129, 1979; M. Blay, "Genèse des couleurs et modèles mécaniques dans l'oeuvre de Hobbes", *Thomas Hobbes. Philosophie première, théorie de la science et politique*, Paris, PUF, 1990. For an accurate reading of the optique of Hobbes: J. Bernhardt, "Hobbes et le mouvement de la lumière", in *Revue d'histoire des sciences*, n°30, 1977; J. Médina, introduction and commentary of the optique of *De Homine*, in the critical edition and French translation, dir. J. Terrel, Parin, Vrin, 2015, and "Hobbes's Geometrical Optics", in *Hobbes Studies*, vol. 29–1, april 2016. See also A. Lupoli, "Optics, Simple Circular Motion and Conatus", *ibid.*; F. Giudice, "Optics in Hobbes's Natural Philosophy", *ibid.*

[3] Concerning these objections, see E. Curley, "Hobbes contre Descartes", in *Descartes. Objecter et répondre*, Paris, PUF, 1994; J. Terrel, "Le matérialisme de Hobbes dans les *Troisièmes Objections*", in *Hobbes et le matérialisme*, dir. J. Berthier, A. Milanese, Paris, Editions matériologiques, 2016.

as regards anthropology, in the medical meaning the term had in the seventeenth century, and as regards the content of *L'Homme*, a treatise which, whether by accident or not, did not offer any chapter devoted to soul.

So, even if Hobbes did not read *L'Homme* while he constructed his anthropology, the question remains what relation there can be between Hobbes and the Cartesian thesis it contained (a content that could also be partly found in *Dioptrics* and *Discours de la méthode*)[4]: a physiology without soul or immaterial spirit, a mechanical explanation of vision, an anatomy of eye and sensation. Hobbes at least studied the *Dioptrics* while he wrote the *Elements of Law* (he finished this treatise in 1640, but he had started to prepare it at least as early as 1636). Now, this physiological matter is not only common to both thinkers, but also crucial for the identity of Hobbesian philosophy: not only mechanism (and some have maintained—wrongly on our account—that this Hobbesian mechanism came from a reception of Cartesian philosophy), but also materialism (and some have also maintained that it could come from a radicalisation of the Cartesian exclusion of any organismal soul). So, we would like to pursue this question: what is the importance of the Cartesian thought for the origin and the development of Hobbesian natural philosophy in general, and anthropology in particular?

15.1 Hobbes and Physiology: A Late Interest

At the beginning of the 1630s, Hobbes was already interested in optics. Sensation was, according to him, his first subject-matter, chronologically speaking. His prose autobiography indeed explained that he studied mathematics in order to understand motions, and motions in order to understand sensation.[5] The correspondence during 1635–36 (the years of his meeting with Mersenne) confirms this point. But this interest in sensation and optics did not immediately mean an intellectual interest for physiology before his exile in France (1641–50), for reasons exposed in a letter to Newcastle, 29 July 1636:

> In things that are not demonstrable, of which kind in the greatest part of Natural Philosophy, as depending upon the motion of bodies so subtile as they are invisible, such as are air and spirits, the most that can be attained unto is to have such opinions, as no certain experience can confute, and from which can be deduced by lawful argumentation, no absurdity.[6]

---

[4] We will not analyse the relation between Hobbes and Descartes in general. See also E. Marquer, "Ce que sa polémique avec Descartes a modifié dans la pensée de Hobbes", and A. Bitbol-Hesperies, "*L'Homme* de Descartes et le *De Homine* de Hobbes", in *Hobbes, Descartes et la métaphysique*, Paris, Vrin, 2005. The second paper gives many elements for an accurate comparison. The author considers however out of discussion that Cartesian thought is anterior and decisive to explain the beginning of Hobbesian physiology and natural philosophy. That is precisely the point we want to question.

[5] J. Terrel, *Hobbes. Vies d'un philosophe*, PUR, 2008.

[6] *Correspondence*, ed. N. Malcolm, Oxford, Clarendon Press, 1994, vol. I, p. 33, we have modernized spelling.

Here, Hobbes defined the largest part of natural philosophy as a field in which there can be no demonstration, understood as the possibility of establishing a cause with certainty, but only probable opinions, because natural phenomena depend on bodies—especially material spirits—too subtle to be perceivable. In this field, the only possible proofs are due to experimental refutations. His account implied that natural history remained a part of natural philosophy (contrary to what Hobbes would think later). When Newcastle claimed, Hobbes wrote, that diverse sensible qualities are nothing but motions in spirits, Hobbes agreed, but he disapproved of the views of Mydorge and Warner when they considered their optics as pure demonstration, because their reasons depended on suppositions. Hobbes then detailed (concerning Warner's *De loco imagines* sent by Warner to Charles Cavendish) the experiments to be made in order to assess the hypotheses of Warner about the place of the image in lenses. In other words, necessity of an experimental control and diffidence towards hypotheses explained his interest in optics *and* his diffidence towards physiology, because there could be no experimental control of the hypotheses concerning material spirits.[7] Later, in *De Corpore* (1655), Hobbes promised to come back to physiology, partly studied in 1655, and to detail the 'fabrica' of the eye, but he never kept his promise. *De Homine* rewrote in Latin the second part of *First Draught of the Optiques* (1645–6),[8] but without any anatomical reference, because the optics of *De Homine* was a geometry of the field of view, and not a physics of light or a physiology of seeing, contrary to what could be expected. We rely here on the study of J. Medina.[9] He recalls that, with Kepler, a change of the meaning of optics had only begun and would culminate in Newton: in Aristotle or Euclid, optics was defined as science of vision; thanks to an objectivation of ocular devices (for instance, the metaphor of the eye as *camera obscura*, in Della Porta or Kepler), optics became a physics of light and its action on bodies. Between Kepler and Newton, both meanings were to be found and combined, and they were clearly distinguished in Hobbes. In the first part of *First Draught of the Optiques* (we now understand the plural), "Of illumination", Hobbes used the new meaning of optics, a physics of light (and this physics was afterwards to be found in *De Corpore*), but, in the second part, "Of vision", we find the first meaning of optics, historically speaking, with the addition (until 1646) of physiological considerations, and that is exactly what was afterwards to be found in *De Homine*.

So Descartes and Hobbes could agree concerning physiology, but Hobbes ventured into it only a little and late. Moreover, he studied physiology in Paris for historical reasons, unrelated to Descartes. Indeed, in Paris, around 1645–6, Hobbes was within a group of English exiles (with Charles Cavendish, William Petty and

---

[7] *Ibid.*

[8] Not the first, because its content was already in *De Corpore*.

[9] Op. cit.

Robert Boyle),[10] who were all corpuscularists[11] and experimenters, far from William Harvey, vitalist, hostile to atomism and to the use of microscopes. This group deviated then from Harvey's works for substantive reasons, but in order to *translate* his thesis into a mechanical conception of physiology. From 1645 to 1646, before the publication of the two treatises of Regius[12] and the one of Hooghelande,[13] which contributed to the diffusion of the Cartesian physiology, Hobbes and Petty studied Vesalian anatomy and chemistry. In 1646, they attended the lessons in chemistry of William Davidson, with whom they realized, for seven months, fifty chemistry experiments on quicksilver, sulphur and distillation of nitre (there are traces of these studies and experiments in *De Homine*, Chap. 1). These years are crucial for both the finalisation of *De Corpore* and the future constitution of Royal Society (Petty founded in 1649 the Buckley Hall group, including Boyle, with Wilkins, and this group became Royal Society), and they testify to the influence of Hobbes on these young physicians. What could be called the program of mechanization of Harvey, Petty presented in 1651, was not as inspired by Descartes, as it was by Hobbes,[14] whose secretary was for a while William Petty (he calligraphed, in 1645, *First Draught*, and drew the figures), who always showed great admiration for Hobbesian natural philosophy, even if after 1648 Hobbes denied the existence of void, whereas Petty, like Boyle, kept on claiming its existence.

So Hobbesian mechanism was already constructed, before he encountered Descartes's books, and had numerous partisans in Oxford,[15] and this was without any doubt the reason why Mersenne esteemed Hobbes as a valorous philosopher in 1634. Detailing what debt Hobbesian anthropology could have to the Cartesianism

---

[10] Also before, without any doubt, but we do not know much about 1644 (Hobbes worked on his *Critique of* De Mundo, till the middle of 1643), because of lack of preserved correspondence. We only know that Hobbes ended his *Cogitata physico-mathematica*, published by Mersenne in 1644, took a close look at Descartes's *Principia philosophiae* and received reactions to his *De Cive*, first published, in a short version, in 1642.

[11] About corpuscularism, long present in England (Thomas Harriot and also Francis Bacon can give testimonies), see R. H. Kargon, *Atomism in England from Hariot to Newton*, Oxford, Clarendon Press, 1966; B. Gemelli, *Aspetti dell'atomismo classico nella filosofia di Francis Bacon et nel seicento*, Leo S. Oschki, 1996. Percy's and Welbeck's circles were not foreign to the transmission of Hariot's works: in 1631, Warner published the *Artis analyticae praxis* of Thomas Hariot, and John Wallis wrote on 20 July 1683 that at least a part of Hariot's papers were once in Hobbes's hands. See K. Schuhmann, *Hobbes. Une chronique*, Paris, Vrin, 1988. Here we have the factual reasons why he was late interested in Harvey: Hobbes found contributors to achieve a project of which he long perceived the difficulty.

[12] *Fundamenta physices*, 1646; *Fundamenta medica*, 1647.

[13] A part of the *Cogitationes*, 1646.

[14] For a study of the manner Hobbes, Petty and Bathurst reappraised and criticized Harvey and Descartes, see J. Medina, "Physiologie mécaniste et movement cardiaque: Hobbes, Harvey, et Descartes", in *Lecture de Hobbes*, Paris, Ellipses, 2013, from whom we borrow many elements. For the idea of a mechanization of the thesis of Harvey by English physicians, after 1645, see R. G. Franck Jr., *Harvey and the Oxford Physiologists*, University of California Press, 1980; "Medecin", in ed. Tyacke, *The History of the University of Oxford*, vol. IV, Oxford University Press, 1997.

[15] See M. Feingold, "The Mathematical Sciences and New Philosophies", in ed. Tyacke, *The History of the University of Oxford*, op. cit., pp. 413sq.

present in *L'Homme*, we now have to (1) say more about the beginning of Hobbesian mechanism, and then (2) detail this program of mechanization of Harvey.

15.2 The Beginning of Hobbesian Mechanism and Physiology

Concerning the first point, we will focus on the statute of the *Short Tract on First Principles*. Like F. Tönnies,[16] J. Bernhardt[17] thought it was written in 1630, the year following Hobbes' legendary Euclidian revelation, in Paris, during his second travel on the continent. In fact, this text was written later in the 1630s, before 1636 (we will give the reasons). The 'revelation' is mostly a legend[18] and this text is not in Hobbes' hand. But, contrary to what is claimed by N. Malcolm,[19] it is obviously mechanistic. It probably was written by Robert Payne,[20] a fellow of an Oxford College, where he taught mathematics, among other things. Both men admired each other and belonged to the Newcastle circle: Hobbes thought that Payne was superior to Warner in a letter to Newcastle in 1635, and approved of him asking Payne to criticize a treatise of Warner concerning burning mirrors. On the other hand, many elements of the text are indisputably Hobbesian,[21] especially the third part about man: the physiological definition of image, along with a cognitive definition, for instance. Such a definition is absent from the optical treatises of Kepler: either Kepler was satisfied with his explanation of the formation of a *pictura*, referring to future works concerning its physiological effects,[22] or he explained the passage from this *pictura* to a mental *imago* by the effect of immaterial spirits operating in the seeing body.[23] Here is another Hobbesian characteristic of the *Short Tract*: the demonstration of determinism on the basis of an identification of a sufficient cause with a necessary cause – this demonstration is that we find in Chap. 9 of *De Corpore*.

In detail, the controversy of the attribution of this tract is complex and implies many non-decisive elements: for instance, the fact that Payne, in England at this moment (Hobbes was in France), translated from Italian to English a work by

[16] *Thomas Hobbes : Leben und Lehre*, 3rd edition, Stuttgart, 1925.

[17] Introduction and commentary to his edition and French translation of *Short Tract*, Paris, PUF, 1988.

[18] See our arguments in A. Milanese, "Sensation et phantasme dans le *De Corpore*: que signifie, chez Hobbes, fonder la philosophie sur la sensation?", in *Lumières*, n°10, 2007.

[19] See "Robert Payne, the Hobbes Manuscripts, and the 'Short Tract'", in *Aspects of Hobbes*, Oxford, Clarendon Press, 2002. He gives a summary of the controversies concerning this text. See also a study to which N. Malcolm also refers, by T. Raylor, "Hobbes, Payne, and *A Short Tract on First Principles*", *The Historical Journal*, 44, 2001, pp. 29–58.

[20] As was shown by Richard Tuck in 1988, and T. Raylor and N. Malcolm agree.

[21] As was already shown by K. Schuhmann.

[22] *Astronomia pars Optica*, 1604.

[23] *Dioptrice*, 1611.

Castelli in 1635, which had significant similarities with *Short Tract*. This point is not conclusive because Hobbes and Payne read Italian and both could have read this text before Payne translated it. The important point, for the question of the beginning of Hobbesian mechanism, is (1) the fact that this tract is a mechanistic one, and (2) the fact that Payne and Hobbes worked together between 1630 and 1634, within the same circle. The source and paternity of many ideas contained in this tract were obviously collective, which is often neglected in this kind of paternity quarrel. Moreover, the way the relation between Payne and Hobbes evolved after 1640 prompts one to think of the influence of Hobbes on Payne rather than the contrary. Payne, whose only works were that of a translator, wrote many notes on *De Corpore*, preserved at Chatsworth, and, in the 1640s, he tried hard to spread Hobbesian ideas and texts in Oxford. For instance, the manuscript of *Human Nature* used for the edition of 1650 was that of Thomas Lockey, a friend of Payne, who also tried by any means, but vainly, to prevent a growing hostility in Oxford toward Hobbes.[24]

Leaving aside this question of paternity, the *Short Tract* is a good testimony of the beginning of Hobbesian mechanism and was written between 1630 and 1635. If it was before 1634, Payne wrote it but including ideas that were forged with Hobbes. If it was written between 1634–5, Payne wrote it (Hobbes was in France) under the influence of Hobbes. But, if it were written after 1635, it would have contradicted the optical theories of Hobbes, in 1636, on an important point: the diffusion of light by the milieu, whereas the *Short Tract* talked about a diffusion by corpuscular sensible species. Now, in this case, it would be surprising to have no traces of a discussion between Hobbes and Payne about this point. In any case, this text indirectly reveals a state of the thought of Hobbes before 1634, because the manuscript is of Payne's hands, Payne remaining in England while Hobbes was on the continent (1634–36), and it was probably during this travel that Hobbes changed his mind or developed his account of the diffusion of light.

So it is clear that Hobbes became a mechanist philosopher before knowing Descartes' works, to which he never acknowledged a significant debt. On the other hand, he did recognize a debt for Galileo, Kepler, and Copernicus. Hobbes sometimes suggested that the new science began with Copernicus because the mechanistic revolution was a consequence of his astronomical discovery, published in *De revolutionibus coelestium* (1543).[25] This early relation to mechanism[26] explains his complicated relation to Harvey and physiology. He probably met him in Bacon's circle, in the 1620s. One of the common points between Hobbes and Descartes was

---

[24] See M. Feingold, "An early Translator of Galileo and a friend of Hobbes: Robert Payne of Oxford", in ed. North and Roche, *The Light of Nature: Essays in the History and Philosophy of Science Presented to A. C. Crombie*, Dordrecht, Martinus Nijhoff, 1985, pp. 265–280; "The Mathematical Sciences and New Philosophies", op. cit., p. 414.

[25] Hobbes often mentioned Copernic. See for instance *Decameron physiologicum* (1678), chap. IX for his statute of beginners of modern physics.

[26] Despite his reputation of late philosopher: accepting mechanism and becoming a philosopher, in the systematic meaning of the term, are different things, even if Hobbes did not make this difference when he wrote in his autobiographies that, in 1634, when he met Mersenne, he was then counted as a "philosopher".

their adhesion to the Harveyian thesis of blood circulation. As far as Descartes is concerned, this is clear and well known. From *Discours de la méthode*, Descartes recognized that Harvey was the first to have stated the circulation of the blood, and Descartes agreed, despite disagreeing with the way in which Harvey changed the roles of diastole and systole (blood is not expulsed with the diastole, but with the systole, for Harvey).

On the other hand, it is difficult to know when Hobbes adhered to the circulation of the blood, considering his misgivings about the possibility of demonstration, and even of probable opinion, in physiology. No text before *Leviathan* (1651) claimed it explicitly. The first treatises of optics even compared the diffusion of light by the sun to diastole and systole, maintaining their ancient and pre-Harveyian meaning. However, the example of Descartes shows that it is possible to admit that the circulation of the blood is due to the heart without accepting the Haveyian meaning of diastole and systole. Anyway, it was only a comparison with the aim of understanding an optical phenomenon (similar to comparisons we find in Descartes' *Dioptrics*), and not an implicit thesis concerning the heart. Furthermore, after 1646, this comparison disappeared, not because Hobbes changed his mind concerning physiology, but because he rejected the void, and so rejected, after 1648, his first explanation of the diffusion of light: that is the reason why the second part of the *First Draught* is altered, in *De Homine*, without the major transformation the elements of the first part underwent when they were included in *De Corpore.*[27]

But chiefly, the explanation of sensation and passions, in *Short Tract* as well as in genuine Hobbesian anthropological treatises, always implied a circular motion of material spirits, associated, in 1651, with the motion of blood, without mentioning Harvey. Indeed Harvey was mentioned only once, in *De Corpore*: physics really started with Copernicus and Galileo, physiology with Harvey and political science with Hobbes (1655). But, in *De Homine* (1658), when Hobbes dealt with blood circulation, Harvey was not mentioned. The single mention of 1655 was permitted, but only circumstantial: circumstantial because, like Hobbes, whom he befriended (they dissected a deer together[28]), Harvey was royalist, and, like him, Hobbes wrote, he had to fight against everybody to make his novelties recognized. This Hobbesian statement is confirmed by the fact that the earlier statement of Descartes concerning Harvey was an exception, in 1637.[29] This mention of 1655 was also permitted because, with Ralph Bathurst, a friend of Petty and Hobbes, he took part in what could be called a program of mechanization of Harvey.[30] Bathurst was an admirer of

[27] See J. Medina, in *De Homine*, op. cit.

[28] See N. Malcolm, "Hobbes and the Royal Society", in *Aspects of Hobbes*, op. cit. This article has also the merit to contribute to demolish the wrong image, partly due to his quarrel with Boyle seen from Boyle's side, of a philosopher opposed to experimentation. See also J. Terrel about the political context of the genesis of this reputation, "Hobbes et Boyle: enjeux d'une polémique", in *La philosophie naturelle de Boyle*, dir. M. Dennehy and C. Ramond, Paris, Vrin, 2009.

[29] See W. Pagel, *William Harvey's Biological Ideas*, Karger Medical and Scientific Publishers, 1967

[30] About Bathurst and this collaboration with Hobbes, see J. Medina, "Physiologie mécaniste et mouvement cardiaque: Hobbes, Harvey, et Descartes", op. cit.

Hobbes, in this young generation of physicians: he wrote a Latin poem in the honour of Hobbes, published as an introduction of the first part of the *Elements of Law*, entitled *Human Nature* in 1650 (without the permission of Hobbes). The first chapter of *De Homine* is one of the results of this program: Hobbes agreed with the mechanistic criticism of Harvey's vitalism, and refused to consider, as Harvey did, blood as a simple element, non-corpuscular, living *in se* and *per se* (for Harvey, the heart is a pump, but receiving its power to move from blood, the real and original source of life). But, for Bathurst and Hobbes,[31] the Cartesian explanation (the heart is not a muscle, but has an innate heat, making it work like a thermal spring) is not sufficient because (1) the speed of blood circulation makes this explanation implausible, (2) it contradicts the muscle structure of heart, obvious thanks to dissection, especially when a still beating heart is extracted from a dead animal, and (3) Cartesian explanation is also insufficiently mechanistic: where does the heat of the heart come from? Hobbes preferred to go back to the pump model (and also the Harveyian inversion of diastole and systole), and even the idea that the motion of the heart is received by blood. The Haveyian explanation was satisfying from an anatomical point of view, even if it implied false principles. The question Harvey did not ask was: where does the motion within the blood—not the circulation of the blood, but the motion contained within the blood and transmitted to heart—come from? The only credible answer for Hobbes was: from respiration. Thanks to respiration, nitre contained by the air is received by blood, which transmits its *conatus* (motions too small to be perceivable) to heart, which becomes able to move blood like a pump. Here was the relation between respiration and the circulation of the blood, thanks to a chemical analysis of the air. This thesis anticipated that of Richard Lower and broke with an Aristotelian thesis, which, with the idea of an innate heat of the heart, Descartes also accepted: respiration is a cooling system.[32]

In sum, Hobbesian mechanism was first developed without Descartes, and his physiological application took more time, not because it was delayed, but because of the Hobbesian epistemological misgivings concerning physiology and of the difficulties of the program of mechanization of Harvey. Furthermore, physiology remained only a small part of *De Homine*, whereas optics was almost half the book, but without the physics of light or physiology of vision (except the affirmation, against Descartes and Kepler, that the optical nerves are not empty but full, an affirmation proved, for Hobbes, by observations with microscope). That is what Hobbes understood by considering rather soul ('*anima*') than body, in *De Homine*.[33]

---

[31] Concerning Hobbes, *De Homine*, chap. 1, and concerning Bathurst, *Praelectiones tres de respiratione* (1655–6), unpublished while Bathurst was alive, and published by T. Warton, in *The Life and literary Remains of Ralph Bathurst*, 1761.

[32] About the heart and respiration, see J. Medina, "Physiologie mécaniste et mouvement cardiaque: Hobbes, Harvey, et Descartes", op. cit. See also R. Garau, "Springs, Nitre, and Conatus. The Role of the Heart in Hobbes's Physiology and Animal Locomotion", in *British Journal for the History of Philosophy*, 2016.

[33] *De Homine*, chap. 1, op. cit., Paris, Vrin, 2015.

15.3 Reconciling Mechanism and Materialism[34]

This last section shows that the role played by Descartes in the formation of Hobbes's thought cannot be direct because Hobbesian anthropology also proceeded from another history, different from that of mechanism, and paying attention to this other history means we must start with Bacon, for whom Hobbes worked in the 1620s. The relation between Bacon and Hobbes is often, in intellectual biographies of Hobbes, a blind spot, because we do not know much about it, but it also allows to explain many elements in Hobbes: the form of its system, the idea that there is a political science, all this can be found in Bacon, as well as the idea of sovereignty.[35] In *De Dignitate et Augmentis Scientiarum* (1623), Bacon distinguished 'rational soul', immaterial, the 'substance and faculties' of which are not known except by sacred theology – philosophy only dealing with 'uses and objects' of these faculties –, and the 'sensitive soul', motion of life and sensation, clearly knowable in its substance as it is material, as already maintained, according to Bacon, by Telesio and Doni.[36] These two references are absent from the English version (*Of the Proficience and Advancement of Learning*, 1605), in which we read only that sacred theology knows the 'substance' of the soul, whereas philosophy knows its 'faculties'. Now, Hobbes took a hand, to an unknown extent, in the composition of *De Augmentis*. So, a physiology deprived from any organic and immaterial soul was present in England at the end of 1620s, Hobbes had access to it, and this source also showed its paternity, which had little relation to mechanism. It has rather a relation to the Renaissance and anti-scholastic tradition of commentary of Aristotle's *De Anima*. These commentaries had indeed two joint developments, especially under the influence of Vives[37] and Melanchthon,[38] and, without any doubt, the anatomy of Vesalius.[39] The first development promoted the idea that the faculties of the soul are functions we exercise and not properties we observe, from which proceeded an enrichment of the commentary with elements taken, for instance, from the *Rhetoric*, and so of the study of *uses* and effects of our mental faculties, in order to know the mind, not thanks to introspection or intellection, but thanks to its 'uses and objects'. The inter-

[34] Most elements of this part of our study are taken from A. Milanese, *Bacon et le gouvernement du savoir. Critique, invention, système: la pensée moderne comme épreuve de l'histoire*, Paris, Classiques Garnier, 2016, part II.

[35] About the relation between Bacon and Hobbes, see J. Bernhardt, "Sur le passage de Fr. Bacon à Th. Hobbes", *Etudes philosophiques*, n°4, 1985; R. Bunce, "Thomas Hobbes' relationship with Francis Bacon – an introduction", *Hobbes Studies*, vol. XVI, 2003; J. Terrel, "Comment Hobbes devient Hobbes", in *Lumières*, n°10, 2007; A. Milanese, "L'histoire de la science et de ses institutions de Bacon à Hobbes: un héritage critique", in *Archives de philosophie*, 77–1, 2014, et "Sur le passage de Bacon à Hobbes: un système et ses tensions", in *Philosophical Enquiries – revue des philosophies anglophones*, n°4, 2015.

[36] See also A. Milanese, "'History as psychology': de quoi est faite une psychologie empiriste chez Bacon?", in *Dix-septième siècle*, n°3, 2014.

[37] *De anima et vita*, 1538.

[38] *Commentarius de anima*, 1540.

[39] *De humani corporis fabrica*, 1543.

est of Hobbes for *Rhetoric* was sufficient to lead him to write and publish a summary of it in 1637. The second development took part in the context of new developments in medecine, thanks to which one reinforced, detailed and corrected Aristotelian theses with more accurate anatomical knowledge. But, the more the anatomical and physiological elements were detailed, the less the adding of an immaterial soul seemed necessary. Then, two kind of commentaries could be distinguished: the one, close to the text, kept an illustrative function to references to Vesalius (for instance in Suarez), the other lead, with Telesio for instance, to the ridding of any reference to the organic soul.[40] This is the line of thought which, starting from Telesio and Doni, Hobbes got through Bacon.

But there remains two points to be explained: (1) what became the immaterial soul of Bacon, in Hobbes?; (2) what is the nature of a matter able to generate life?

(1) On the first question, Telesio and Bacon had in common the idea of maintaining an immaterial soul, but mainly because of sacred theology. Telesio also justified the fact that philosophers believe what can be read in Scriptures about an immaterial soul because of the superiority of human beings over the other animals.[41] The materialist anthropology of Hobbes fits well in this history: there is no physiological soul and there is no immaterial soul, but only because there is nothing about immateriality in Scriptures. What is called '*anima*' could be nothing else but motions of material spirits. The structure of its argumentation, including the need to explain the superiority of human cognitive power with a transformation of animal cognitive power (see for instance the first five chapters of *Elements of Law*), was well adjusted to a critical reappraisal of Baconian analyses, but not to those of Descartes. This explains why Hobbes never justified the inexistence of immaterial substances, except by arguing for their incomprehensibility.

(2) Second point. The vivifying efficiency of matter, because this idea implied also something like a vitalism (for instance in Campanella[42] or Harvey), and, from this point of view, talking about a mechanistic and materialist physiology seemed to imply to reconcile the irreconcilable: the idea of a spontaneous matter, implied in a materialism, and the rejection of the same, implied in a mechanism. The concept of *conatus* in Hobbes reconciled these, and the lack of such a concept in Descartes meant, for Hobbes, that he was unable to understand that '*anima*' is nothing but motions in material spirits inducing vital motion and mental powers, cognitive as well as affective. This concept allowed one to conceive something close to the spontaneity of a finite body, explaining the apparent spontaneity of life, or, more generally, of matter. In his criticism of Descartes' *Dioptrics*, Hobbes thought, like Fermat for instance, that the Cartesian distinction between a motion and an inclination to

[40] See K. Park, "The organic soul", in *The Cambridge History of Renaissance Philosophy*, Cambridge University Press, 2008.

[41] The difference is that, according to Bacon, this superiority is a sign that what Scriptures tell on that subject is believable, whereas, according to Telesio, this superiority is a way, for philosophy, to demonstrate that there is an immaterial soul.

[42] About Campanella and Hobbes, see E. Sergio, "Hobbes lecteur de Campanella: autour des sources cachées du matérialisme hobbésien", in *Hobbes et le matérialisme*, op. cit.

motion could not be understood in a mechanistic natural philosophy.[43] The mechanisation of Harvey was then a part of a general mechanisation of materialism. In this way, we can explain the importance of the phenomena of reaction and resistance in Hobbesian natural philosophy, and Hobbes made a great use of these concepts in his criticism of Descartes, especially of the Cartesian concept of rest: his criticism not only of the Cartesian explanation of hardness, but also, more generally, of the 4th law of collision, in *Principia philosophiae*, their common point being the opposition, unfounded for Hobbes, between motion and rest, whereas rest is rather, for him, a balance of opposing motions. We can also think about another reproach made by Hobbes to Descartes: the substitution of a geometrical scheme to a '*ratio physica*' (an explanation thanks to the action of matter) – for instance, the demonstration proposed by Hobbes of the sine law.[44] Extension is not naturally anterior to motion, according to Hobbes – motion is co-eternal to matter, and the late affirmation of the corporality of God (1668) is decisive. Certainly, the existence of a first cause remained hypothetical: *in fine*, according to Hobbes, not only there is no proof of the existence of a God, but also, in *De Corpore* (1655), reason alone cannot come to the idea that there is a God (before, Hobbes seemed to write the contrary). But, the idea that, if there is a first cause, then it is material, reconciles mechanism and the spontaneity of matter: even if no finite body can move oneself, the entire matter has its own motion, if a supposed first cause is also corporeal.[45]

15.4 Conclusions: Cartesian Hobbes and Hobbesian Descartes

From a Cartesian point of view, it remains difficult to admit what Hobbes claimed in *De Homine*: Hobbes considered, in this text, the soul ('*anima*') rather than the body (this single use of the concept of *anima*, in Hobbes, had without doubt a polemical function). But this sentence is meaningful within the history we have reconstructed. In Hobbes, mind became a corporeal power, but a power that cannot be fully explained, and most of all the very fact of appearance.[46] The distinctions made by Bacon are maintained, but translated: the immaterial soul is rejected on exegetical grounds; sensation itself is partly known thanks to its 'uses and objects'.

[43] See J. Bernhardt, "La polémique contre la *Dioptrique* dans le *Tractatus Opticus II*", op. cit.

[44] Correspondence shows an early interest in this subject, around Hobbes: on 17 October 1634, Warner asked Payne to deliver to Hobbes the problem of refraction. He also wanted what Mydorge had written on this subtect-matter, and he thought that Hobbes could obtain it, which shows that, in 1634, Hobbes knew well Mydorge and this kind of question. And, on 13 june 1636, Hobbes wrote (from Lyon) to Newcastle that Mydorge just sent to Charles Cavendish his treatise on refraction (*Prodomi*, books III and IV, which were to be published in 1639).

[45] On this point, see A. Milanese, "L'accès intellectual à Dieu chez Hobbes", in *Hobbes et la religion*, Bordeaux, PUB, 2012.

[46] "Among every available phenomenon, the most admirable is *to phaïnesthaï*", *De Corpore*, IV, 25, we translate.

Hobbesian anthropology linked two points of view: that of the physiological basis and that of the phrase 'read thyself'.[47] Now, within myself, acts, and so relations, are read: cognitive and affective relations between me and the world around me (from this reading proceeded a subjective optics and an anthropology of desires[48]), and between me and other men (from which proceeded an analysis of sociality[49]). Here is the reason why political history and science are essential to such an anthropology of *anima*, or, more often, *mind*, *mens* or *animus*. Here is the essentially non-Cartesian character of Hobbesian anthropology.

On this basis, we can understand how the Cartesian prism could impose a bias in the way we read Hobbes. This bias leads to build a materialistic anthropology on the basis of Cartesian dualism (without considering union), in taking away thinking substance. This way to think of materialism by taking away is very common and diverse,[50] and implies an ignoring of the knowledge of mental powers by their uses. The anthropological result cannot avoid being inconsistent. Nevertheless, we find it in the readings that dissociate natural philosophy and political philosophy in Hobbes (whereas anthropology is, according to him, what gathers 'both sides of a precipice'[51]), and in the readings that interpret recourse to the history of human affairs as a break in the continuum of its philosophy (whereas this recourse is a consequence of the essence of his materialism).[52] The readings of Leo Strauss, for instance, despite their differences, perfectly match this sketch.[53]

But the reverse prism also exists, especially in English philosophy. The beginning of the *Divine Dialogues* by Henry More (1668), published the year Hobbes published the thesis of the corporality of God, showed how Descartes could be substituted for Hobbes as exemplar of materialism. Sophron is worried when Cuphophron announces the arrival of Hylobares, a materialist thinker:

> Who is this Hylobares? Is it he who is so much famed for holding, That there is nothing but body or matter in the world; that there is nothing just or unjust in its own nature; That all pleasures are alike honest, tho' it be never so unaccountable a satisfaction of either a man's cruelty or his lust?

Here is precisely the way critics of Hobbes read his philosophy at this moment. But Cuphophron reassures him: it is not him, because it is possible to discuss with

---

[47] *Elements of Law*, I, 5, 9. In this occurrence, the phrase "read thyself" is the basis of the whole philosophy, because sensation is a mental act thanks to which the whole world appears.

[48] *De Homine*, chap. 2–9, and then 11–13.

[49] *Ibid.*, chap. 11–15. And in *Leviathan*, introduction, the phrase "read thyself" is the basis of anthropology and political science.

[50] See for instance J.-L. Marion, "Hobbes et Descartes: l'étant comme corps", in *Hobbes, Descartes et la métaphysique*, op. cit.: he reconstructed Hobbesian theory of knowledge on the basis of the Cartesian list of simple natures, but restricted to material simple natures.

[51] *De Homine*, dedicate.

[52] For an interesting political implementation of this idea of a Cartesian prism in the way Hobbes is read, and for the reasons why we should read more accurately Hobbesian anthropology today, see S. Frost, *Lessons from a Materialist Thinker*, Stanford University Press, 2008.

[53] *The Political Philosophy of Hobbes* (1936) and *Natural Right and History* (1953).

Hylobares, and More already reassured the reader in the presentation of the characters: Hylobares is 'a young, witty, and well moralized Materialist', so not Hobbes. During the dialogue, Hylobares develops Cartesian thesis, similar to what Berkeley lent to his Hylas in *Three Dialogues between Hylas and Philonous* (1713), at the beginning of the second dialogue. And it is no coincidence if Cuphophron is the character who reassures Sophron: he is also described as a Cartesian, but a Cartesian who centres his thought on the immaterial thinking substance – he is 'Platonist or Cartesian' in the presentation of characters. In a different manner, when in his *Disquisition about the Final Causes of Natural Things* (1688), Boyle reproached Descartes for not having shown that mechanism allowed a cosmological proof of the existence and the immateriality of God, because mechanistic principles imply that every body, including the whole matter, *receives* its motion, so that the first mover can be nothing but immaterial, such an argument was not brought against Descartes, but against the fact that he forgot what he should have written to refute the materialist interpretation of mechanism. For the same reason, this argument is also aimed at Hobbes: mechanism cannot be materialist because it implies the immateriality of God, contrary to what Hobbes wrote in 1668. Descartes did not, for Boyle, make use of all the resources available in his philosophy against a Hobbesian reading of Cartesian thesis. Boyle did not *Hobbesianize* Descartes, but identified this way of reading Descartes as a risk that Descartes had not done enough to prevent.

So, there can be a Cartesian Hobbes: dualism, minus thinking substance – there remains a passive matter and no specific reflection about the way in which we can know mental functions. And there can be a Hobbesian Descartes: an active matter, able to produce every property – including the observable consequences of thought –, to which a useless soul is added, as a mere 'Ghost in the Machine'.[54] Seeing through its posterity, Hobbesian anthropology has at least one debt to Cartesian physiology, the same that Descartes has sometimes to Hobbes: a biased and restrictive reading.

[54] G. Ryle, "Descartes' myth", in *The Concept of Mind*, Hutchinson, London, 1949.

Chapter 16
Enlightenment Criticisms of Descartes' Anthropology

Stephen Gaukroger

Abstract Descartes took the notion of the cultivation of the self seriously, drawing on the physiology of *L'Homme* as well as ethical precepts drawn from writers such as Seneca. Enlightenment thinkers such as Diderot were engaged in the same anthropological project, but they rejected Descartes' account as being too individualistic.

I shall take anthropology, for the purposes of this chapter, to be a study of human cognitive and affective states, the relation between them, an account of what determines them, and an account of human well-being. Descartes does not offer an anthropology in this comprehensive sense, but he does offer accounts that his Enlightenment successors criticized on anthropological grounds. There are two sets of questions on which the issues turned. The first was the extent to which cognitive and affective states could be characterized in terms of physiology, a project which *l'Homme* takes furthest. Here there seems to be some degree of conflict between *l'Homme* and metaphysical works such as the *Meditationes*. The second is whether human well being is to be conceived of and secured in individual terms. Although this latter question is not broached as such in *l'Homme*, one might ask whether physiology could engage questions of human well-being. As we shall see, once physiology is medicalized, as it was in the Enlightenment, social and other questions are opened up.

At first sight, there looks to be a lack of fit between works like the *Meditationes* and *L'Homme*. In the *Meditationes*, what we might broadly refer to as 'states of mind' stand on one side of a sharp metaphysical divide between the mental and corporeal realms. This is particularly the case with cognitive states, which fall squarely under 'reason' or the 'intellect'. Towards the end of the *Meditationes* the divide has begun to look so exclusive that Descartes finds it necessary to warn the reader that the mind or soul is not like a pilot in a ship: the interaction between the mind and the body is more intimate than that. Nevertheless, we are not given any

S. Gaukroger (✉)
Unit for the History and Philosophy of Science, University of Sydney,
Level 3, Carslaw Building (F07), Eastern Avenue, Sydney, NSW 2006, Australia
e-mail: Stephen.gaukroger@sydney.edu.au

D. Antoine-Mahut, S. Gaukroger (eds.), Descartes' *Treatise on Man* and its Reception, Studies in History and Philosophy of Science 43,
DOI 10.1007/978-3-319-46989-8_16

account of how this interaction takes place, and are left with the impression that, as Gilbert Ryle put it, the mind is 'the ghost in the machine.'[1] But this is quite at odds with the account in *L'Homme*, which discusses various cognitive operations, such as perceptual judgement of distance, in corporeal terms. There is no doubt that Descartes rejected any reductionist (materialist) account of the mind—the soul is essentially disembodiable and has an identity in its disembodied state on his account—but how it functions in an embodied state clearly depends on bodily operations, and a great deal of *L'Homme* is devoted to a physiological account of cognitive functions.

The *Meditationes* and *L'Homme* do nevertheless have a connection: in Descartes' original plan for his *Principia*. In this original plan, the *Principia* was to be a six-part account. Part I sets out to provide the general principles of knowledge, offering a recapitulation of the *Meditationes*. Parts II to IV cover the nature of matter and the laws of motion, celestial and planetary motions and the transmission of light, and various terrestrial phenomena. These Parts draw extensively on *Le Monde*, while revising its account of some topics. The fifth part was to deal with 'living things', or animals and plants, and a sixth, 'concerning man'.[2] The structure of the scholastic textbooks that the *Principia* is in explicit competition with indicates that the Part V would have relied heavily on *L'Homme*. *L'Homme* does not mention plants, but Descartes held there to be a continuity between animal and plant physiology, treating the circulation of blood and the movement of sap was being cognate phenomena, for example. As for the final part, *Les Passions de l'âme* takes us from the physiology and psycho-physiology of the projected Part V into the intricacies of the relation between mind and body as they relate to affective states. The project of Part VI, in line with the coverage of topics in the late scholastic textbooks, is psychology and the nature of morality.

If we are to understand the relationship between *L'Homme* and the *Meditationes*, we need to understand not only the connection between Part I and the projected Part V of the *Principia*, but also that between Part I, Part V and the projected Part VI.

On the first question, the crucial thing to note is that Part I is in some way separate from the main treatment; Parts II, III, IV, and V follow the sequence of *Le Monde* and *L'Homme*. In what way Part I is separate is the difficult question. Descartes told Burman that one should not devote much time to the *Meditationes*: one should read it through once and then turn one's attention to empirical matters. Some of Descartes' early followers, notably Regis and Rohault, dispensed with the metaphysical foundations as unnecessary, but Descartes himself insisted on them.[3] They established his method of proceeding—differing as it did from those of scholastics, Gassendi, and others—as the only secure one, and this was crucial for its legitimacy, which was an issue for Descartes in a way that it was not for his immediate successors. The metaphysical foundations of Part I were not dispensable for

[1] Gilbert Ryle, *The Concept of Mind* (London, 1949).

[2] See Stephen Gaukroger, *Descartes' System of Natural Philosophy* (Cambridge, 2002).

[3] See Theo Verbeeck, 'The Invention of Nature: Descartes and Regius', in S. Gaukroger, J. Schuster, and J. Sutton et al., *Descartes' Natural Philosophy* (London, 2000), 149–67.

Descartes, but they do not seem to bear on questions of content. The revisions of *Le Monde* incorporated into Parts II, III, and IV do not depend in any way on metaphysical issues. The minor revisions to Descartes' physiology in the writings on developmental physiology and the passions of the soul that postdate *l'Homme* are likewise not at all motivated by metaphysical considerations, and we have every reason to believe that this is the physiological material that would have appeared in Part V.

This is crucial to understanding *L'Homme* in the general scheme of Descartes' philosophy. Consider the question of visual cognition: knowing and understanding the world on the basis of what we see. The metaphysics of Part I offers a picture in which sense perception is unreliable, and in which there is a problem of how the mind and the corporeal world (which includes the sense organs) might be connected. The problems are all generated epistemologically in the first place, and metaphysically in the second. It would not matter what our sense organs are actually like. By contrast, on the account in *L'Homme* and the projected Part V, understanding visual cognition, for example, means knowing about physiology (including brain physiology) and optics, and it is this that provides the key to understanding various forms of perceptual judgement, such as visual judgement of distance and colour. These are not the kind of things that can be dealt with in terms of epistemology or metaphysics. In short, if we want to understand Descartes' account of visual cognition it is here, rather than the *Meditationes* or Part I of the *Principia*, that we need to look.

The second connection, that between Part I, Part V, and the projected VI of the *Principia*, takes us to the question of anthropology. There are a number of contexts within which anthropological questions arise in Descartes. His metaphysics, if taken by itself, has little to offer. Mind body dualism, to the extent to which it insulates one's mental life from the body and the physical environment, is antithetical to anthropological projects. But when Descartes comes to discuss the passions of the soul and questions in moral psychology, metaphysical questions are ignored. Rather, what we get is a mixture of considerations from psycho-physiology and considerations from traditional humanist thinking. The two are closely related because Descartes' concern with ethical questions is exclusively with moral psychology. When he dealt with ethical questions in the last decade of his life, in the letters to Princess Elizabeth and in *Les Passions de l'âme*, the context was not that of determining which actions were moral and which were not, or whether there was some general principle (such as universalizability) that we were compelled to follow in genuinely moral judgements. Rather, it was a question of what might be termed a theory of psychological preparation for moral agency: how to reshape one's life so that the anxieties, melancholia, and uncontrolled passions that prevent one taking full control of one's life can be overcome. It is to the degree that one overcomes these, and as a consequence is able to act freely, that one can be said to be a morally responsible agent.[4] Descartes attempted to provide Princess Elizabeth with advice on what to do to transform herself from someone subject to dispersed sensations and

[4] See Stephen Gaukroger, *Descartes, An Intellectual Biography*, ch. 10.

affections, with a reduced sense of self, into a unified locus of subjectivity, one that could exercise genuine agency, above all genuine moral agency. His primary focus is with the conditions under which one can make oneself a fully responsible moral agent, and what measures one can take to overcome the various obstacles to this: most notably, in Elizabeth's case, melancholia. Only someone who is fully psychologically robust, as it were, and self-aware, is capable of taking complete responsibility for their life, that is, of attaining full moral agency.

Such a conception of ethics has many attractions, exploring the psychological conditions for moral agency in a particularly instructive way, rather than just ascribing, on a priori grounds, the ability to exercise unhindered moral agency to anyone. But it comes as part of an aristocratic, paternalistic package which is antithetical to the democratic conceptions of personhood that were to come to the fore in the eighteenth century. Descartes' approach is reinforced in his metaphysical conception of subjectivity, which is effectively insulated from social and environmental questions. The 'self' remains purely intellectual and separate from the world, and insofar as it is the self that exercises moral agency, we seem to be in an almost solipsistic realm.

This was precisely the criticism made by the Enlightenment *philosophes*, especially Diderot. In his *Letter on The Blind*,[5] Diderot focussed on the question of how the 'mentality' of a blind person, not just his perceptual states, differs from that of a sighted person, and what this tells us about sensibility in general. His interest focused on the case of Nicholas Saunderson, the blind Lucasian Professor of Mathematics at Cambridge and author of a large posthumously-published two-volume *Elements of Algebra*. In effect, what Diderot does is to use the case of Saunderson to pit unity of sensibility against a Cartesian unity of subjectivity, arguing that the unity of sensibility, properly construed, is essentially something socially responsible that encourages a well-formed *persona*, whereas the Cartesian is insensible to the world and works merely in abstractions. It is Saunderson's very blindness that in effect denies him a fully-developed unity of sensibility. A deficient sensibility is primarily a question of an emotional, aesthetic, and moral challenge for Diderot. Because of their impoverished sensibilities, the blind turn their minds inwards and are drawn to thinking in terms of abstractions. As Jessica Riskin notes, 'this made them natural mathematicians and rationalists: in a word, Cartesians. Conversely, Cartesians' abstract, inward focus made them insensible to the world outside their minds: philosophically blind.' This leads Diderot to suggest that both the blind and Cartesians, because of their solipsistic cast of mind, were inhumane.[6] The blind offer a crucial case study for Diderot because he believes that their abstract manner of experiencing pain in others weakens their sense of sympathy for the suffering of others.[7] The situation is in effect the analogue of what in the Cartesian case would be someone—lacking the ability to unify their mental life (perhaps because

[5] *Lettre sur les aveugles*, in Denis Diderot, *Oeuvres Philosophiques*, ed. P. Vernière (Paris: Garnier, 1961), 81–146.

[6] Jessica Riskin, *Science in the Age of Sensibility* (Chicago: University of Chicago Press, 2002), 21.

[7] See, for example, Diderot, *Oeuvres Philosophiques*, 92–3, where he suggests that the reaction to a man urinating and spurting blood is effectively on a par in the blind.

of melancholia or what we would now think of as various forms of neuroses), and thereby failing to shape them satisfactorily into a moral *persona*—whose moral agency, and humanity, would be deficient in comparison with someone (an *honnête homme*) who had achieved this.

What thinkers such as Diderot and Rousseau rejected in this account was the individualistic solution, as if one could simply reshape one's own psychological resources. In its place they offered a developmental account in which the social and cultural context, especially throughout childhood and adolescence, is crucial for the shaping of sensibility. As Anne Vila puts it, 'Rousseau's morale sensitive would provide the key for making people virtuous not by training their minds, but rather, by carefully controlling the impressions made upon their bodies—or more precisely, upon their sensitive systems'.[8] As far as Diderot is concerned, the difference between him and Descartes on this question is that Descartes believes that the ideas that regulate our lives—our moral, emotional, social, political and intellectual lives—are generated in the private realm, whereas Diderot argues that they can only come from public instruction, above all from secularization and reform of the educational system.

The difference here seems to be clear-cut. Diderot thinks that shaping of character, including shaping of one's moral *persona*, is a social issue, whereas Descartes thinks it is a personal one. But in fact matters are not so straightforward. A link between the two appears with the medicalization of physiology. In the work of Haller and others in the first half of the eighteenth century, the physiology of nervous sensitivity had been explored in detail. There had always been a connection between nervous sensibility and the more general phenomenon of sensibility, and by mid-century a particularly strong link between the two began to be established in terms of medicalization. In mid to late French eighteenth-century medicine, a revival of the Galenic doctrine of 'non-naturals' played a crucial role. Factors relating to health are divided into the naturals, the non-naturals, and the contra-naturals. The naturals were structural and functional elements innate in each body such as the temperaments, humours, parts of the body, faculties, and functions. The non-naturals were those factors that determined the state of the body without being controlled by the natural functioning of the body: ambient air, food and drink, movement or exercise and rest, sleep and waking, excretion and retention, and the passions of the soul. The contra-naturals comprised diseases, and these could result from an internal imbalance in the naturals or from an imbalance between naturals and non-naturals.[9] Health on this account was the result of a proper ordering of the naturals and a proper regimen of the non-naturals, brought under the general notion of 'hygiene'.

The idea of a harmonious state was generally perceived as a 'natural' state, both in the case of the correct balance of the naturals, and in the case of the balance of the naturals and the non-naturals. The task was that of guiding individuals to the

---

[8] Anne Vila, *Enlightenment and Pathology* (Baltimore, 1998), 183.

[9] See Peter H. Niebyl, 'The Non-Naturals', *Bulletin of the History of Medicine* 45 (1971), 486–92; and Chester R. Burns, 'Nonnaturals: A Paradox in the Western Concept of Health', *Journal of Medicine and Philosophy* 1 (1976), 202–11.

achievement of a natural state. For the Enlightenment *médecins philosophes*, the subject matter of medicine included every aspect of human experience, and anything external to the body that affected physical, mental, intellectual, or emotional states was included within the scope of medicine, for it had an effect on the well-being of the person. The traditional Christian idea that medicine should confine itself to the body while the Church dealt with the soul was no longer viable. Central to this approach was the idea that physical, mental, and moral questions formed an intimately linked whole, and could not be considered in isolation from one another.

It is of interest here that some in the eighteenth century assimilated Descartes to the materialist cause on the basis of *L'Homme*, which La Mettrie in particular saw as his 'true' account of human beings. As far as metaphysical issues are concerned, there is no doubt that Descartes was not a materialist. I have suggested that we read *L'Homme* as providing the material for Part V of the *Principia*, and so leading to Part VI, the material for which can be provided by *Les passions de l'âme* where the move from physiology to psychology is made, and by the correspondence with Elizabeth, where the move from psychology to morality is made. In this context, *L'Homme* fits in rather well, not with materialism, but with a medical exploration of the conditions which shape the individual *persona*. The *médecins philosophes* believed that medicine, suitably broadened, had come to replace metaphysics, and the kind of philosophically-motivated exploration of physiology that Descartes initiated—especially given the place of this in *Les Passions de l'âme* and the *Principia*, where it forms a prelude to the study of the psychology of the human being—can be seen as a starting point on this path.

Part IV
L'Homme Today

Chapter 17
L'Homme in Psychology and Neuroscience

Gary Hatfield

Abstract *L'Homme* presents what has been termed Descartes' "physiological psychology." It envisions and seeks to explain how the brain and nerves might yield situationally appropriate behavior through mechanical means. On occasion in the past 150 years, this aim has been recognized, described, and praised. Still, acknowledgement of this aspect of Descartes' writing has been spotty in histories of neuroscience and histories of psychology. In recent years, there has been something of a resurgence. This chapter argues that Descartes ascribed a range of active functions to the brain acting on its own in seeking to explain psychological functions such as sense perception, attention, memory, and emotional response.

Within the history of philosophy, interest in *L'Homme* has recently revived, which is to say that, after a hiatus, there is once again interest in and recognition of this work. Going back a century for purposes of comparison, standard histories of philosophy from around 1900 typically included, in the chapter on Descartes, a section on his natural philosophy, and within this some mention of his mechanistic physiology and animal-machine hypothesis.[1] If we look more closely into specialist works on Cartesian philosophy, there was at about this time a considerable appreciation of Descartes' stature as a scientist. Kemp Smith in 1902 felt a need to revive interest in Descartes' metaphysics as opposed to his "scientific works."[2] Among physiologists and psychologists, there was an appreciation of Descartes' mechanistic physiology and his efforts to explain all animal and some human capacities through processes

[1]As an example: Harald Höffding, *A History of Modern Philosophy*, trans. B. E. Meyer, 2 vols. (London: Macmillan, 1908), mentions *L'Homme* and describes the physiology of animal spirits (1: 232–5). Kuno Fischer, *Descartes and His School*, trans. J. P. Gordy, ed. Noah Porter (London: Unwin, 1890), is more focused and offers two chapters on Descartes' "Philosophy of Nature"; but he treats pineal physiology in the chapter on soul–body union, citing the *Passions*. He is unusual for his time in not mentioning *L'Homme*.

[2]Norman Kemp Smith, *Studies in the Cartesian Philosophy* (London: Macmillan, 1902), v–vi. Earlier, Francisque Bouillier, *Histoire et critique de la révolution Cartésienne* (Lyon: Boitel, 1842; 3rd edn., 1868) gave prominence to Descartes' scientific work and animal-machine hypothesis.

G. Hatfield (✉)
Department of Philosophy, University of Pennsylvania, Philadelphia, PA 19104-6304, USA
e-mail: hatfield@sas.upenn.edu

D. Antoine-Mahut, S. Gaukroger (eds.), Descartes' *Treatise on Man* and its Reception, Studies in History and Philosophy of Science 43,
DOI 10.1007/978-3-319-46989-8_17

in the brain. This appreciation is also found in the histories produced within the specialized scientific disciplines, including physiology and psychology.

The situation changed in the decades after the 1950s. Treatments of Descartes in histories of philosophy emphasized metaphysical and epistemological aspects of his work.[3] This can in part be put down to changes in the conception of philosophy. In central parts of the Anglophone world, a conception of philosophy held sway that saw little need and little value in engaging science.[4] Philosophy and science, and so also philosophy and psychology, were distinct and mutually irrelevant enterprises. One can see the product of these attitudes in three books on Descartes' philosophy that appeared in 1978, by Margaret Wilson, Bernard Williams, and Edwin Curley. Wilson mentions Descartes' role as a scientist, but says little else on the topic. Williams and Curley devote some attention to the foundations of Descartes' physics, and Williams fruitfully surveys the role of experience in Descartes' natural philosophy; but neither engages in any detail the physiology and psychology in Descartes.[5]

From the 1970s on, interest in the relations between the history of science and the history of philosophy has revived. And within philosophy in general, a "naturalism" has been in play that values the relevance of scientific findings for various areas of philosophy, including metaphysics and epistemology. Furthermore, the resurgence of interest in perceptual and cognitive psychology, the subsequent rise of cognitive science, and the even more recent growth of cognitive neuroscience has invigorated the connections between those fields and the philosophies of mind and psychology. These developments have engendered a new interest in Descartes' *neuroscience* – or, in more traditional terms, his *physiological psychology* – among historians of philosophy as well as historians of psychology and neuroscience, and hence a renewed interest in *L'Homme*.

This chapter will first document and characterize the appreciation of Descartes as a scientist, and especially as a contributor to neuroscience and psychology, as manifest in the period from 1850 to 1950 by active scientists and in specialist disciplinary histories, focusing on references to *L'Homme* and related texts, especially the *Passions*. We shall find that one dominant narrative portrayed Descartes as a vanquisher of vitalism through his mechanistic physiology and animal-machine hypothesis, illustrating this aspect of his work through the notion of reflexive motion

[3] Etienne Gilson and Thomas Langan, *Modern Philosophy: Descartes to Kant* (New York: Random House, 1963), give five sentences to animal mechanism (ch. 6), with no citation of *L'Homme*; Anthony Kenny, *Descartes: A Study of His Philosophy* (New York: Random House, 1968), has a chapter on matter and motion (ch. 9) which mentions animal mechanism, without citing *L'Homme*. He repeats a frequent (and questionable) assertion that mere fragments of *Le Monde* (presumably including *L'Homme*) were published after Descartes' death (7); *L'Homme* seems complete, unless we count as missing from it the intended third part of *Le Monde* on the rational soul.

[4] See Bertrand Russell, "The Cult of 'Common Usage'," *British Journal for the Philosophy of Science*, 3 (1953): 303–7, for a contemporary diagnosis of this trend in British philosophy.

[5] Margaret Wilson, *Descartes* (London: Routledge and Kegan Paul, 1978), 3; Edwin M. Curley, *Descartes against the Skeptics* (Cambridge: Harvard University Press, 1978), ch. 8; Bernard Williams, *Descartes: The Project of Pure Enquiry* (Harmondsworth: Penguin, 1978), chs. 8–9.

as automatic and mechanistic, a notion attributed to Descartes. The chapter then considers historiographical issues arising from this narrative and attribution. It goes on to note the renewed interest in *L'Homme* within the history and philosophy of psychology and neuroscience. It characterizes the types of work being done and suggests areas that need more attention. Many specialized histories retain the narrative of Descartes as hero in promoting mechanism over vitalism. Greater attention should be given to the types of mechanistic explanations Descartes gave in neuroscience and psychology. Reflecting the emphasis of *L'Homme* itself, close reading of that work is needed in order to understand the specific mechanisms by which Descartes sought to account for sensorimotor psychological functions. These reflections lead to consideration of the notions of psychology brought to bear on seventeenth-century texts such as *L'Homme* by historians of philosophy and historians of science, as opposed to specialist historians of psychology.

17.1 *L'Homme* and Its Reception, 1850–1950: Mechanistic Theory and Reflex Motion

We have already noted that in the decades around 1900 the commonplace assessment of Descartes' place in the history of philosophy included, as a major portion, his efforts to establish a new scientific picture of the world. Within the nineteenth-century scientific community there was appreciation of his efforts and debate over exactly how exalted a figure Descartes should cut in a narrative of scientific progress.[6]

Descartes received widespread praise for two achievements. On a grand scale, he was credited with initiating or greatly promoting a mechanistic attitude toward the organism and the action of the nervous system. While it was generally conceded that the details were wrong, both anatomically and in terms of basic physics, there was a general appreciation that if one substituted "nervous vibrations" or "nerve impulses" for Descartes' "animal spirits," it became clear that he had foreseen the doctrine that nerve action is physicochemical and not dependent on a sensitive soul or other vital principle. On a more specific scale, Descartes was credited with discovering the notion of reflex action.

Emil Dubois-Reymond, in a memorial lecture for Johannes Müller in 1858, credited Descartes with a "fully correct" description of reflex motion, pushing back the date of that discovery by a century and a half. He cited two passages from the *Passions*, one in which Descartes describes an involuntary eye-blink when a hand is thrust toward someone's face (even if the person consciously knows that a friend is doing this "in fun"), and one in which he speaks of animal spirits being "reflected"

[6] For a brief indication of such debates, see John Theodore Merz, *A History of European Thought in the Nineteenth Century, Vol. 1: Introduction, Scientific Thought, Part 1* (Edinburgh: Blackwood, 1896), 292–3.

from sensory processes in the gland into the nerves that move the muscles.[7] His citations on the physiology of animal spirits extended only to the *Passions*, a work that was itself deeply informed by *L'Homme*. He understood Descartes' account to be a mechanistic one.[8]

Thomas Huxley, in talks from 1870 and 1874, emphasized these same two points about Descartes' legacy – his mechanization of physiology and description of the reflex – while heaping even greater praise on him. In an address celebrating the *Discourse*, he describes Descartes as a scientific visionary, as belonging among those "who attain greatness because they embody the potentiality of their own day, and magically reflect the future. They express the thoughts which will be everybody's two or three centuries after them."[9] Huxley had several different aspects of Descartes' thought in mind, including his description of the subjectivity of sensory qualities such as "blueness," "roundness," or "hardness."[10] But he takes Descartes' central achievement to have been his attempt "to resolve all the phenomena of the universe into matter and motion, or forces operating according to law": a "grand conception" that was, as Huxley recounts, sketched in the *Discourse* and "more fully developed" in the *Principles* and *L'Homme*. According to Huxley, in *L'Homme* Descartes arrived "at that purely mechanical view of vital phenomena towards which modern physiology is striving."[11] Huxley provides extensive extracts from *L'Homme* on sensory motor control and quotes the final passage of that work, which reviews Descartes' claim to have mechanized the vital and sensitive souls.[12]

In a subsequent talk on animal automatism, Huxley again glorified Descartes, championing him "as a physiologist of the first rank" who brought to prominence the notion that animal behavior is produced via purely mechanical causes. Accordingly, Descartes opened up the "mechanical theory" of the processes of "motion and sensation."[13] Citing the *Principles*, *Passions*, *Dioptrics*, and *L'Homme*, he formulates five proposals that comprise Descartes' development of the animal-machine hypothesis: that "the brain is the organ of sensation, thought, and emotion"; that animal motion is the result of muscle shortening, induced by motions in the nerves; that "the sensations of animals are due to a motion of the substance of the [sensory] nerves"; that motion of the sensory nerves may be transmitted through the brain to the motor nerves, as a basis for reflex action; and that nerve motion

---

[7] AT xi. 339, 356.

[8] Emil du Bois-Reymond, "Gedächtnisrede auf Johannes Müller, Gehalten in der Leibniz-Sitzung der Akademie der Wissenschaften am 8. Juli 1858," in *Reden von Emil du Bois-Reymond*, ed. Estelle du Bois-Reymond (Leipzig: Veit, 1912), 1: 135–317, on 300–1.

[9] Thomas H. Huxley, "On Descartes' 'Discourse Touching the Method of Using One's Reason Rightly and of Seeking Scientific Truth'," in *Methods and Results: Essays* (New York: Appleton, 1896), 166–98, on 167.

[10] "Descartes' 'Discourse'," 173–4.

[11] "Descartes' 'Discourse'," 181.

[12] AT xi. 130–2, 201–2.

[13] "On the Hypothesis that Animals Are Automata, and Its History," *Methods and Results*, 199–250, on 201.

causes changes in the brain that constitute "the physical basis of memory."[14] He avers that if Descartes had become aware of spinal preparations and the shorter-loop "reflex" of the spinal cord, he would have embraced those results as favorable to his hypothesis.[15]

The neurophysiologist Sherrington was equally fulsome with praise for Descartes. In his own scientific work, Sherrington offers accolades to Descartes for the notion of antagonistic muscle action in the control of the eye muscles, reproducing a diagram from *L'Homme*.[16] In a later book of historical reflections on the natural history of human beings, he discusses the formation of the concept of the reflex. He finds the concept in the sixteenth-century physiologist Jean Fernel. But Sherrington credits Descartes with making the notion prominent: "Descartes, developing the doctrine of mindless motor acts in man and animals, put it forward with a force and clearness which caught the abiding attention of the world."[17] More generally, he finds that "Descartes' biology still possesses for the biologist an interest not merely antiquarian. Its primary touch is with anatomy, yet with anatomy for the light it can throw on function."[18]

---

[14] "Animal Automatism," 203, 206, 208, 211, 213. Note that, although Descartes did not distinguish sensory and motor nerves as independent nerve fibers, he did distinguish sensory and motor functions within individual nerves. Motor function consists in animal spirits flowing down the nerves; sensory function consists in fibrils inside the nerve tubes that operate independently until they reach the brain, where they can influence the flow of animal spirits. See Hall, *Treatise*, 24, n. 27.

[15] "Animal Automatism," 219–26. John Sutton, *Philosophy and Memory Traces: Descartes to Connectionism* (Cambridge: Cambridge University Press, 1998), 77, reads *L'Homme* as envisioning a short-loop reflex circuit that does not involve animal spirits flowing out of the pineal gland (but remains located in the brain). Accordingly, the mechanism behind the boy pulling his foot away from the fire (AT xi. 142) and the man withdrawing his hand (191–3) is a simple loop in which the sensory fiber opens the motor tubule that encases it, causing animal spirits to flow not from the pineal gland but from the ventricle. I find Descartes making the pineal gland, which is located in the ventricle, the font of all animal spirits (130, mentioning a "certain little gland" without naming it), so that, even if only one sensorimotor nerve is involved, the spirits originate from the gland. Moreover, context shows that the pineal gland is involved in the second example (the hand), as Descartes introduces "the gland H" as the source of animal spirits (170) and then analyzes in detail the conditions on spirit flow (from the gland). With the boy, the fact that all spirits originate from the gland might affect whether competing stimuli are noticed but otherwise play little role in differentiating the sensorimotor response to the fire. With the man, more nervous structure is involved and the force of pineal flow is crucial. This example shows the need for more systematic readings of *L'Homme*. For instance: the example of the boy (141–2) occurs with the first treatment of the external senses (141–63), but the example of the hand occurs in the middle of a second, more extensive account of the sense of vision in relation to motor movements and muscle sense (174–6, 181–8). Beginning at 170, there is a deeper account of brain physiology and, at 174–97, of sensorimotor function. How do the later accounts of vision and sensorimotor processes relate to the earlier (textually, and for what they could suggest about the stages in which the work was composed)?

[16] Charles Sherrington, *The Integrative Action of the Nervous System* (New York: Scribners, 1906; as re-issued, Cambridge: Cambridge University Press, 1948), 286–7.

[17] Charles Sherrington, *Man on His Nature* (Cambridge: Cambridge University Press, 1963; first published, 1940), 152.

[18] *Man*, 153.

Among psychologists, William James wrote in the *Principles*: "To Descartes belongs the credit of having first been bold enough to conceive of a completely self-sufficing nervous mechanism which should be able to perform complicated and apparently intelligent acts."[19] James chides Descartes for only partly extending his mechanistic account to human beings, while also noting that the denial of consciousness to animals was "too paradoxical" to gain wide acceptance. Some years later, no doubt responding to Descartes' reputation among physiologists, the physiological psychologist Ivan Pavlov noted:

> Three hundred years ago Descartes evolved the idea of the reflex. Starting from the assumption that animals behaved simply as machines, he regarded every activity of the organism as a *necessary* reaction to some external stimulus, the connection between stimulus and response being made through a definite nervous path: and this connection, he stated, was the fundamental purpose of the nervous system in the animal body.[20]

Neither James nor Pavlov mentions *L'Homme*, but that work was widely recognized among physiologists and historians of physiology and of psychology as Descartes' central work on the physiology of animal behavior and his most extensive development of the animal-machine hypothesis.

Turning now to more detailed histories of physiology and psychology, Descartes was again treated as a hero for his role in mechanization and often was credited with the concept of the reflex (perhaps while allowing that Galen had described the pupillary reflex). His prominence in histories of physiology, as in the lectures of Michael Foster from the 1890s, is well-documented.[21] Less attention has been payed to the somewhat surprising portrayals of Descartes in histories of psychology in the decades after 1900, where, perhaps through the influence of Huxley or James, the animal-machine hypothesis and the concept of reflex were noticed.

Some psychologist-historians, such as George Brett, offered a comparatively circumspect portrayal. While anointing Descartes as a world-historical figure, Brett is more critical and more historically nuanced than our physiologists, or even James. He doubts that Descartes' reflex concept is the same as that expressed in late eighteenth and nineteenth-century physiology. He cites *L'Homme* and recounts the animal-machine hypothesis, but is critical of it, favoring a view that accords moral qualities to animals.[22] Boring's *History of Experimental Psychology* (New York: Appleton-Century, 1929) acknowledges Descartes' conception of the body as a machine, describes some brain physiology, and quotes from the *Passions* but doesn't

---

[19] James, *Principles of Psychology*, 2 vols. (New York: Holt, 1890), 1: 130.

[20] I. P. Pavlov, *Conditioned Reflexes: An Investigation of the Physiological Activity of the Cerebral Cortex*, trans. G. V. Anrep (Oxford: Oxford University Press, 1927), 4. Pavlov considered himself to be a physiologist, but his work was received as foundational in American psychology. The back-cover blurb on the Dover reprint (New York, 1960) portrays the book as a contribution to "experimental psychology."

[21] C. U. M. Smith, "Descartes and Modern Neuroscience," *Perspectives in Biology and Medicine*, 42.3 (1999), 356–71, on 356.

[22] George Sidney Brett, *A History of Psychology, Vol. 2: Medieval and Early Modern Period* (London: Allen and Unwin, 1921), 199, 204, 214–17.

cite *L'Homme*. The second edition (1950) mentions *L'Homme* as the main work for Descartes' "physiological psychology."[23] In the meantime, the psychologist Franklin Fearing had published an especially thorough study of Descartes' contribution to the notion of reflex, in his 1930 book *Reflex Action: A Study in the History of Physiological Psychology* (Baltimore: Williams and Wilkins, 1930). It includes numerous quotations from *L'Homme* and the *Passions*, and reproduces the now familiar diagram from *L'Homme* of the boy with his foot near a fire. Fearing granted to Descartes an important role in developing the concept of mechanistic explanations for animal actions and in enunciating a physiological principle of reflex action. But he agreed with Brett that Descartes had lumped together a wide range of animal responses under the notion of an automatic or reflex response, including "activities such as breathing, singing, walking, swallowing, yawning, bodily accompaniments of emotion, eye movements, intra-ocular adjustments, excretory actions, protective responses to external stimuli, postural responses, etc."[24]

These citations show that the general project of treating the body as a machine, including rejection of vegetative and sensitive souls, was well known to physiologists and psychologists in the period from 1850 to 1950, and that *L'Homme* was frequently referenced. Those authors who quoted extensively from that work tended to cite statements of the general program of mechanizing the vegetative and sensitive souls and the basics of pineal hydraulics. The latter included the sensorimotor loop formed by sensory stimulation that opens tubes, causing spirit flow, which may then be directed to the muscles – a loop that can produce those behaviors that were later identified as exhibiting the reflex concept. There is not much by way of detailed study of the specific mechanisms that realize functions which are more advanced psychologically than the reflex.[25] These specific mechanisms pertain to associative memory, attentional mechanisms, binocular focusing of the eyes, accommodation

---

[23] *History of Experimental Psychology*, 161.

[24] *Reflex Action*, 27. Fearing's book was reissued in paperback by the MIT Press in 1970, with a new introduction by the physiological psychologist Richard Held. Fearing (25) quotes a passage naming a certain gland as the source of animal spirits, but doesn't go into the details of Descartes' envisioned mechanisms, beyond sensorimotor loops and animal spirits inflating muscles.

[25] Norman Kemp Smith, *New Studies in the Philosophy of Descartes: Descartes as Pioneer* (London: Macmillan, 1953), devotes Chapter 5 to the physiological principles in *L'Homme* and refers to that work in Chapter 6, on the "embodied self" in Descartes. The physiological principles pertain to mechanized vital processes, flowing animal spirits, and the "monstrous" thesis (136) of animal insentience. The chapter on embodiment mentions Descartes' account of distance perception but refers only to the *Dioptrics*; that chapter also discusses memory, but without noting the mechanical account of associative memory found in *L'Homme* (AT xi. 177–9). Interestingly, despite his touting of *L'Homme* in his two books on Descartes, Smith did not include an excerpt in his *Descartes: Philosophical Writings* (New York: Modern Library, 1958), but did include other works of physiological and psychological import (*Dioptrics*; *Discourse*, but missing the middle of Part V; and large segments of the *Passions*). By contrast, Ralph M. Eaton, in his earlier *Descartes: Selections* (New York: Scribner's, 1927), excerpted *L'Homme* (5 pages), along with large sections of the *Principles* (from all four parts, 45 pages); *Passions* (43 pages); and *Le Monde: Lumière* (38 pages).

of the lens, and brain states that co-vary with the distance to seen objects, as well as the more elaborate sensorimotor loops controlling the search for food and the avoidance of danger, which are at least suggested in *L'Homme*.

17.2 Historiography of the Reflex Concept

I have mentioned notes of disagreement on whether Descartes actually invented the concept of sensorimotor reflex action. This question was investigated with special thoroughness by Georges Canguilhem in *La Formation du concept de réflexe aux XVIIe et XVIIIe siècles* (Paris: Puf, 1955). Canguilhem suspected that nineteenth and twentieth-century authors, upon reading the famous passages in Descartes on the eye-blink or hands extended in a fall,[26] had projected a later concept of reflex action back onto the writings of Descartes.

Amidst other agendas, Canguilhem found it important to distinguish among "descriptions of the phenomena of neuro-muscular automatism"; experimental studies of the functions of anatomical structures; and the formulation of the reflex concept and its generalization in a theory.[27] He was willing to grant that Descartes had expressed the notion of automatism but unwilling to ascribe to him the notion of reflex, which he finds to be first properly articulated in the latter part of the seventeenth century by Thomas Willis.[28]

There are two aspects of Canguilhem's argument that I wish to highlight. First, he treats the notion of the reflex as that of an integrated neural subsystem and then denies that Descartes, because of his mechanistic outlook, can describe the body as an integrated whole independently of its relation to the soul, which means that he can't have integrated subsystems independently of that relation, either.[29] Second, he maintains that Descartes did not really form the notion of an automated loop in which peripheral sensory processes elicit peripheral motor actions, but that he privileged central control by the pineal gland of motor processes, without a true sensorimotor loop.[30]

The question of whether Descartes' animal machines, and the machine of the human body conceived apart from the soul, can legitimately be ascribed an integrated, functional unity within the constraints of his metaphysics and natural philosophy, remains open. Some interpreters adopt the line of Canguilhem (who cites Martial Guéroult in support), Rodis-Lewis, and Des Chene, which denies that Descartes can legitimately ascribe a functional unity to a body without taking into

---

[26] *Passions*, AT xi. 338–9; *Meditations*, vii. 230.

[27] Canguilhem, *Formation*, 3.

[28] Canguilhem, *Formation*, 1, 77, was also intent upon establishing that the concept of reflex did not require a materialistically mechanist framework but that blind reflex might also be posited in a vitalist framework.

[29] *Formation*, 53–4.

[30] *Formation*, 45–6.

account its relation to a soul or mind. Others find that within his natural philosophy Descartes has the resources to ascribe functional unity to animal machines.[31]

The issues can't be settled here. Let us, therefore, suspend judgment on this aspect of Canguilhem's argument and proceed to consider what resources Descartes had for making functional distinctions among the various "automatic" responses as listed above by Fearing. Careful attention to the text shows that Descartes lists four factors that condition the sensorimotor flow of spirits, apart from the mind: current sensory stimulation; innate plumbing of the brain ("instincts"); alterations to the plumbing due to previous stimulation ("memory"); and the character of spirits arriving from the heart (as influenced by recent ingestions and by bodily temperament).[32] Applying these notions, we can separate singing, which would clearly depend upon memory – in this case, associative memory, such as is described in *L'Homme* – from the instinctual eye-blink, or the protective extension of the hands. Indeed, all of the other items listed, including bodily accompaniments of emotion, would be considered by Descartes as instinctual in the first instance and thus as initially effected by bodily mechanisms without mental intervention – even if some of them, such as walking, eye-movements, and intra-ocular adjustments,[33] are subject to voluntary mental control. Thus, Descartes need not consider all "automatic" or "mechanistic" control of behavior to be of a single, undifferentiated kind.

Be that as it may, Canguilhem's second argument, if it could be generalized to all cases, would exclude the reflex notion from Descartes by showing that he didn't really have the conception of automated sensorimotor loops that operate independently of the soul. This argument focuses on the interpretation of the text that accompanies the diagram in *L'Homme* showing a woman pointing at an arrow, with the side of her head rendered transparent to allow inspection of the pineal gland and the animal spirits flowing from it (Fig. 17.1). In the diagram, the woman looks at arrow *ABC*, and she points where she looks. The pointing (the position of the arm)

---

[31] Geneviève Rodis-Lewis, "La Conception de *L'Homme* dans le cartésianisme," in Rodis-Lewis, *L'Anthropologie cartésienne* (Paris: Puf, 1990), 19–38, on 29, makes organic unity depend on mind–body union, denying such unity to animal bodies; Dennis Des Chene, *Spirits and Clocks: Machine and Organism in Descartes* (Ithaca: Cornell University Press, 2001), 11, has the unity depend on human projection or else divine intention. Others find that Descartes develops a notion of functional integrity (whether teleological or systematic) independent of mind for both animal and human bodily machines: Gary Hatfield, "Animals," in Janet Broughton and John Carriero, eds., *Companion to Descartes* (Oxford: Blackwell, 2008), 404–25, on 411–17; Delphine Kolesnik-Antoine, *L'homme cartesien: La "force qu'a l'âme de mouvoir le corps": Descartes, Malebranche* (Rennes: Presses universitaires de Rennes, 2009), 27–30; Deborah J. Brown, "Cartesian Functional Analysis," *Australasian Journal of Philosophy*, 90.1 (2012): 75–92; Barnaby R. Hutchins, "Descartes, Corpuscles and Reductionism: Mechanism and Systems in Descartes' Physiology," *Philosophical Quarterly* 65.4 (2015): 669–89.

[32] AT xi. 166, 192–193. At 190, he notes other global conditions on animal motion: the current location of the limbs, and the interconnections among concurrent sensory stimulations. Of course, in human beings, the mind can also influence the flow of spirits, as when a person wills to move an arm.

[33] *Passions*, AT xi. 361–2.

Fig. 17.1 Woman (as bodily machine) pointing, by the physician Gerard van Gutschoven for the 1664 edition of *L'Homme*. The machine is looking at arrow *ABC*. The diagram shows the pattern of causal influence on the eyes, neural transmission into the brain, and a pattern of animal spirits flowing from the pineal gland (*H*). Spirits also flow from the gland to nerve *8*, controlling the muscles of the arm. See the text for further explanation (Source: 1677 edition of *L'Homme* (author's collection))

is under the control of the animal spirits, in particular, of the manner in which they flow from the gland into the motor nerve that controls the muscles of the arm.

Canguilhem contends that Descartes has only a notion of central control of the spirits, and hence invokes no sensorimotor loop of the kind that other interpreters treat as reflexive. He puts great weight on passages such as the following, which describes the flow of spirits from point *b* on the gland toward the motor nerve, tube *8*: "one can suppose that what makes tube *8* turn toward point *b* rather than toward some other point is merely that the spirits leaving point *b* tend with greater force toward *8* than do any other spirits."[34] This and other passages do suggest that the

[34] Hall, *Treatise*, 92; AT xi. 91.

flow of the spirits from the gland determines how the arm is pointed. But that raises a question: what determines the flow of spirits? Above, we saw four factors. One of these is current sensory stimulation, or, as Descartes puts it, "the action of the objects that impinge on the senses."[35] In another diagram featuring the same arrow, Descartes notes that "when the degree of openness of tubules *2*, *4*, and *6*, for example, is increased by the action of object *ABC*, the spirits, which commence at once to flow toward them more freely than they did before, draw the gland after themselves a little."[36] Thus, sensory activity controls spirit flow. In the schema of the first diagram, the fact that the eyes are drawn to point *B* on the object might then cause the arm to point there through purely mechanical causes. Conversely, if the arm is forced by whatever cause to point toward *C*, the eyes follow. The movements of the eyes and the arm are linked reciprocally, as in a reflex action. The pineal flow cannot simply be classified as a "central" controlling factor, since it is subject to causal control by stimulation of the sensory nerves.

We may conclude, then, that the question of reflex action in Descartes awaits further study, including more sustained interpretive scrutiny of the relevant texts.

17.3 *L'Homme* in Recent Histories of Neuroscience and Psychology

Having seen the prominent discussions of Descartes' mechanizing explanations of behavior and role in developing the reflex concept, we would expect to find these themes in subsequent histories of neuroscience and psychology. This expectation is often but not always met.[37] But we may now also ask, what advances have been

[35] *Treatise*, 96; AT xi. 185.

[36] *Treatise*, 97; AT xi. 185.

[37] The psychologist-historian Erwin A. Esper, *A History of Psychology* (Philadelphia: Saunders, 1964), 221–3, shows a good appreciation of Descartes' mechanistic theory, its hydraulic character, and even of his dissecting activities. Comparative psychologist Robert Boakes, *From Darwin to Behaviourism: Psychology and the Minds of Animals* (Cambridge: Cambridge University Press, 1984), 85–9, references *L'Homme* liberally and reproduces two diagrams, while offering a critical assessment of Descartes' reflex doctrine (suggesting that he must have known about the behavior of decapitated animals, undermining his exclusive emphasis on central-brain reflex processes). Thomas Steele Hall, *Ideas of Life and Matter: Studies in the History of General Physiology, 600 B.C.–A. D. 1900*, 2 vols. (Chicago: University of Chicago Press, 1969), 1: 250–64, especially 263, portrays Descartes as a leader in micromechanization, while focusing on his basic physiology and not his physiological psychology. In 1972, Hall published his translation of the *Treatise* with extensive notes, taking more notice of Descartes' "physiological psychology," which he finds to be the dominant topic of the work (xxxvii). By contrast, John G. Benjafield, *History of Psychology*, 2nd edn. (Oxford: Oxford University Press, 2005), 24–6, gives no notice of *L'Homme* or Descartes' physiological psychology. Stanley Finger, *Origins of Neuroscience: A History of Explorations into Brain Function* (New York: Oxford University Press, 1994), holds that Descartes' "metaphysical ideas about the visual system were entirely speculative and inadequate" (77), while admitting the impact of his "ideas about the brain as a reflexive machine" (26), without giving details.

made more recently in interpreting the significance of *L'Homme* within the histories of neuroscience and psychology?

One advance has been the development of a more sophisticated understanding of what Descartes was doing in developing his "mechanized" account of animal and (some) human behavior. One gets the feeling, in the descriptions of Huxley, Sherrington, and others, that Descartes is being used as a hero figure in an ongoing struggle to establish a mechanistic physiology, or a physicochemical account of how behavior is produced. The vision of a mechanized body thus becomes the main contribution of Descartes' *L'Homme*. But this portrayal leaves out what Sherrington, without elaborating, called Descartes' contribution to the analysis of functions.

Some historians of psychology have appreciated that Descartes was not just mechanizing behavior but intended to mechanize behavior under a functional description. In an insightful passage, the psychologist-historian Thomas Hardy Leahey has written, of Descartes' mechanistic psychology:

> What is important is Descartes's view of the human body as a machine incorporating many faculties previously assigned to the soul. Like Aristotle and the medieval faculty psychologists, Descartes wrote about memory, imagination, and common sense. However, unlike them, Descartes assigned these faculties to the body, implying that although they appear to be mental activities, they may be explained as bodily activities. Therefore Descartes sought to account for as much of the mind as possible on materialist, mechanical terms within the sphere of science, reserving only self-consciousness at most to philosophy.[38]

Leahey is not the first to make this point, but he understands that Descartes is not simply pushing mechanical explanation, but mechanical explanations of psychologically characterized activities, such as memory and imagination. As Richard Lowry puts it, Descartes sought to render the functions of the vegetative and sensitive souls as "functions of the body."[39] Quite recently, John D. Greenwood, quoting *L'Homme*, shows that Descartes extended his mechanistic explanations to learned behaviors dependent on memory, without invoking "mentality or consciousness."[40] Raymond E. Fancher has long appreciated Descartes' physiological psychology or

---

Antonio Damasio, *Descartes' Error: Emotion, Reason, and the Human Brain* (New York: Putnam, 1994), does not cite *L'Homme* and cites the *Passions* only to make the erroneous point that Descartes held that controlling the passions by reason and the will is what makes us human (124). Reason and will are unique to Descartes' humans, but so is the feeling of the passions. The passions are not subject to direct control by the will, Descartes' framework does not exclude empirical investigation of the mind, etc. See Geir Kirkebøen, "Descartes' Embodied Psychology: Descartes' or Damasio's Error?" *Journal of the History of the Neurosciences*, 10.2 (2001): 173–91, and Hatfield, "The *Passions of the Soul* and Descartes's Machine Psychology," *Studies in History and Philosophy of Science*, 38.1 (2007), 1–35.

[38] Leahey, *A History of Psychology: Main Currents in Psychological Thought*, 2nd edn. (Englewood Cliffs, NJ: Prentice-Hall, 1987), 95.

[39] Lowry, *The Evolution of Psychological Theory: A Critical History of Concepts and Presuppositions*, 2nd edn. (New York: Aldine, 1982), 9; he accepts (85–6), on the basis of John Cottingham, "'A Brute to the Brutes?': Descartes' Treatment of Animals," *Philosophy*, 53 (1978): 551–9, that Descartes attributed sentience to animals.

[40] Greenwood, *A Conceptual History of Psychology: Exploring the Tangled Web*, 2nd edn. (Cambridge: Cambridge University Press, 2015), 82–3.

"neuropsychology," including his mechanization of the sensitive soul and proffered explanations for both simple reflexes and learned behaviors.[41]

These psychologist-historians show a deepened appreciation of Descartes' project. They recognize the functional significance of his attempted mechanization of memory processes, but do not go into the details of his proposed mechanisms. On this topic, historians of philosophy have gone further in showing the richness of Descartes' conception of the brain mechanisms of memory, including John Sutton, *Philosophy and Memory Traces: Descartes to Connectionism* (Cambridge: Cambridge University Press, 1998), and Stephen Gaukroger, *Descartes' System of Natural Philosophy* (Cambridge: Cambridge University Press, 2002), Chaps. 7 and 8.

Within our second targeted group, of neuroscientists and historians of neuroscience, recent results have been mixed. C. U. M. Smith has undertaken historical studies on Descartes as anatomist and brain physiologist, even reproducing one of Descartes' own anatomical drawings.[42] He finds that although Descartes' program of mechanization has been a fruitful influence, "it is time to move on."[43] His criticism targets a perceived lack of dynamism in Descartes' conception of brain function and his supposed conception of the brain as a mere passive transmitter of sensory stimulation. As we shall see, there are ample grounds for denying both charges.

By contrast, I. M. L. Donaldson has delved into the details of *L'Homme* on the perception of eye position. He finds that Descartes had an "outflow" theory, according to which we are able to perceive the positions of our limbs and our eyes because the mind is affected by the brain state that controls the muscles that determine those positions.[44] Relying on Hall's commentary,[45] he correctly suggests that the very flow of spirits which causes the eyes to be held in certain positions simultaneously causes the mind to perceive those positions. Donaldson does not draw out the implication that awareness of eye position might be yoked to convergence of the eyes as a source of information for distance perception.

Descartes himself makes the link between eye position and distance perception, with his celebrated "natural geometry."[46] Natural geometry as a means for perceiving distance relies on the triangle of convergence formed by the distance between the eyes and their angles of convergence, with each set of angles being correlated with a specific distance to the focal object (Fig. 17.2). The interpretation of this natural geometry, and especially the question of whether it involves informational processes of calculation, has been fraught. The most common position postulates

---

[41] Fancher, *Pioneers of Psychology* (New York: Norton, 1979), 21–27, 31–41; with several citations to Hall's *Treatise* and to the *Passions*.

[42] Smith, "Descartes and Neuroscience," 358.

[43] "Descartes and Neuroscience," 370.

[44] Donaldson, "The Functions of the Proprioceptors of the Eye Muscles," *Philosophical Transactions: Biological Sciences*, 355, No. 1404 (2000): 1685–1754, on 1688, 1729.

[45] Hall, *Treatise*, 62–3, n. 107.

[46] *Dioptrics*, AT vi. 137; *L'Homme*, xi. 160.

Fig. 17.2 The triangle of convergence. The converging eyes fixate point *N*. Inter-ocular line segment *LM*, together with the angles of rotation for the eyes in a given fixation, form a specific triangle. Because segments *LN* and *MN* are intended to converge on the focal point of the eyes, the lines should be straight, not broken as in the diagram, and eye *M* should be rotated slightly left. Produced by van Gutschoven (Source: 1677 edition of *L'Homme* (author's collection))

cognitive or at least informational processes that represent inter-ocular distance, the angles of convergence, and consequently the distance from the baseline between the eyes to a focal object. These cognitive operations may be of the same sort as ordinary calculation, but go unnoticed because they are habitual or rapid.[47] Or they may involve informational content of the sort that some investigators believe Descartes attributed to both animals and humans.[48] In either case, "natural geometry" involves representations and calculations over them.

Another interpretation was put forward some 30 years ago by Hatfield and Epstein, who argued, from passages in *L'Homme*, that Descartes can plausibly be read as offering a purely psychophysical account of distance perception (in addition to other explanations involving cognition or representation, such as calculating distance from known size and visual angle). Accordingly, Descartes' natural geometry is realized by brain mechanisms that control convergence (and accommodation as well), which operate like mechanical compasses to "calculate" the distance to the focal objects – but do so mechanically, without requiring representations.[49]

[47] Jody L. Graham, "The Intellect's Burden: Geometrical Inferences in Descartes's Theory of Vision," *Theoria*, 64.1 (1998): 55–83, provides an overview.

[48] Gaukroger, *Descartes: An Intellectual Biography* (Oxford: Clarendon Press, 1995), 276–90; Sutton, *Memory Traces*, ch. 3, and 293–7.

[49] Gary Hatfield and William Epstein, "The Sensory Core and the Medieval Foundations of Early Modern Perceptual Theory," *Isis* 70.3 (1979): 363–84. For the comparison to Descartes' mechanical compasses, see Hatfield, "On Natural Geometry and Seeing Distance Directly in Descartes," in Vincenzo de Risi, ed., *Mathematizing Space: The Objects of Geometry from Antiquity to the Early Modern Age* (Berlin: Birkhäuser, 2015), 157–92. For the relation between this interpretation and "outflow" theories of eye position, see Hatfield, "Natural Geometry in Descartes and Kepler," *Res Philosophica*, 92.1 (2015): 117–48.

As Descartes explains in *L'Homme*, the convergence of the eyes on an object at a certain distance and the accommodation of the lens of the eye for that distance are controlled by a central brain state, consisting of the pineal gland and the flow of spirits from it to the nerve tubes controlling the ocular muscles. This mechanism is set up such that the gland leans forward by one amount or another depending on the distance for which the eyes converge and the lenses accommodate. The lean of the gland thus realizes a corporeal (non-mental) "idea of distance,"[50] which causes the mind to experience the distance to the object. This system is a mechanical realization of the physiological operations underlying the psychological function of distance perception.

The system is dynamic and tracks changing relations between the eyes and objects in the world. The arrangement does not fit Smith's notion of the brain as a passive conduit. What's more, in describing this system, Descartes portrays a mechanism of accommodation according to which spirit flow weakens when the retinal image becomes blurry and strengthens when the image is sharply focused, setting up a steady-state system. The target of this system is a well-focused image, a steady state that it "seeks" by mechanical means alone.[51] Descartes' mechanization of associative memory has already been mentioned. He also describes a corporeal mechanism to explain aspects of attentional shifts and limitations on the scope of attention.[52] *L'Homme* is not to be valued simply for mechanization as such, but for its imagining of physiological mechanisms for realizing psychological functions. Its brand of mechanization was something new.

17.4 *L'Homme* and Psychology

L'Homme as a work in physiology devotes 12 of its 83 pages (in the AT edition) to vital (or vegetative) functions. It devotes 71 pages to sensorimotor functions, about a third of which are concerned with vision. There is one page of summary at the end, where Descartes reviews his claim to have mechanized the offices of the vegetative and sensitive souls.

The recently renewed interest in *L'Homme* has thus far been disproportionately focused on basic physiology, the vegetative functions.[53] This is consistent with

---

50 AT xi. 183–8. On corporeal ideas, with references to *L'Homme*, see Emily Michael and Fred S. Michael, "Corporeal Ideas in Seventeenth-Century Psychology," *Journal of the History of Ideas*, 50.1 (1989): 31–48.

51 AT xi. 187–9.

52 Hatfield, "L'attention chez Descartes: l'aspect mental et l'aspect physiologique," *Études philosophiques* (forthcoming). Human attention includes dynamic interaction between mental experience and brain processes.

53 Des Chene, *Spirits and Clocks*, importantly focuses on questions of organic unity and function, along with basic physiology. However, he strays in suggesting that philosophers have been too concerned with "the cognitive faculties" and not with basic physiology in Descartes and in *L'Homme*; for that work is primarily concerned with the physiological psychology of sensorimotor processes. Des Chene neglects this material. Psychological topics also are not emphasized in the

Hall's original approach to the work as primarily a work of physiology. But, as our psychologist-historians have reminded us, the primary thrust of *L'Homme* is to provide a physiological psychology. The accounts of memory function provided by Sutton and distance perception by Hatfield and Epstein are examples of detailed work that brings out the psychological subject matter of *L'Homme*. We can hope for additional work showing how psychological functions are realized mechanistically.

Why has this aspect of the work remained in comparative neglect? An answer to this question would tell us more about the attitudes of philosophers and historians of science toward psychology than about the actual history of psychology (and neuroscience). Among the reasons for neglect is an earlier conception that psychology is not relevant to philosophy, perhaps tied to a notion that, prior to the nineteenth century, when a major figure such as Descartes engaged in psychological topics it was because he mistakenly approached an epistemological topic psychologically. Accordingly, any "psychology" of the seventeenth century is either epistemology manqué or collapses into philosophy of mind. This view is, however, mistaken.[54]

If we look to major surveys as bellwethers, there is good evidence that the notion of a seventeenth-century psychology has not been incorporated into the disciplinary self-conceptions of the history of science and the history of philosophy. In recent Cambridge histories, psychology in the seventeenth century has received even less attention than biology, which, until the past 15 years or so, was woefully neglected. The *Cambridge History of Seventeenth-Century Philosophy*[55] contains no systematic treatment of biological or psychological topics as such. In examining the relation between philosophy and the new science, it in effect adopts a late eighteenth-century conception of "physics" as mechanics (matter in motion). By contrast, in the seventeenth century, "physics" was understood to be the science of nature in general, or of all natural phenomena, and so included what we would call biological, chemical, and psychological phenomena. Even the recent and

---

apparatus to Annie Bitbol-Hespériès and Jean-Pierre Verdet, eds., *Le Monde, L'Homme* (Paris: Editions du Seuil, 1996), nor in Delphine Antoine-Mahut, "La machine du corps," in Frédéric de Buzon, Èlodie Cassan, and Denis Kambouchner, eds., *Lectures de Descartes* (Paris: Ellipses, 2015), 229–52, or Frédéric de Buzon and Denis Kambouchner, "L'âme avec le corps," *Lectures de Descartes*, 279–328. A good number of chapters in Gaukroger, John Schuster, and John Sutton, eds., *Descartes' Natural Philosophy* (London: Routledge, 2000), discuss *L'Homme* in relation to psychological topics, especially the senses (chs. 16, 17, 21, 23, 26, 28).

[54] On the existence and character of psychology in the seventeenth century, see Fernando Vidal, *The Sciences of the Soul: The Early Modern Origins of Psychology*, trans. Saskia Brown (Chicago: University of Chicago Press, 2011), and Hatfield, "Rationalist Roots of Modern Psychology," in John Symons and Paco Calvo, eds., *Routledge Companion to Philosophy of Psychology* (London: Routledge, 2009), 3–21. On distinguishing psychology from epistemology and metaphysics of mind, see Hatfield, "The Workings of the Intellect: Mind and Psychology," in Patricia Easton, ed., *Logic and the Workings of the Mind: The Logic of Ideas and Faculty Psychology in Early Modern Philosophy* (Atascadero, CA: Ridgeview, 1997), 21–45.

[55] Michael Ayers and Daniel Garber, eds., *Cambridge History of Seventeenth Century Philosophy* (Cambridge: Cambridge University Press, 1998).

innovative *Cambridge History of Early Modern Science*,[56] while acknowledging this contextual meaning of "physics," relegates biological topics to the chapter on medicine and gives virtually no attention to psychological topics.

L'Homme is primarily a work that combines physiology with the psychology of sensorimotor response. It portrays an organism devoid of mind. This organism is nonetheless attributed systems that include senses, imagination, and memory, all realized as bare arrangements of matter. This mindless machine adjusts to its environment, pulling its hand from the fire, roving about to find food when its stomach is empty, and forming associations in its corporeal memory. More generally, it will pursue beneficial and avoid harmful things.[57] These topics direct us toward the mechanical neuroscience, or physiological psychology, of *L'Homme*. Along with the novel account of distance perception and the steady-state mechanisms of accommodation, these topics are worthy of closer investigation. *L'Homme* is a remarkable work, whose depths remain to be plumbed.

[56] Katharine Park and Lorraine Daston, eds., *The Cambridge History of Science, Volume 3: Early Modern Science* (Cambridge: Cambridge University Press, 2006).

[57] AT xi. 179, 192–5.

Chapter 18
The Embodied Descartes: Contemporary Readings of *L'Homme*

Barnaby R. Hutchins, Christoffer Basse Eriksen, and Charles T. Wolfe

Abstract A certain reading of Descartes, which we refer to as 'the embodied Descartes', is emerging from recent scholarship on *L'Homme*, in keeping with the interpretive trend which emphasizes Descartes's identity as a natural philosopher. This reading complicates our understanding of Descartes's philosophical project: far from strictly separating human minds from bodies, the embodied Descartes keeps them tightly integrated, while animal bodies behave in ways quite distinct from those of other pieces of extended substance. Here, we identify three categories of embodiment in contemporary readings of Descartes's physiology: (1) bodily health and function, (2) embodied reflex and memory, and (3) embodied cognition. All present more or less strong versions of the embodied Descartes. Together, they constitute a compelling reading of a Cartesian natural philosophy that, if not expressly antidualist, is an awfully long way from the canonical picture.

18.1 Introduction

The canonical understanding of Descartes portrays him as radically devaluing the body in favour of the mind. After all, according to the Second Meditation, "in the strict sense," each of us is "*only* a thing that thinks."[1] By emphasising that the mind is essential while the body can be doubted away, Descartes's cogito appears to make the body disposable. And given that the cogito's separation of the mind from the

[1] AT VII, 27/CSM II, 18 (our emphasis).

B.R. Hutchins
Department of Philosophy, Bar-Ilan University, Ramat Gan 5290002, Israel

C.B. Eriksen
Department of Philosophy and History of Ideas, Aarhus University, Jens Chr. Skous Vej 7, 8000 Aarhus C, Denmark

C.T. Wolfe (✉)
Department of Philosophy and Moral Sciences, Ghent University, Blandijnberg 2, Ghent B-9000, Belgium
e-mail: ctwolfe1@gmail.com

D. Antoine-Mahut, S. Gaukroger (eds.), Descartes' *Treatise on Man* and its Reception, Studies in History and Philosophy of Science 43,
DOI 10.1007/978-3-319-46989-8_18

body is Descartes's solution to the epistemological problems of the (bodily) senses, it is no wonder that the mind looks like the proper topic of Cartesian philosophy, while the body looks like little more than a tiresome, forgettable impediment. It is precisely this that allows for celebrated reconstructions of Descartes as conceiving, inter alia, of the mind as simply a "ghost in the machine" (Ryle), the machine being the extended body.[2] This position has recently been updated by Hacking, for whom the Cartesian standpoint means positing the body as 'other', wholly "an assemblage of replaceable parts."[3] In that regard, not the *Treatise on Man*—hereafter *L'Homme*—but its summary in the *Discourse* is already something of a counter-argument, since Descartes explicitly warns there, and then, later, in the *Meditations*, that we should not conceive of the mind in the body like a pilot in a ship,[4] thus defusing one giant and persistent phenomenological objection against him in advance.[5]

Nevertheless, on the received view, Descartes is hopelessly naïve in regarding the human body as simply a mechanical automaton hosting a disembodied mind, and by extension, he is often seen in more historical disciplines, but also throughout the humanities, as the classic 'objectifier of nature', as in Sawday's otherwise elegant study of early modern anatomy: "As a machine, the body became objectified; a focus of intense curiosity, but entirely divorced from the world of the speaking and thinking subject."[6] Sutton nicely observes that here,

> [Descartes's] objectification of the human body is … but one symptom of the mechanistic violation of an earlier enchanted world. Where once holistic herbalists and natural magicians embraced analogy and sympathy over representation and intervention, coupling earthy bodily realism with organicist ecologism, the Cartesian birth of modernity enforced divisions of philosophy from biology, science from history, power-mongering manipulators of nature from the dead ecology which they exploit, and of active rational male observers from passive fragmented female bodies.[7]

---

[2] Ryle, *The Concept of Mind*; Williams, *Descartes*. For a critical discussion of this caricature, see Reiss, "Denying the Body?". A related, and influential, caricature of Descartes (at least in the later decades of the twentieth century) was that of Richard Rorty, for whom Descartes committed the "original sin of epistemology," by introducing the modern idea of representation through clear and distinct ideas, thus installing epistemology at the foreground of early modern philosophy while bracketing off matters such as the relation between the body and soul as "not something for philosophy" (Rorty, *Philosophy and the mirror of nature*, 60–61; see discussion in Sutton, *Philosophy and Memory Traces*, 50–55 and in Introduction to Brown, *Descartes and the Passionate Mind*).

[3] Hacking, "The Cartesian Body," 13.

[4] "It is not sufficient for [the rational soul] to be lodged in the human body like a helmsman in his ship . . . but that it must be more closely joined and united with the body in order to … constitute a real human being" (AT VI 59; CSM I 141); cf. Kirkebøen, "Descartes' Embodied Psychology", 181.

[5] For a reconstruction of an 'embodied Descartes' based not on *L'Homme* but (primarily) on the Sixth Meditation, see Brown, *Descartes and the Passionate Mind*.

[6] Sawday, *The Body Emblazoned*, 29, cf. also 22, 37. For a critique of this narrative, see Snider, "Cartesian Bodies."

[7] Sutton, "The Body and the Brain," 697–698, elaborating on Sutton, *Philosophy and Memory Traces*, 82f.

But a spate of recent literature has started to take the Cartesian body seriously as a subject in its own right, reversing the trend that privileges the foundationalist metaphysics found notably in the *Meditations*. Recall that, in the prefatory letter to the *Passions of the Soul*, Descartes wrote, "My intention was to explain the passions only as a natural philosopher [*physicien*], and not as a rhetorician or even as a moral philosopher."[8] As James has put it, "[b]y treating the *Meditations on First Philosophy* as Descartes' philosophical testament, scholars have created a one-sided interpretation of Cartesianism in which the division between body and soul is overemphasized and sometimes misunderstood."[9] But for some, James included, *L'Homme* is at least as central to Descartes's philosophy as the first two Meditations, and the body is at least as indispensable as the mind.

In this chapter, we distinguish three main ways in which the scholarship has reunited Cartesian philosophy with the body: (i) health and function, (ii) memory and reflex, and (iii) embodied cognition.[10] Of those, (iii) applies to bodies united with minds, while (i) and (ii) also apply to Cartesian bodies in themselves, in the absence of minds (whether the body–machines of *L'Homme*, described in isolation from the minds with which God could unite them, or animals and plants which, for Descartes, would always lack minds). The literature has tended to focus on each of these aspects independently: each highlights a certain way in which human and animal (and, largely implicitly, plant) bodies look a little odd, a little different, compared to the rest of the material world. It would also be possible to focus on the passions, which are transversal here, as animal passions match points (i) and (ii) while properly human passions belong in addition to point (iii). This chapter is not an exhaustive review of the literature, but points to conceptually coherent components in recent approaches to Descartes's treatment of bodies, and aims to show how they have built up a picture at odds with the one-sided, mind-privileging view of Descartes's philosophy. We refer to this picture as 'the embodied Descartes'—a reading of Cartesian philosophy in which bodies are not understood as alienated bundles of spare parts (*pace* Hacking and many others) but are truly inhabited; and in which they are not simply lumps of extended substance but have their own peculiar properties that do not seem to be shared with the rest of the material world.

In the following, we first address the scholarship's treatment of the machine–bodies of *L'Homme* considered in isolation from any mind or soul. Then, we turn to the subject of the missing section of *L'Homme*: the soul and its union with the body. In the absence of this section from the extant text, the literature has looked to its recapitulation in the *Meditations* and *Treatise on the Passions of the Soul*, and has found an account of truly embodied cognition, in which the mind, while in principle

[8] AT XI, 326/CSM I, 327.

[9] Susan James, *Passion and Action: The Emotions in Seventeenth-Century Philosophy*, 106.

[10] 'Life' does not rate its own independent category here, notably because Descartes adopts such a deflationary attitude towards it. Further, our analysis is different from more standard treatments such as Duchesneau's (*Modèles du vivant*), which discusses Descartes's approach to organic life, while we are concerned with a revision of the historiographic and philosophical categories with which Descartes has been interpreted.

ontologically distinct and separable from the body, is utterly bound up in their union.

18.2 "Living" Bodies

Before addressing the ways in which the literature has picked out the peculiarities of human and animal bodies, it is worth saying a little about what kind of bodies we are dealing with. There are, of course, two major senses of 'body' in Descartes's philosophy. There are human and animal (and perhaps even automaton[11]) bodies—bodies, that is, with some autonomy; bodies that we might well call 'living'. These are the bodies we are concerned with here. And then there are bodies in general—that is, material bodies, or, rather, bits of extended substance. The latter have enjoyed a somewhat longer period of attention, reflecting an emphasis on Descartes as natural philosopher.[12]

For Descartes, 'living' bodies[13] are a special class of material bodies. He is quite clear that all physiology is purely material, and thus that a human or animal (or plant) body is nothing but a piece of matter, albeit a mechanically rather complex one. In the opening of *L'Homme*, Descartes describes the body as "just a statue or a machine made of earth."[14] In that particular case, the body he describes is an analogue of the human body; like *Le Monde*, *L'Homme* is built around the rhetorical conceit of a thought experiment in which God creates a mechanical world, and (in the latter) mechanical humans to populate it. His aim is to show that a mechanical, material human body is metaphysically possible—that it is the kind of thing that God could create—without explicitly committing himself to the position that our own bodies really are purely material and mechanical.

That machine made of earth, he goes on to write, contains "inside it all the parts needed to make it walk, eat, breathe, and imitate all those functions we have which

---

[11] The human bodies that are the subject of *L'Homme* are ostensibly 'just' automata that God is capable of creating. Given that, automata (or, at least, certain kinds of automata) clearly belong in the same category as human bodies. In addition, it seems unlikely that Descartes has any principled means of distinguishing between animal bodies and automata (see Hutchins, "Descartes and the Dissolution of Life").

[12] Thus prominent scholars such as Gaukroger (*Descartes' System of Natural Philosophy*) and Garber (*Descartes Embodied*) have called attention to the treatment of material bodies in Descartes's natural philosophy, stressing that Descartes's work should be understood as that of a natural philosopher (see also the essays collected in the volume *Descartes' Natural Philosophy*).

[13] Exactly what life consists in for Descartes, or if, indeed, it consists in anything at all, is an ongoing problem in the literature; see most recently, Detlefsen, "Descartes on the Theory of Life" and Hutchins, "Descartes and the Dissolution of Life." Here, we use the term 'living' only to provide a convenient means of distinguishing between human, animal, and plant bodies and other bits of extended substance.

[14] AT X, 120/TM 99. As Gaukroger points out in his translation (TM 99, n. 3), 'earth' here refers to Descartes's third element, rather than to macro-scale dirt, mud, clay, etc.

can be imagined to proceed from matter and to depend solely on the disposition of our organs."[15] The prevailing conception of a living body at the time was an Aristotelian one, in which the body's operations depended on immaterial souls. The hypothetical machine of *L'Homme*, though, does everything a human can, and yet it is nothing but matter. Later (by the time of writing the *Meditations*, *Principles of Philosophy*, *Treatise on the Passions of the Soul*, and *Description of the Human Body*), Descartes drops the pretence that the mechanical body is merely an analogue, and he explicitly upholds that actual human and animal bodies really are nothing but extended substance.

So, for Descartes, living bodies are pieces of matter, just like rocks, or bits of metal, or any other body in general. But, living bodies are not *just* like any other body. As Sutton puts it, "[t]he earthen machines described in *L'Homme* are importantly unlike the clocks and simple automata with which they are conceptually analogous, for their capacities far outstrip those we usually imagine or ascribe to them."[16] Yes, we are machines made of earth, like statues, but "these 'statues' are animated, these machines dream."[17] Some object that, if animals are just machines, how can they possess memory? And the answer is that there is a vast conceptual space between the automatic and the non-physical,[18] as we discuss in Sect. 18.3.2. Minimally, "The difference between an animal as traditionally conceived and a Cartesian automaton is not a difference between soft, fleshy organic entities and clockwork robots, but a conceptual difference between how physiological processes are to be modelled."[19]

We aim in this chapter to examine how Descartes's recognition of how living bodies differ from 'mere' bodies yields a (cumulatively developed) interpretation of the 'embodied Descartes'. In the case of humans, they are bodies united with minds; the union, it has been argued in the literature, makes a significant difference to both the mind and the body. Human cognition is, it turns out, affected by the body. And unlike minds in themselves, which cannot be divided (and so cannot be destroyed or harmed), unions of mind with bodies are capable of health and disease. They also have sensations. But such properties are not (or at least seem not to be) restricted to bodies united with minds. Animal bodies, and human bodies considered in isolation, present us with properties peculiar to them and absent from the rest of the material world—properties that make them worthy of separate study. Animals seem just as capable of health and disease as humans, they appear to be functional, and they exhibit the bodily aspects of both memory and a certain kind of sensation.

---

[15] TM 99.

[16] Sutton, "Body and Brain," 700.

[17] Sutton, *Philosophy and Memory Traces*, 56. For more on the fruitfulness of the body-machine analogy see Wolfe, "Le mécanique face au vivant."

[18] Sutton, *Philosophy and Memory Traces*, 81, critiquing Marjorie Grene's claims in her study, *Descartes*, 47-48.

[19] Gaukroger, "The resources of a mechanist physiology," 386–387.

18.3 Living Bodies in Isolation

Embodiment discussion in the humanities (including cultural studies, gender studies, and history) tends to focus on 'my own sense' of my body, and, equally, of its historicity and constructed status. It does not focus specifically on embodied cognition, as in the field of that name within cognitive science. Yet in the Cartesian context, talk of embodiment does mean embodied cognition, albeit in a very broad sense, i.e. that thought, specifically, is not separate from body, and conversely, that many sensorimotor functions do not involve thought, for Descartes. Strictly speaking, the bodily machines of *L'Homme* are incapable of cognition in themselves. Thoughts are modifications of thinking substance (i.e. soul or mind), and thus bodies (i.e. extended substances) cannot, by definition, think. So, whatever those machines do themselves, without the input of a soul, cannot be cognitive. In addition, given that Descartes is consistently explicit that animals are nothing but machines, and never have souls, this also means that there is no such thing as animal cognition for Descartes. And yet, the machines of *L'Homme* are capable of "movements . . . just like the movements in us that testify to malice, timidity, inconstancy, tardiness, and ruthlessness"[20] (amongst many other things). Cartesian bodies, by themselves, are also perfectly capable of self-preservation, raising their arms to break a fall, with no intervention from the mind; and soulless sheep can identify a wolf and flee from it.[21] While all these activities are non-cognitive for Descartes, they bear more of a resemblance to the behaviour of humans than to that of water, or salt crystals, or magnets. And this, after all, is Descartes's entire point in *L'Homme*: such complex, human-like behaviour can be produced by purely material bodies, in the total absence of any soul.

While no more immaterial than any other piece of extended substance, then, Cartesian living bodies behave in ways that other bits of matter cannot—ways that others might want to identify with cognitively-influenced behaviour. The scholarship has picked out a range of such ways of behaving. We categorise them here under two main categories: living bodies are healthy and functional (or unhealthy and dysfunctional)[22], and they have memory and reflexes.

18.3.1 Health and Function

Bodies seem to present Descartes with a significant problem of teleology. A large part of his opposition to Aristotelianism is his rejection of teleology in the natural world. The natural world is the world of extended substance, and teleology is the

[20] AT XI, 167/TM 141.

[21] Fourth Replies (AT VI, 230/CSM II, 161). Cf. Rorty, "Descartes on thinking with the body," 377–379.

[22] Health and function might appear to be an outlying category here, in that it seems less obviously pseudo-cognitive than, say, memory. But health and function are goal-directed processes, and, for Descartes, only minds can be goal-directed.

preserve of minds. Because Descartes takes extended substance to be incapable of self-direction, and because he takes whatever intentions God might have had in creating the world to be utterly inscrutable to us,[23] ends are off-limits to Cartesian natural philosophy. And yet, his descriptions of living bodies are full of teleological terms, such as *usus/usage* and *officio/office*, as in *L'Homme*, where he speaks about the "the functions of the waking state,"[24] the "normal bodily function,"[25] and the "functions that I have attributed to this machine,"[26] or when, in a letter to Elisabeth, he speaks of the "office" of the liver,[27] and, in the *Passions*, the "office" of the stomach (which CSM render as "function").[28]

This apparent teleology is a problem for purely material Cartesian bodies. The machines of *L'Homme*, considered in isolation from any soul, are precisely the kinds of things that cannot have intrinsic ends. But, as it turns out, living bodies are very difficult to explain (at least for Descartes), without recourse to teleology. Most obviously, it is very difficult for him to discuss health and illness without involving ends in the matter—the resources of extended substances seem to allow little prospect for explaining just what it is that makes a living body either healthy or ill. And health was an important part of Descartes's own understanding of his natural-philosophical project. In a letter to the Marquess of Newcastle from October 1645, he wrote that "the preservation of health has always been the principal end of my studies,"[29] and medicine takes up a prominent place in the tree of knowledge described in the preface to the French edition of the *Principles of Philosophy*, as one of three principal branches of philosophy. This commitment to medicine makes it problematic that Descartes has such difficulty establishing a way to distinguish the normal, or healthy, from the pathological, or unhealthy. As Shapiro puts it, "[t]he problem stems from the conflict of two principles: first, the natural world is to be conceived non-teleologically; and second, the norms that constitute our concept of health are essentially teleological."[30]

Some scholars have identified this problem in Descartes's medical thought without attempting to save him from his own apparent inconsistency[31], whereas others have tried to give explanations for how Descartes can maintain a concept of health

---

[23] See Meditation Four (AT VII, 55/CSM II, 39) and the Fifth Replies (AT VII, 374–375/CSM I, 258).

[24] AT X, 197/TM 165.

[25] AT X, 144/TM 119.

[26] AT X, 201/TM 169. See also *Discourse*, AT VI, 53/CSM I, 138 (the function of respiration).

[27] Descartes to Elisabeth, May 1646, AT IV, 407.

[28] *Passions* II.98, AT XI, 402/CSM I, 363. For further discussion of teleological concepts in Descartes, see Simmons, "Sensible Ends" (on sensation), Brown, "Cartesian Functional Analysis" and Distelzweig, "The Uses of *Usus*" (on biology and medicine).

[29] AT IV, 329/CSMK 275.

[30] Shapiro, "Health of the Body-Machine," 424.

[31] E.g. Distelzweig, "The Uses of *Usus*." On the charge of inconsistency overall see Sutton, McIlwain et al., "Applying intelligence to the reflexes," 99n., referring to Grosholz, *Cartesian Method*, Shapin, "Descartes the Doctor," and Des Chene, *Spirits and Clocks*.

without unsettling his mechanical system of natural philosophy. Here, we will examine three possible solutions, namely intrinsic structure, reciprocal dependency, and projectionism.

Shapiro proposes an account of Cartesian health that attributes a non-teleological form of health to living bodies themselves. The bodies of animals and plants, and the machines in *L'Homme*, she argues, each have a "stable intrinsic structure" which makes that particular machine the particular machine it is, without appealing to any extrinsic purposes (such as those of the body's creator).[32] Shapiro characterizes a stable structure as a structure that must be maintained in order for the machine to carry out some consistent kind of work. To use her example—and Descartes's—a clock is characterized by its parts being organized in such a way that it shows the time. The clock maintains that activity as long as it maintains the structure that enables that activity, and there is no need to refer to the intensions of its designer for that to be the case. The same goes for living bodies: the idea here is that health is a structural norm; it is simply the stable persistence of the particular structure that allows the body to continue operating. This notion of a stable structure, Shapiro argues, allows Descartes to speak of health without involving teleology. Or, rather, it would have allowed Descartes to do so, had this been the position he upheld. For Shapiro, stable intrinsic structure is a "way out" of the antinomy of mechanism and teleology, but she shows through textual evidence that it is not the way out that Descartes chooses. Rather, his notion of health is not purely mechanical: according to the Sixth Meditation, it is based on the union of the soul and body.[33]

In a 2012 paper, Brown, like Shapiro, also aims to provide Descartes with an objective, naturalized account of an apparently teleological aspect of living bodies—in this case, function rather than health.[34] Brown focuses on Descartes's

[32] Shapiro, "Health of the Body-Machine," 435. In part, Shapiro is responding to Des Chene (*Spirits & Clocks*, 125ff.), who floats the idea of a dispositional unity of the body (in which its unity is given by the arrangement of its parts), but concludes that, for Descartes, living bodies themselves cannot be fully understood without appealing to teleology, and "[e]nds cannot be entirely supplanted by dispositions, even in animals" (ibid., 140). Shapiro argues that stable intrinsic structure circumvents this, although it is not clear that the stability required really is definable without reference to ends. One can also take elements from Des Chene's "dispositional unity" and Shapiro's "structural integrity" to construe the mechanistically understood body as a form of emergent unity contained in the parts. The heart, lungs or liver do not function on the basis of a predetermined end but in accordance with an almost morphogenetically understood law of spatial disposition. In a formula, *the mechanical is the functional*: a material arrangement of parts (a clock, a bodily organ, a body) is *disposed* to act in a certain manner. Descartes speaks—to be sure, in strongly spatialist terms—of the "disposition of our organs," but precisely in order to derive "functions" from them (Descartes, *L'Homme*, AT XI, 120, 202 and *Description du corps humain*, AT XI, 226); cf. Des Chene, *Spirits and Clocks*, 116, 120-121, and Wolfe, "Teleomechanism redux."

[33] Shapiro, "Health of the Body-Machine," 437. This "way out" is the projectionist reading discussed below.

[34] Brown, "Cartesian Functional Analysis."

attempt to produce a fully mechanical account of embryogenesis.[35] The problem implicit in embryogenesis was how to explain the consistent production of consistently-formed foetuses from formless seed through nothing more than efficient causation between pieces of matter, and thus without recourse to final causes. Brown notes that Descartes's account, spurious though it may be, relies on reciprocal dependencies[36] between different parts of the developing foetus in order to avoid teleology: "*nothing intends anything in this process*—but the formation of the brain is necessary for the persistence of the heart and the formation of the heart a necessary precondition for the formation of the brain."[37] In this way, intrinsic function is grounded in interdependence: the heart has a function in embryogenesis and development not because it intends to help produce a foetus or because it has been assigned that function by a designer, but because it depends on other parts of the foetus that also depend on it.

Other scholars have sought to explain the possibility of Descartes's speaking of the health of the living body through a projectionist reading of natural teleology.[38] Human bodies are unique, for Descartes, in that they are united with minds, and since minds are capable of having intrinsic ends, the union can confer teleology upon the human body. On the projectionist reading, when Descartes attributes health and function to animals, and to human bodies considered in isolation from their minds, this is a projection of the human case onto machines similar to that of the human body. In recent papers, Manning has argued that animals and body–machines exist as intermediary substances between ensouled humans and bodies in general.[39] These intermediary substances can be ascribed health by human minds, by comparison with human bodies. In distinction to Shapiro's and Brown's accounts, this does not place health (and function) within the machine itself, but makes it an extrinsic denomination projected from the human case onto the living body. Indeed, on this reading, attempts to objectify health and function invert the situation, by trying to locate human health and function within the machine itself: as Manning puts it, "[t]he health of the machine is parasitic on the health of the human being, not the other way around."[40] In this way, the projectionist reading operates with an anthropology where natural-philosophical explanations are based on the sense that we humans have of our own, living bodies. On this reading, it is the experience of being in a body which is primary for Descartes, and the basis on which he can make physiological arguments about bodies regarded naturalistically or mechanically.

---

[35] The account is found in the final section of the *Description of the Human Body*, a late manuscript in which Descartes reworks the explanations of *L'Homme*.

[36] Brown restricts her analysis on the account of embryogenesis, but interdependence is also central to the physiology of the developed body. See Hutchins, "Descartes, Corpuscles, and Reductionism."

[37] Brown, "Cartesian Functional Analysis," 86.

[38] See Des Chene, "Life and Health"; Manning, "Descartes' Healthy Machines," "Descartes' Metaphysical Biology"; Hutchins, *Obscurity and Confusion*, Ch. 6.

[39] Manning, "Descartes' Healthy Machines," "Descartes' Metaphysical Biology."

[40] Manning, "Descartes' Healthy Machines," 261.

Despite its anthropocentrism, the projectionist reading still points to a particularity of living bodies: a body–machine lends itself to the projection of health and function, while a stone or a lump of metal does not. What projectionism might appear to risk, however, is antirealism about health and function in non-ensouled bodies. If health and function are nothing but projection from the human situation—if, that is, they are extrinsic to living bodies themselves—they might start to look like anthropocentric fantasies. In other words, if health and function are not to be found in living bodies themselves, it is not at all clear that we are talking about the body itself when we talk of its health; it is not clear that we are talking about the heart itself when we talk of its function. If Shapiro's and Brown's accounts do not hold, and if Descartes has no means of allowing for intrinsic bodily health and function, there appears to be nothing about those bodies themselves that can constitute their health and function; in which case, isn't the ascription of health and function to bodies simply an illusion? Hutchins has recently suggested that teleology in the natural world might constitute an epistemic gap in Descartes's system—that is, there are some phenomena that cannot be accounted for within Descartes's system, but this does not imply that Descartes takes them to be unreal.[41] On this reading, we are capable of recognizing health and function in living bodies even if we cannot identify a particular material structure in which it consists. Consequently, bodily health and function is more than an anthropocentric fantasy: it is something intrinsic to living bodies, albeit something inexplicable.

18.3.2 Memory and Reflex

Descartes's neurophysiology of memory and sensation is a key moment in 'revisionist' reconstructions of his thought, not least given that most discussions of the 'embodied mind', in the wake of Varela et al.'s influential 1991 work, reiterate in yet another form the 'whipping boy' role for Descartes.[42] Consider the now-classic presentation of the situation in Canguilhem's 1955 study of the origins and development of reflex action, which is partly a polemic against Descartes and in favour of Thomas Willis, including the latter's account of animal spirits.[43] In contrast, Sutton does not proceed according to the same somewhat 'Manichean' method of opposition, but seeks to inscribe Cartesian neurophysiology into a more fluid, dynamic narrative, notably in his groundbreaking 1998 book *Philosophy and memory traces*, and subsequent papers. In a complementary vein, Kirkebøen in his detailed 2001 paper on "Descartes' Embodied Psychology," which takes Antonio Damasio's *Descartes' Error* as its foil, supplies a detailed account of the complexities of a

[41] Hutchins, *Obscurity and Confusion*, Ch. 6.

[42] Namely, Varela, Thompson and Rosch, *The Embodied Mind.* This is further discussed in Roux's "L'ennemi cartésien. Cartésianisme et anti-cartésianisme en philosophie de l'esprit et en sciences cognitives."

[43] Canguilhem, *Formation du concept de réflexe*, 30, 33.

Cartesian mechanistic psychology and indeed its posterity, via William James, Pavlov and (unintentionally) Müller's law of "specific nervous energy" and Fechner's psychophysics.

How much can a machine do? Differently put, how much can a mechanist physiology achieve? Without challenging substance dualism *per se*, the answer is quite a lot: "although there can be no Cartesian science of the self-conscious mind, there can and must be sciences of memory, imagination, dreaming, and so on."[44] Descartes's neurophysiology of spirits, pores and memory traces is not as easily brushed aside as one might think, given that it seems to present a kind of archaic survival of items that have no place in a mechanistic ontology, such as animal spirits. While some more positivist historians of science have found the presence of animal spirits in Descartes's neurophysiology to be something of an embarrassment, or at least an inconsistency, Sutton has consistently maintained that it is on the contrary an indicator of the fertility and plasticity of the Cartesian machine model of life and the nervous system, and in addition, a tenet which is consistent with other core components of what we are terming the 'embodied Descartes' here. *L'Homme*, on this reading, gives a detailed account of "animal spirits roaming through the pores and traces of body and brain, which is entirely consistent with his scattered remarks elsewhere, through to *The Passions of the Soul*, on corporeal memory and the dynamics of embodied cognition."[45]

Indeed, as Sutton has insisted, fluid dynamics forms the explicit basis for Descartes's physiology, and moreover, "the picture of a static, rigid body … is entirely foreign to Descartes's physiology."[46] An act of remembering is then more about reconstructing "patterns of motion in the animal spirits" than about disembodied representation, in a context of what Sutton refers to as "causal holism"[47]: "Every trace in a brain region affects any episode of processing, so every memory is composite, just as every sensation dangerously carries the perceptual history of the perceiver. This is how 'chimeras and hypogryphs are formed in the imaginations of those who daydream', who neglect the twin direction offered by external objects and by reason."[48] This is where the importance of Descartes's treatment of memory becomes clearer. It is not just that he had a neurophysiology of memory explained in terms of cerebral folds and animal spirits.[49] Rather, it helps bring out the lack of

[44] Sutton, "Body and brain," 708; *Philosophy and Memory Traces*, 75f.; Sutton, McIlwain et al., "Applying intelligence to the reflexes," 84f. (for the point that the treatment of memory in *L'Homme* directly rebuts the caricature of the deterministic, input-output model of the Cartesian automaton).

[45] Sutton, McIlwain et al., "Applying intelligence to the reflexes," 84.

[46] Sutton, "Body and brain," 716.

[47] Sutton, *Philosophy and Memory Traces*, 55; cf. 58, 61, 86f.

[48] Sutton, McIlwain et al., "Applying intelligence to the reflexes," 85, citing *L'Homme*, AT XI, 185.

[49] For more on animal spirits, see Sutton, *Philosophy and Memory Traces*, 102f. For a more developmental perspective, see the chapters on Descartes in Emanuela Scribano, *Macchine con la mente* (Roma: Carocci, 2015). Scribano argues that Descartes modified his theory of knowledge and perception over time, in search of greater 'scientific' coherence, particularly in a neurophysiological vein (an emphasis already present in Sutton's 1998 book), but she has a less naturalistic reading of the union than Sutton does. Thanks to Claudia Matteini for discussion of this text.

a kind of crude, Pavlovian (stimulus-response) determinism of the Cartesian automaton. Automatic behaviour is not non-deterministic, but it is not a strict linear determinism, because "the corporeal causes act holistically."[50] Contrasting with the idea that all bodily functions are explainable in terms of reflex arcs, Sutton and others highlight Descartes's concerns with flexible, non-hardwired cerebral mechanisms in *L'Homme* (but also in the *Dioptrics* and the *Passions*).

On the Cartesian view, Sutton writes, "there is no reason to accept that hardwiring or biology, on the one hand, and current stimuli, on the other, must be the sole determinants of machine behavior."[51] In the case of the memory processes of the automaton,

> the effects of experience are transmitted over long temporal gaps, and are causally involved in behaviour mediated by complex internal processes... To put it another way, memory shows that an automaton's physiology changes over time. Automata with different histories, different 'experiences' marking their brains and bodies, will ... respond differently, and one automaton will respond differently to the same stimulus at different times, after new experience has modified the pores and folds of its brain.[52]

For the body and the brain to be treated 'just as machines' turns out to mean: treated in a dynamical and detailed way. Rather than reducing neurophysiology and passions to a "linear biophysics of barren matter," Cartesian mechanism leaves quite a lot in, also as regards what we might expect in terms of reductionism: "mechanism did not require the elimination of puzzling and complex natural phenomena. Indeed, Descartes accepts some of the stranger facts of the organicist world: he rejects not the baffling phenomena (the bleeding of wounds on the approach of the murderer, the weapon salve, sympathies, the maternal imagination imprinting on the foetus), but only certain candidate explanations of these phenomena which attribute thought or free will to corpuscles."[53]

If we take this revised picture of a Cartesian neurophysiology together with Descartes's focus on the passions, we can imagine with Timothy Reiss that "Descartes's final achievement would have been to bring back the particular body with its specific passions produced by particular perceptions and controlled by a

---

[50] Sutton, "Body and brain," 709.

[51] Ibid.

[52] Sutton, "Body and brain," 709. Sutton notes that Amélie Rorty still divides "informational and maintenance systems" in her otherwise embodied vision of how intertwined "epistemology and physiological homeostasis are" in Descartes's "thinking with the body" (Sutton, *Philosophy and Memory Traces*, 92). For Rorty, "The criteria for identifying a medically sound body might sometimes vary, depending on whether the body is considered primarily and solely as a homeostatic machine, or as a homeostatic machine designed to serve an epistemically sound information system" ("Descartes on thinking with the body," 385). Sutton notes that Rorty does not address the way that animal spirits, themselves information-bearing entities, are generated by non-cognitive bodily processes.

[53] Sutton, McIlwain et al., "Applying intelligence to the reflexes," 99n; Sutton, "Body and Brain," 701-702, referring also to *L'Homme* AT XI, 177.

reason and a will, specific but yet common to all humanity".[54] But for that, we need to discuss Cartesian psychosomatics, as we do below in Sect.18.4.

18.4 Embodied Cognition

In the available text of *L'Homme*, whenever Descartes refers to the relation between the body–machine and the soul, he seems to treat the latter as simply superadded to the former. At one point, after having compared the body to the intricate fountains at Saint-Germain-en-Laye[55], he writes,

> when a rational soul is present in this machine it will have its principal seat in the brain and will reside there like the fountaineer, who must be stationed at the tanks to which the fountain's pipes return if he wants to initiate, impede, or in some way alter their movements.[56]

As Gaukroger notes, this "comes dangerously close to the idea of the mind as being like a pilot guiding a ship"[57]—that is, from this passage, as well as from the few other references to the soul in *L'Homme* itself, it is easy to get the impression that the Cartesian mind really is divorced from its body. But, it is not incidental that what we have of *L'Homme* is not the whole story: in the summary of the treatise he provides in the *Discourse*, Descartes describes a later (now missing, and possibly never-written) section of the text, in which he

> showed how it is not sufficient for [the soul] to be lodged in the human body like a helmsman in his ship, except perhaps to move its limbs, but that it must be more closely joined and united with the body in order to have, besides this power of movement, feelings and appetites like ours and so constitute a real man.[58]

Evidently, Descartes had intended to nuance his fountaineer analogy, to show that the fountaineer of our body is not merely stationed at the tanks, but is actually within the pipes and fluids themselves—the true fountaineer is, indeed, not separate from but embodied within the fountain. In order to understand how the body–machine is united with its soul, then, the existing text of *L'Homme* is somewhat lacking; the scholarship has had to look elsewhere—principally, to the *Meditations* and the *Treatise on the Passions of the Soul*—to reconstitute the full story.

In a paper expressly targeted against Damasio's (admittedly rhetorical) use of the 'Descartes as father of the disembodied mind' trope, Kirkebøen gives a broad over-

---

[54] Reiss, "Denying the Body?", 603.

[55] Descartes specifies only "grottoes and fountains in the royal gardens" (AT XI, 130/TM, 107), but the description fits those at Saint-Germain. See Gaukroger, *Descartes*, 63–64. On Descartes's use of the fountain analogy, see Des Chene, *Spirits & Clocks*, Ch. 6; Sutton, *Philosophy and Memory Traces*, 94f.

[56] AT XI, 131–132/TM 107.

[57] TM 107, n. 19.

[58] AT VI, 59/CSM I, 141. Descartes famously repeats the point in the Sixth Meditation (AT VII, 81/CSM I, 56).

view, in which he wants to show that, from the *Rules* to the *Passions*, Descartes consistently adopts an embodied mind position. Descartes's "explanations of psychological phenomena," Kirkebøen argues, "are *all* embodied."[59] Rorty emphasizes a similar point in her 1992 paper, "Thinking with the Body." Addressing the issue of whether it matters for the mind how the body is structured, she argues that the Cartesian body actively and necessarily contributes to thought in general, and to thought pertaining to kinematics, the biological sciences, and morals, especially.[60] It is the influence of the body, and its union with the mind, that drives us to seek those things we take to be good for us (i.e. healthy), and to avoid harmful things—in both our practical activities and cognitive processes.[61]

This underlying concern for the health of the union is a recurring element in the scholarship on Cartesian embodied cognition. In an influential 2001 paper, Simmons shows how the mind's experience of sensation is inextricably tied up with the health of the union:

> sensation occurs only in a mind that is united to a body; because it is essential to such a mind that its place of residence be preserved, it needs to be alerted to any damage to the body so that it can initiate body-preserving behavior … and thereby aid in the preservation of the body and so of the mind-body union. The biological function of the senses … clearly sets the backdrop for Descartes' treatment of the production of sensations throughout his career in natural philosophy.[62]

Similarly, drawing partly on the work of Grene, Rorty argues that, for example, the judgement that one piece of matter has moved from the vicinity of one set of bodies to that of another,[63] cannot be established by pure thought alone, but needs sensory observation and experiment to be determined.[64] So, when it comes to general ideas, such as that of extension, they are found through intellectual analysis, but when it comes to particular ideas about things in motion, i.e. the behaviour of natural phenomena, they come from bodily-sourced perceptual ideas[65], which, if Simmons is correct, all (more or less implicitly) must refer to the mind's union with its body. This means that, in contrast to the picture of Descartes's philosophy in which all knowledge [66] is due to pure cognition that is fundamentally divorced from the body, it turns out that all our empirical knowledge is bound up with the health of our bodies.

---

[59] Kirkebøen, "Descartes' Embodied Psychology," 174 (our emphasis). Kirkebøen also provides a wealth of details concerning Descartes's influence on experimental psychology, in figures such as Pavlov, Sherrington and McCulloch (179, 188); Hatfield, "Descartes' Machine Psychology," also notes Descartes's influence on William James and Pavlov.

[60] Rorty, "Descartes on thinking with the body," 372.

[61] Ibid.

[62] Simmons, "Sensible Ends," 60. See also "Re-Humanizing Descartes," 63.

[63] Cf. *Principles*, Part II, 25.

[64] Rorty, "Descartes on thinking with the body," 376; cf. Grene, *Descartes*.

[65] Rorty, "Descartes on thinking with the body," 377.

[66] See, e.g., Alston, "Foundationalism," 384.

And it is this embodiment at the heart of sensation that prevents our being fountaineers stationed at the water tanks of our own bodies—as Simmons puts it, "[t]hese sensations, [Descartes] insists, cannot be made sense of on the pilot-in-a-ship or, equivalently, the angel-in-a-machine [or fountaineer-in-a-fountain] model. Why not? Because sensations are not purely intellectual thoughts of the sort a pure intellect has."[67] The non-embodied fountaineer receives explicit information about the status of the fountain, and manipulates its pipes explicitly to produce the intended effects.[68] But that is not how the Cartesian mind interacts with its body: Descartes is well aware that we *feel* thirst and simply lift the water glass to our mouths in response; our minds do not receive explicit information about our bodies' dehydration and then deliberately tug on the correct strings in our brains to make the arm reach out for the glass.

All this is not to deny, however, that Descartes does take the mind and the body to be radically distinct—and to be fully separable, at least in principle. Brown provides a way to make sense of the apparent conflict between Descartes's account of an embodied mind and the real distinction between thinking and extended substances that he propounds elsewhere.[69] She argues that Descartes has two separate ways of identifying the self. On the one hand, I am a "minimal self"—that is, the version of myself that consists in only what indubitably belongs to me. My minimal self is nothing but a thinking thing. But, on the other hand, I am a "maximal self": a union of mind and body—an embodied mind.[70] My maximal, embodied self is my natural state, and it takes "an extraordinary and unsustainable effort"[71] to shift to seeing myself as a minimal self.[72]

One significant upshot of this tight integration between the Cartesian mind and the Cartesian body has, perhaps not surprisingly, to do with health: the embodiment of the mind has practical implications for Cartesian medicine. For Des Chene, Descartes addresses health in two ways: biomechanics and psychosomatics.[73] The former treats the mechanism of the body itself; the latter treats the body in union with the mind—Descartes takes it that the mind can intervene in the health of the body through their union, and that somatic and psychological, and even moral, pathologies are interdependent.[74] The reception of Descartes's work has taken biomechanics to be Cartesian medicine per se—this is fully consistent with the long-received picture of the Cartesian body as a set of interchangeable parts that can be

[67] Simmons, "Re-Humanizing Descartes," 57.

[68] Cf. AT VII, 81/CSM II, 56.

[69] See also Simmons, "Re-Humanizing Descartes."

[70] Brown, "Descartes and the Embodied Self," 245.

[71] Ibid., 240.

[72] Cf. Hutchins (*Obscurity and Confusion*, Ch. 4), who argues that our natural, embodied state has a certain epistemic priority over our minimal state.

[73] Des Chene, "Life and health in Cartesian natural philosophy," 723. Note that psychosomatic medicine involves altering bodily memory, on which see Sect. 18.3.1 here. See also Sutton, "Body and brain," 715, on Descartes's infamous psychosomatic self-cure for his heterotropia fetish.

[74] On the moral side, see Brown, *Descartes and the Passionate Mind*, Ch. 8.

worked on as a mechanic works on any other mechanism. But, as Des Chene notes,[75] Descartes himself favoured psychosomatics over biomechanics. He took the greatest medical benefit to be found not in mechanical intervention with the body-machine itself (which, as we have discussed in Sect. 18.3, was never "just" a machine anyway), but in the embodied treatment possible only between a mind and a body that are united.

18.5 Conclusion

The recent literature (of which we have presented only a sampling here) builds up a picture of Descartes's philosophy at odds with the traditional, received view in which the mind is at its centre, and in which the body, if it is to be considered at all, is dispensable and alienated. This picture of an embodied Descartes radically re-evaluates the Cartesian body. Human and animal bodies are, ontologically, still fully continuous with any other lump of extended substance—there is no ontological difference between them, of course. But, human and animal bodies are, nevertheless, different. They are, the literature has suggested, "embodied" differently—unlike rocks or bits of metal, human and animal bodies are, in some sense, inhabited.

In the human case, bodies are inhabited by minds with which they are fully combined. As Kirkebøen shows, this means that the cognition of the Cartesian mind is influenced by the passions of the body to which it is united; and, as Rorty and Simmons show, all empirical knowledge available to the mind is fundamentally mixed up with a concern for the health of its union with its body. And even medicine, for Descartes, is, in its most efficacious form, an embodied medicine, as set out in Des Chene's account of psychosomatics. In the animal case, on the other hand, bodies are inhabited not by minds as such, but by processes that bear more than a passing resemblance to what others take to be psychological. They are inhabited, as we see in Canguilhem's and Sutton's analyses, by their reflexes and memories, and perhaps also, as Brown and Shapiro have argued, by a material interest in their own health. The posterity of *L'Homme*, even if it is overshadowed by biomechanics, has many other nuances, whether it leaves us with absolute consistency or not. That is, the scholarship should not just be understood as presenting either an 'embodied Descartes' or its contrary. Faced with complexity, some readers could also charge him with inconsistency: how can a mechanist have such an account of health? of the passions? of animal spirits? And so on. We hope to have sketched a different presentation, both of Descartes and of the literature, offering a form of coherence.[76]

[75] Des Chene, "Life and health in Cartesian natural philosophy," 724.

[76] A further option which goes beyond the boundaries of this chapter is to explore whether Descartes actually "*requires* us to contract full intimacy with our own body and our own peculiar past," so that "paradoxically, Descartes himself could hint at the possibilities and the perils of what's become known as 'post-Cartesian agency'" (Sutton, "Body and brain," 699, 700).

The picture of the embodied Descartes that emerges from this literature is, if not quite that of a phenomenologist of the intimacy of the lived body, of its joys and its pains, nevertheless quite far removed from the old caricature of the brute mechanist with his pile of disposable limbs.

Acknowledgement Partial support for this research was provided by the Israel Science Foundation grant 469/13.

References

Alston W. Foundationalism. In: Dancy J, Sosa E, Steup M, editors. A companion to epistemology. Oxford: Blackwell; 1992. p. 382–5.

Brown D. Descartes and the passionate mind. Cambridge: Cambridge University Press; 2006.

Brown D. Cartesian functional analysis. Australas J Philos. 2012;90(1):75–92.

Brown D. The sixth meditation: Descartes and the embodied self. In: Cunning D, editor. The Cambridge companion to Descartes' meditations. Cambridge: Cambridge University Press; 2014. p. 240–57.

Canguilhem G. La formation du concept de réflexe aux XVIIe et XVIIIe siècles, 2nd revised ed. Paris: Vrin; 1977. (First published 1955).

Damasio A. Descartes' error. Emotion, reason, and the human brain. New York: Putnam; 1994.

Des Chene D. Life and health in Cartesian natural philosophy. In: Gaukroger S, Sutton J, Schuster J, editors. Descartes' natural philosophy. London: Routledge; 2000. p. 723–5.

Des Chene D. Spirits and clocks: machine & organism in Descartes. Ithaca/London: Cornell University Press; 2001.

Descartes R. Œuvres, Adam C, Tannery P, editors, new presentation by B. Rochot and P. Costabel. Paris: Vrin-CNRS; 1964–1974. (AT).

Descartes R. The world and other writings. (trans. and editor Gaukroger S). Cambridge: Cambridge University Press; 1998.

Distelzweig PM. The use of *Usus* and the function of *Functio*: teleology and its limits in Descartes's physiology. J Hist Philos. 2015;53(3):377–99.

Duchesneau F. Les modèles du vivant de Descartes à Leibniz. Paris: Vrin; 1998.

Garber D. Descartes embodied. Reading Cartesian philosophy through Cartesian science. Cambridge: Cambridge University Press; 2001.

Gaukroger S. Descartes: an intellectual biography. Oxford: Oxford University Press; 1995.

Gaukroger S. The resources of a mechanist physiology and the problem of goal-directed processes. In: Gaukroger S, Sutton J, Schuster J, editors. Descartes' natural philosophy. London: Routledge; 2000a. p. 383–400.

Gaukroger S, Schuster J, Sutton J, editors. Descartes' natural philosophy. London: Routledge; 2000b.

Gaukroger S. Descartes' system of natural philosophy. Cambridge: Cambridge University Press; 2002.

Grene M. Descartes. Brighton: Harvester Press; 1985.

Grosholz E. Cartesian method and the problem of reduction. Oxford: Clarendon Press; 1991.

Hacking I. The Cartesian body. BioSocieties. 2006;1:13–5.

Hatfield G. The *Passions of the soul* and Descartes's machine psychology. Stud Hist Phil Sci. 2007;38:1–35.

Hutchins BR. Descartes, corpuscles and reductionism: mechanism and systems in Descartes's physiology. Philos Q. 2015;65(261):669–89.

Hutchins BR. Descartes and the dissolution of life. South J Philos. 2016a;54(2):155–73.

Hutchins BR. Obscurity and confusion: nonreductionism in Descartes's biology and philosophy. PhD dissertation. Ghent: Department of Philosophy and Moral Sciences, Ghent University; 2016b.
James S. Passion and action: the emotions in seventeenth-century philosophy. Oxford: Clarendon Press; 1997.
Kirkebøen G. Descartes' embodied psychology: Descartes or Damasio's error? J Hist Neurosci. 2001;10(2):173–91.
Manning G. Descartes' healthy machines and the human exception. In: Garber D, Roux S, editors. The mechanization of natural philosophy. Dordrecht: Springer; 2013. p. 237–62.
Manning G. Descartes's metaphysical biology. HOPOS J Int Soc Hist Philos Sci. 2015;5(2):209–39.
Reiss T. Denying the body? Memory and the dilemmas of history in Descartes. J Hist Ideas. 1996;57(4):587–607.
Rorty R. Philosophy and the mirror of nature. Princeton: Princeton University Press; 1979.
Rorty AO. Descartes on thinking with the body. In: Cottingham J, editor. The Cambridge companion to Descartes. Cambridge: Cambridge University Press; 1992. p. 371–92.
Roux S. L'ennemi cartésien. Cartésianisme et anti-cartésianisme en philosophie de l'esprit et en sciences cognitives. Astérion 11. 2013. http://asterion.revues.org/2419
Roux S, editor. Le corps et l'esprit, problèmes cartésiens, problèmes contemporains. Paris: Editions des archives contemporaines; 2015.
Ryle G. The concept of mind. London: Hutchinson; 1949.
Sawday J. The body emblazoned: dissection and the human body in Renaissance culture. London: Routledge; 1995.
Scribano E. Macchine con la mente. Fisiologia e metafsica tra Cartesio e Spinoza. Roma: Carocci; 2015.
Shapin S. Descartes the doctor: rationalism and its therapies. Br J Hist Sci. 2000;33:131–54.
Shapiro L. The health of the body-machine? Or seventeenth century mechanism and the concept of health. Perspect Sci. 2003;11(4):421–42.
Simmons A. Sensible ends: latent teleology in Descartes' account of sensation. J Hist Philos. 2001;39(1):49–75.
Simmons A. Re-humanizing Descartes. Philos Exch. 2013;41(1):53–71. http://digitalcommons.brockport.edu/phil_ex/vol41/iss1/2
Snider A. Cartesian bodies. Mod Philol. 2000;98(2):299–319.
Sutton J. Philosophy and memory traces: Descartes to connectionism. Cambridge: Cambridge University Press; 1998.
Sutton J. The body and the brain. In: Gaukroger S, Sutton J, Schuster J, editors. Descartes' natural philosophy. London: Routledge; 2000. p. 697–722.
Sutton J, McIlwain D, Christensen W, Geeves A. Applying intelligence to the reflexes: embodied skills and habit between Dreyfus and Descartes. J Br Soc Phenomenol. 2011;42(1):78–103.
Varela F, Thompson E, Rosch E. The embodied mind. cognitive science and human experience. Cambridge: MIT Press; 1991.
Williams B. Descartes: the project of pure enquiry. London: Pelican; 1978.
Wolfe CT. Le mécanique face au vivant. In: Gaillard A, Roukhomovsky B, Roux S, editors. L'automate: modèle, machine, merveille. Bordeaux: Presses Universitaires de Bordeaux; 2012. p. 115–38.
Wolfe CT. Teleomechanism redux? Functional physiology and hybrid models of Life in early modern natural philosophy. Gesnerus Special issue: Teleology. Mech Early Mod Med. 2014;71(2):290–307.

The manufacturer's authorised representative in the EU is Springer Nature Customer Service Centre GmbH, Europaplatz 3, 69115 Heidelberg, Germany. If you have any concerns regarding our products, please contact ProductSafety@springernature.com

Printed and bound by CPI Group (UK) Ltd, Croydon, CR0 4YY

25/09/2026

02222869-0014